NONLINEAR PHYSICAL SCIENCE

NONLINEAR PHYSICAL SCIENCE

Nonlinear Physical Science focuses on recent advances of fundamental theories and principles, analytical and symbolic approaches, as well as computational techniques in nonlinear physical science and nonlinear mathematics with engineering applications.

Topics of interest in *Nonlinear Physical Science* include but are not limited to:

- New findings and discoveries in nonlinear physics and mathematics
- Nonlinearity, complexity and mathematical structures in nonlinear physics
- Nonlinear phenomena and observations in nature and engineering
- Computational methods and theories in complex systems
- Lie group analysis, new theories and principles in mathematical modeling
- Stability, bifurcation, chaos and fractals in physical science and engineering
- Nonlinear chemical and biological physics
- Discontinuity, synchronization and natural complexity in the physical sciences

SERIES EDITORS

Albert C.J. Luo
Department of Mechanical and Industrial Engineering
Southern Illinois University Edwardsville
Edwardsville, IL 62026-1805, USA
Email: aluo@siue.edu

Nail H. Ibragimov
Department of Mathematics and Science
Blekinge Institute of Technology
S-371 79 Karlskrona, Sweden
Email: nib@bth.se

INTERNATIONAL ADVISORY BOARD

Ping Ao, University of Washington, USA; Email: aoping@u.washington.edu
Jan Awrejcewicz, The Technical University of Lodz, Poland; Email: awrejcew@p.lodz.pl
Eugene Benilov, University of Limerick, Ireland; Email: Eugene.Benilov@ul.ie
Eshel Ben-Jacob, Tel Aviv University, Israel; Email: eshel@tamar.tau.ac.il
Maurice Courbage, Université Paris 7, France; Email: maurice.courbage@univ-paris-diderot.fr
Marian Gidea, Northeastern Illinois University, USA; Email: mgidea@neiu.edu
James A. Glazier, Indiana University, USA; Email: glazier@indiana.edu
Shijun Liao, Shanghai Jiaotong University, China; Email: sjliao@sjtu.edu.cn
Jose Antonio Tenreiro Machado, ISEP-Institute of Engineering of Porto, Portugal; Email: jtm@dee.isep.ipp.pt
Nikolai A. Magnitskii, Russian Academy of Sciences, Russia; Email: nmag@isa.ru
Josep J. Masdemont, Universitat Politecnica de Catalunya (UPC), Spain; Email: josep@barquins.upc.edu
Dmitry E. Pelinovsky, McMaster University, Canada; Email: dmpeli@math.mcmaster.ca
Sergey Prants, V.I.Il'ichev Pacific Oceanological Institute of the Russian Academy of Sciences, Russia; Email: prants@poi.dvo.ru
Victor I. Shrira, Keele University, UK; Email: v.i.shrira@keele.ac.uk
Jian Qiao Sun, University of California, USA; Email: jqsun@ucmerced.edu
Abdul-Majid Wazwaz, Saint Xavier University, USA; Email: wazwaz@sxu.edu
Pei Yu, The University of Western Ontario, Canada; Email: pyu@uwo.ca

Albert C.J. Luo

Discontinuous Dynamical Systems on Time-varying Domains

With 96 figures, 4 of them in color

Author
Albert C.J. Luo
Department of Mechanical and Industrial Engineering
Southern Illinois University Edwardsville
Edwardsville, IL 62026-1805, USA
Email: aluo@siue.edu

ISSN 1867-8440   e-ISSN 1867-8459
Nonlinear Physical Science

ISBN 978-7-04-025759-5
Higher Education Press, Beijing

ISBN 978-3-642-00252-6   e-ISBN 978-3-642-00253-3
Springer Dordrecht Heidelberg London New York

Library of Congress Control Number: 2009920463

© Higher Education Press, Beijing and Springer-Verlag Berlin Heidelberg 2009
This work is subject to copyright. All rights are reserved, whether the whole or part of the material is concerned, specifically the rights of translation, reprinting, reuse of illustrations, recitation, broadcasting, reproduction on microfilm or in any other way, and storage in data banks. Duplication of this publication or parts thereof is permitted only under the provisions of the German Copyright Law of September 9, 1965, in its current version, and permission for use must always be obtained from Springer. Violations are liable to prosecution under the German Copyright Law.
The use of general descriptive names, registered names, trademarks, etc. in this publication does not imply, even in the absence of a specific statement, that such names are exempt from the relevant protective laws and regulations and therefore free for general use.

Cover design: Frido Steinen-Broo, EStudio Calamar, Spain

Printed on acid-free paper

Springer is part of Springer Science+Business Media (www.springer.com)

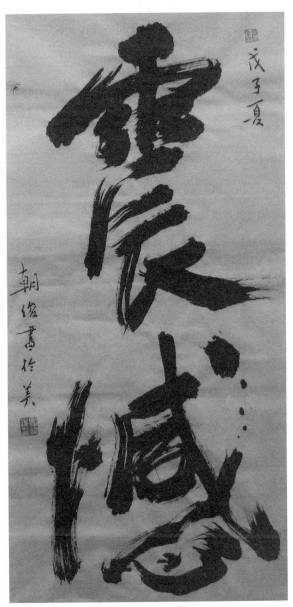

Earthquake and Heartbreak
*that shocks one's passion and spirit,
and recalls everyone to a sense of duty.*

Preface

This book is about discontinuous dynamical systems on time-varying domains. I had not planed to write this book originally. As a scientist working on dynamics and vibration, the 5.12 earthquake of Wenchuan (Sichuan province, China) shocked my heart and made me feel guilty because my research cannot make any direct contributions to help them. Therefore, I would like to write two words "Zhenhan" in Chinese callig-raphy on the dedication page to express my passion. The meaning of "Zhenhan" is "Earthquake and Heartbreak" that shocks one's passion and spirit, and recalls everyone to a sense of duty. Herein, I would like to accumulate recent research developments of discontinuous dynamical systems on time-varying domains. One likes to use continuous models for discontinuous dynamical systems. However, sometimes such con-tinuous modeling cannot provide adequate descriptions of discontinuous dynamical systems. Recently, researchers have gradually realized that discontinuous modeling may provide an adequate and acceptable predi-cation of engineering systems. Currently, most research still focuses on discontinuous dynamical systems on time-invariant domains. To better describe practical problems, some research on discontinuous systems on time-varying domains is scattered here and there but without a systemati-cal theory. The purpose of this book is to systematically present a theory of discontinuous dynamical systems on time-varying domains for univer-sity students and researchers.

This book mainly focuses on the switchability of discontinuous dy-namical systems on time-varying domains. Based on such concepts, prin-ciples of dynamical system interactions without any connections are pre-sented. This book consists of seven chapters. Chapter 1 discusses two examples to show where discontinuous dynamical systems exist. Chapter 2 presents a basic theory for the switchability of a flow to the separation boundary in discontinuous dynamical systems, and switching bifurca-tions are also addressed. In Chapter 3, transversality and sliding phenomena for a controlled dynamical system to an inclined line boundary of control logic are presented to show how to apply such a new theory. In Chapter 4, dynamics of a frictional oscillator on a traveling belt with time-varying speeds is presented, which is a simple example of discon-tinuous dynamical systems on the time-varying domains. Chapter 5 pre-sents the dynamics mechanism of impacting chatter and stick phenomena in two dynamical systems with impact laws. In Chapter 6, dynamical behaviors of two systems connected with friction are presented. In Chap-ters 3 6, the similar writing styles and formats are adopted to show how to apply the new theory to different problems. In Chapter 7, a generalized theory for the interaction principle of dynamical systems is developed from discontinuous dynamical systems on time-varying domains. I hope the information presented in this book may stimulate more research in the area of discontinuous dynamical systems.

Finally, I would like to appreciate my students (Brandon M. Rapp, Brandon C. Gegg, Dennis O'Connor and Sagun Thapa) for applying the new concepts to mechanical systems and completing all numerical com-putations. This book is dedicated to the people who died during the earthquake of Wenchuan (Sichuan province, China) on May 12, 2008. Finally, I would like to thank my wife (Sherry X. Huang) and my chil-dren (Yanyi Luo, Robin Ruo-Bing Luo, and Robert Zong-Yuan Luo) for their tolerance, patience, understanding and support.

Albert C. J. Luo
Edwardsville, Illinois
August, 2008

Contents

1 **Introduction** 1
   1.1 Discontinuous systems 1
   1.2 Book layout 5
   References 6

2 **Flow Switchability** 9
   2.1 Discontinuous dynamic systems 9
   2.2 *G*-functions 11
   2.3 Passable flows 15
   2.4 Non-passable flows 19
   2.5 Tangential flows 28
   2.6 Switching bifurcations 39
   References 52

3 **Transversality and Sliding Phenomena** 55
   3.1 A controlled system 55
   3.2 Transversality conditions 57
   3.3 Mappings and predictions 60
   3.4 Periodic and chaotic motions 67
   References 75

4 **A Frictional Oscillator on Time-varying Belt** 77
   4.1 Mechanical model 77
   4.2 Analytical conditions 80
      4.2.1 Equations of motion 80
      4.2.2 Passable flows to boundary 83
      4.2.3 Sliding flows on boundary 85
      4.2.4 Grazing flows to boundary 89
   4.3 Generic mappings and force product criteria 91
      4.3.1 Generic mappings 91

		4.3.2	Sliding flows and fragmentation 93

        4.3.2 Sliding flows and fragmentation 93
        4.3.3 Grazing flows 96
   4.4 Periodic motions 98
        4.4.1 Mapping structures 98
        4.4.2 Illustrations 100
   4.5 Numerical simulations 112
   References 113

5 Two Oscillators with Impacts and Stick 115
   5.1 Physical problem 115
        5.1.1 Introduction to problem 115
        5.1.2 Equations of motion 117
   5.2 Domains and vector fields 119
        5.2.1 Absolute motion description 119
        5.2.2 Relative motion description 125
   5.3 Mechanism of stick and grazing 127
        5.3.1 Analytical conditions 128
        5.3.2 Physical interpretation 133
   5.4 Mapping structures and motions 134
        5.4.1 Switching sets and basic mappings.................... 134
        5.4.2 Mapping equations 137
        5.4.3 Mapping structures 141
        5.4.4 Bifurcation scenario 144
   5.5 Periodic motion prediction 144
        5.5.1 Approach 145
        5.5.2 Impacting chatter 147
        5.5.3 Impacting chatter with stick 150
        5.5.4 Parameter maps 152
   5.6 Numerical illustrations 153
        5.6.1 Impacting chatter 153
        5.6.2 Impacting chatter with stick 155
        5.6.3 Further illustrations 158
   References 160

6 Dynamical Systems with Frictions............................... 161
   6.1 Problem statement 161
   6.2 Switching and stick motions 164
        6.2.1 Equations of motion 164
        6.2.2 Analytical conditions 166
   6.3 Periodic motions 172
        6.3.1 Switching planes and mappings 172
        6.3.2 Mapping structures and motions 175
        6.3.3 Bifurcation scenario 178
   6.4 Numerical illustrations 181
        6.4.1 Periodic motion without stick 181

		6.4.2	Periodic motion with stick 184
		6.4.3	Periodic motion with stick only 189
	References 189

7   Principles for System Interactions 191
   7.1   Two dynamical systems 191
      7.1.1   Dynamical systems with interactions 191
      7.1.2   Discontinuous description 195
      7.1.3   Resultant dynamical systems 196
   7.2   Fundamental interactions 199
   7.3   Interactions with singularity 206
   7.4   Interactions with corner singularity 210
   References 215

Appendix 217
   A.1   Basic solution 217
   A.2   Stability and bifurcation 219

Index 221

Chapter 1
Introduction

In engineering, discontinuous dynamical systems exist everywhere. One usually uses continuous models to describe discontinuous dynamical systems. However such continuous models cannot provide suitable predictions of discontinuous dynamical systems. To better understand discontinuous systems, one should realize that discontinuous models will provide an adequate and real predication of engineering systems. Thus, one considers a global discontinuous system consisting of many continuous sub-systems in different domains. For each continuous subsystem, it possesses dynamical properties different from the adjacent continuous subsystems. Because of such difference between two adjacent subsystems, the switch ability and/or transport laws on their boundaries should be addressed. The investigation on such discontinuous systems mainly focused on the time-independent boundary between two dynamical systems. In fact, the boundary relative to time is more popular. In this book, discontinuous dynamical systems on time-varying domains will be of great interest. A brief survey will be given through two practical examples. Finally, the book layout will be presented, and the summarization of all chapters of the main body of this book will be given.

1.1 Discontinuous systems

In mechanical engineering, there are two common and important contacts between two dynamical systems, i.e., impact and friction. For example, gear transmission systems possess both impact and friction. Such gear transmission systems are used to transmit power between parallel shafts or to change direction. During the power transmission, a pair of two gears forms a resultant dynamical system. Each gear has its own dynamical system connected with shafts and bearings. Because two subsystems are not connected directly, the power transmission is completed through the impact and friction. Because both of sub-systems are independent each other except for impacting and sliding together, such two dynamical systems have a common time-varying boundary for impacting, which causes domains for the two dynamical systems to be time-varying.

In the early investigation, a piecewise stiffness model was used to investigate dynamics of gear transmission systems. Although such a dynamical system is discontinuous, the corresponding domains for vector fields of the dynamical system are time-independent. For instance, den Hartog and Mikina (1932) used a piecewise linear system without damping to model gear transmission systems, and the symmetric periodic motion in such a system was investigated. For low-speed gear systems, such a linear model gave a reasonable prediction of gear-tooth vibrations. With increasing rotation speed in gear transmission systems, vibrations and noise become serious. Ozguven and Houser (1988) gave a survey on the mathematical models of gear transmission systems. The piecewise linear model and the impact model were two of the main mechanical models to investigate the origin of vibration and noise in gear transmission systems. Natsiavas (1998) investigated a piecewise linear system with a symmetric tri-linear spring, and the stability and bifurcation of periodic motions in such a system were investigated through the variation of initial conditions. Based on a piecewise linear model, the dynamics of gear transmission systems was investigated by Comparin and Singh (1989), and Theodossiades and Natsiavas (2000). Pfeiffer (1984) presented an impact model of gear transmissions, and the theoretical and experimental investigations on regular and chaotic motions in the gear box were later carried out by Karagiannis and Pfeiffer (1991).

To model vibrations in gear transmission systems, Luo and Chen (2005) gave an analytical prediction of the simplest periodic motion through a piecewise linear impacting system. In addition, the corresponding grazing of periodic motions was observed, and chaotic motions were simulated numerically through such a piecewise linear system. From the local singularity theory in Luo (2005), the grazing mechanism of the strange fragmentation of the piecewise linear system was discussed by Luo and Chen (2006). Luo and Chen (2007) used the mapping structure technique to analytically predict arbitrary periodic motions of such a piecewise linear system. In that piecewise linear model, it was assumed that impact locations were fixed, and the perfectly plastic impact was considered. The separation of the two gears occurred at the same location as the gears impact. Compared with the existing models, this model can give a better prediction of periodic motions in gear transmission systems, but the relative assumptions may not be realistic to practical transmission systems because all the aforementioned investigations are based on a time-independent boundary or a given motion boundary. To consider the dynamical systems with the time-varying boundary, Luo and O'Connor (2007a,b) proposed a mechanical model to determine the mechanism of the impacting chatter and stick in gear transmission systems. The corresponding analytical conditions for such impacting chatter and stick were developed.

In mechanical engineering, the friction contact between surfaces of two systems is important for motion transmissions (e.g., clutch systems, brake systems, etc.). In addition, the two systems are independent except for the friction contact. Such problem will have time-varying boundary and domains. For such a friction problem, it should return back to the 1930's. den Hartog (1931) investigated the peri-

1.1 Discontinuous systems

odic motion of a forced, damped, linear oscillator contacting a surface with friction. Levitan (1961) investigated the existence of periodic motions in a friction oscillator with a periodically driven base. Filippov (1964) investigated the motion existence of a Coulomb friction oscillator, and presented a differential equation theory with discontinuous right-hand sides. The differential inclusion was introduced via the set-valued analysis for the sliding motion along the discontinuous boundary. The investigations of discontinuous differential equations were summarized by Filippov (1988). However, the Filippov's theory mainly focused on the existence and uniqueness of solutions for non-smooth dynamical systems. Such a differential equation theory with discontinuity is difficult to apply to practical problems. Luo (2005a) developed a general theory to handle the local singularity of discontinuous dynamical systems. To determine the sliding and source motions in discontinuous dynamical systems, the imaginary, sink and source flows were introduced by Luo (2005b). The detailed discussions can be referred to Luo (2006).

On the other hand, Hundal (1979) used a periodic, continuous function to investigate the frequency-amplitude response of a friction oscillator. Shaw (1986) investigated non-stick, periodic motions of a friction oscillator through Poincaré mapping. Feeny (1992) analytically investigated the non-smooth of the Coulomb friction oscillator. To verify the analytic results, Feeny and Moon (1994) investigated chaotic dynamics of a dry-friction oscillator experimentally and numerically. Feeny (1996) gave a systematic discussion of the nonlinear dynamical mechanism of stick-slip motion of friction oscillators. Hinrichs, Oestreich and Popp (1997) investigated the nonlinear phenomena in an impact and friction oscillator under external excitations (also see, Hinrichs, Oestreich and Popp (1998)). Natsiavas (1998) developed an algorithm to numerically determine the periodic motion and the corresponding stability of piecewise linear oscillators with viscous and dry friction damping (also see, Natsiavas and Verros (1999)). Ko, Taponat and Pfaifer (2001) investigated the friction-induced vibrations with and without external excitations. Andreaus and Casini (2002) gave a closed form solution of a Coulomb friction-impact model without external excitations. Thomsen and Fidlin (2003) gave an approximate estimation of response amplitude for the stick-slip motion in a nonlinear friction oscillator. Kim and Perkins (2003) investigated stick-slip motions in a friction oscillator via the harmonic balance or Galerkin method. Li and Feng (2004) investigated the bifurcation and chaos in a friction-induced oscillator with a nonlinear friction model. Pilipchuk and Tan (2004) investigated the dynamical behaviors of a 2DOF mass-damper-spring system contacting on a decelerating rigid strip with friction. Awrejcewicz and Pyryev (2004) gave an investigation on frictional periodic processes by accelerating or braking a shaft-pad system. In 2007, Hetzler, Schwarzer and Seemann (2007) considered a nonlinear friction model to analytically investigate the Hopf-bifurcation in a sliding friction oscillator with applications to the low frequency disk brake noise.

In the aforementioned investigations, the conditions for motion switchability to the discontinuous boundary were not considered enough. Luo and Gegg (2005a) used the local singularity theory of Luo (2005a, 2006) to develop the force criteria for motion switchability on the velocity boundary in a harmonically driven linear oscillator with dry-friction (also see, Luo and Gegg (2005b)). Through such an investigation, the traditional eigenvalue analysis may not be useful for motion switching at the discontinuous boundary. Lu (2007) mathematically proved the existence of such a periodic motion in a friction oscillator. Luo and Gegg(2006a,b) investigated the dynamics of a friction-induced oscillator contacting on time-varying belts with friction. Recently, to model the disk brake system, many researchers still considered the mechanical model as in Hetzler, Schwarzer and Seemann (2007). Luo and Thapa (2007) proposed a new method to model the brake system which consists of two oscillators, and the two oscillators are connected through a contacting surface with friction. Based on this model, the nonlinear dynamical behaviors of a brake system under a periodical excitation were investigated.

The other developments on non-smooth dynamical systems in recent decades will be addressed as well. Feigin (1970) investigated the C-bifurcation in piecewise-continuous systems via the Floquet theory of mappings, and the motion complexity was classified by the eigenvalues of mappings, which can be referred to recent publications (e.g., Feigin 1995; di Bernnado et al 1999). The C-bifurcation is also termed the grazing bifurcation by many researchers. Nordmark (1991) used "grazing" terminology to describe the grazing phenomena in a simple impact oscillator. No strict mathematical description was given, but the grazing condition (i.e., the velocity $dx/dt = 0$ for displacement x) in such an impact oscillator was obtained. From Luo (2005a, 2006), such a grazing condition is a necessary condition only. The grazing is the tangency between an n-D flow of dynamical systems and the discontinuous boundary surface. From differential geometry points of view, Luo (2005a) gave the strict mathematic definition of the "grazing", and the necessary and sufficient conditions of the general discontinuous boundary were presented (also see, Luo 2006). Nordmark's result is a special case. Nusse and Yorke (1992) used the simple discrete mapping from Nordmark's impact oscillator and showed the bifurcation phenomena numerically. Based on the numerical observation, the sudden change bifurcation in the numerical simulation is called the *border-collision bifurcation*. So, the similar discrete mappings in discontinuous dynamical system were further developed. Especially, Nordmark and Dankowicz (1999) developed a discontinuous mapping from a general way to investigate the grazing bifurcation, and the discontinuous mapping is based on the Taylor series expansion in the neighborhood of the discontinuous boundary. Following the same idea, di Bernardo et al (2001a,b; 2002) developed a normal form to describe the grazing bifurcation. In addition, di Bernardo et al (2001c) used the normal form to obtain the discontinuous mapping and numerically observed such a border-collision bifurcation through a discontinuous mapping. From the discontinuous mapping and its normal form, the aforementioned bifurcation theory structure was developed for the so-called, codimension 1 dynamical system.

The discontinuous mapping and normal forms on the discontinuous boundary were developed from the Taylor series expansion in the neighborhood of the discontinuous boundary. However, the normal form requires the vector field with the C^r-continuity and the corresponding convergence, where the order r is the highest order of the total power numbers in each term of normal form. For piecewise linear and nonlinear systems, the C^1-continuity of the vector field cannot provide an enough mathematical base to develop the normal form. The normal form also cannot be used to investigate global periodic motions in the discontinuous system. Leine et al (2000) used the Filippov theory to investigate bifurcations in nonlinear discontinuous systems. However, the discontinuous mapping techniques were employed to determine the bifurcation via the Floquet multiplier. The more discussion about the traditional analysis of bifurcation in non-smooth dynamical systems can be referred to Zhusubaliyev and Mosekilde (2003). Based on the recent research, the Floquet multiplier also may not be adequate for periodic motions involved with the grazing and sliding motions in non-smooth dynamical systems. Therefore, Luo (2005a) proposed a general theory for the local singularity of non-smooth dynamical systems on connectable domains (also see, Luo (2006)). To resolve the difficulty, Luo (2007b) developed a general theory for the switching possibilities of flow on the boundary from the passable to non-passable one, and so on. In this book, from recent developments in Luo (2007a; 2008a,b), a generalized theory for discontinuous systems on time-varying domains will be presented.

1.2 Book layout

To help readers easily read this book, the main contents in this book are summarized as follows.

In Chapter 2, from Luo (2008a,b), a switchability theory for a flow to the boundary in discontinuous dynamical systems will be presented. The G-functions for discontinuous dynamical systems will be introduced to investigate singularity in discontinuous dynamical systems. Based on the new G-function, the full and half sink and source, non-passable flows to the separation boundary in discontinuous dynamical systems will be discussed, and the switchability of a flow from a domain to an adjacent one will be addressed. Finally, the switching bifurcations between the passable and non-passable flows will be presented.

In Chapter 3, the switchability of a flow from one domain into another in discontinuous dynamical systems will be presented through a periodically forced, discontinuous dynamical system with a time-invariant boundary. The normal vector-field for flow switching on the separation boundary will be introduced and the passability condition of a flow to the separation boundary will be given through such normal vector fields. The sliding and grazing conditions to the separation boundary will be presented as well. This investigation may help one better understand the sliding mode control.

In Chapter 4, an oscillator moving on the periodically traveling belt with dry friction is investigated as a dynamical system with a time-varying boundary. The conditions of stick and non-stick motions in such an frictional oscillator will be developed, and the periodic motions in the oscillator will be investigated as well. The grazing and stick (or sliding) bifurcations will be investigated. The significance of this investigation is to show how to control motions in such friction-induced oscillators in industry.

In Chapter 5, impact and stick motions of two oscillators at the time-varying boundary and domains will be presented under an impact law. The dynamics mechanism of the impacting chatter with stick at the moving boundary will be investigated from the local singularity theory of discontinuous dynamical systems. The analytical conditions for the onset and vanishing of stick motions will be obtained, and the condition for maintaining stick motion is achieved as well. This chapter will show how two dynamical systems interact.

In Chapter 6, two dynamical systems connected with friction will be presented, and the motion switchability on the moving discontinuous boundary will be discussed through the theory of discontinuous dynamical systems. The onset and vanishing of motions will be discussed through the bifurcation and grazing analyses. It will be observed that two different systems can stick together forever.

In Chapter 7, a theory for two dynamical systems with a general interaction will be presented. Such an interaction occurs at a time-varying boundary. From the discontinuous theory in Chapter 2, a general methodology for two system interactions will be presented in order to determine the complex motion of the two systems, which is caused by the interaction between two systems.

References

Andreaus,U. and Casini, P. (2002), Friction oscillator excited by moving base and colliding with a rigid or deformable obstacle, *International Journal of Non-Linear Mechanics,* **37**, pp.117-133.

Awrejcewicz, J. and Pyryev, Y. (2004), Tribological periodical processes exhibited by acceleration or braking of a shaft-pad system, *Communications in Nonlinear Science and Numerical Simulation,* **9**, pp.603-614.

Comparin, R. J. and Singh, R. (1989), Nonlinear frequency response characteristics of an impact pair, *Journal of Sound and Vibration,* **134**, pp.259-290.

Dankowicz, H. and Nordmark A. B. (2000), On the origin and bifurcations of stick-slip oscillations, *Physica D,* **136**, pp.280-302.

den Hartog, J. P. (1931), Forced vibrations with Coulomb and viscous damping, *Transactions of the American Society of Mechanical Engineers,* **53**, pp.107-115.

den Hartog, J. P. and Mikina, S. J. (1932), Forced vibrations with non-linear spring constants, *ASME Journal of Applied Mechanics,* **58**, pp.157-164.

di Bernaedo, M., Budd, C. J. and Champneys, A. R. (2001a), Grazing and border-collision in piecewise-smooth systems: a unified analytical framework, *Physical Review Letters,* **86**, pp.2553-2556.

di Bernardo, M., Budd. C. J. and Champneys, A. R. (2001b), Normal form maps for grazing bifurcation in n-dimensional piecewise-smooth dynamical systems, *Physica D,* **160**, pp.222-254.

di Bernardo, M., Budd. C. J. and Champneys, A. R. (2001c), Corner-collision implies border-collision bifurcation, *Physica D,* **154**, pp.171-194.

di Bernardo, M., Feigin, M. I., Hogan, S. J. and Homer, M. E. (1999), Local analysis of C-bifurcations in n-dimensional piecewise-smooth dynamical systems, *Chaos, Solitons & Fractals,* **10**, pp.1881-1908.

di Bernardo, M., Kowalczyk, P. and Nordmark, A. B.(2002), Bifurcation of dynamical systems with sliding: derivation of normal form mappings, *Physica D,* **170**, pp.175-205.

References

Feeny, B. F. (1992), A non-smooth Coulomb friction oscillator, *Physics D,* **59**, pp.25-38.

Feeny, B. F.(1994), The nonlinear dynamics of oscillators with stick-slip friction, in: A. Guran, F. Pfeiffer and K. Popp (eds), *Dynamics with Friction*, River Edge: World Scientific, pp.36-92.

Feeny, B. F. and Moon, F. C. (1994), Chaos in a forced dry-friction oscillator: experiments and numerical modeling, *Journal of Sound and Vibration,* **170**, pp.303-323.

Feigin, M. I. (1970), Doubling of the oscillation period with C-bifurcation in piecewise-continuous systems, *PMM,* **34**, pp.861-869.

Feigin, M. I. (1995), The increasingly complex structure of the bifurcation tree of a piecewise-smooth system, *Journal of Applied Mathematics and Mechanics,* **59**, pp.853-863.

Filippov, A. F. (1964), Differential equations with discontinuous right-hand side, *American Mathematical Society Translations, Series 2,* **42**, pp.199-231.

Filippov, A. F. (1988), *Differential Equations with Discontinuous Righthand Sides*, Dordrecht: Kluwer Academic Publishers.

Hetzler, H., Schwarzer, D. and Seemann, W. (2007), Analytical investigation of steady-state stability and Hopf-bifurcation occurring in sliding friction oscillators with application to low-frequency disc brake noise, *Communications in Nonlinear Science and Numerical Simulation,* **12**, pp.83-99.

Hinrichs, N., Oestreich, M. and Popp, K.(1997), Dynamics of oscillators with impact and friction, *Chaos, Solitons and Fractals,* **8**, pp. 535-558.

Hinrichs, N., Oestreich, M. and Popp, K. (1998), On the modeling of friction oscillators, *Journal of Sound and Vibration,* **216**, pp.435-459.

Hundal, M. S.(1979), Response of a base excited system with Coulomb and viscous friction, *Journal of Sound and Vibration,* **64**, pp.371-378.

Karagiannis, K. and Pfeiffer, F. (1991), Theoretical and experimental investigations of gear box, *Nonlinear Dynamics,* **2**, pp.367-387.

Kim, W. J. and Perkins, N. C. (2003), Harmonic balance/Galerkin method for non-smooth dynamical system, *Journal of Sound and Vibration,* **261**, pp.213-224.

Ko, P. L., Taponat, M. C. and Pfaifer, R. (2001), Friction-induced vibration–with and without external disturbance, *Tribology International,* **34**, pp.7-24.

Leine, R. I., van Campen, D. H. and van de Vrande (2000), Bifurcations in nonlinear discontinuous systems, *Nonlinear Dynamics,* **23**, pp.105-164.

Levitan, E. S. (1960), Forced oscillation of a spring-mass system having combined Coulomb and viscous damping, *Journal of the Acoustical Society of America,* **32**, pp.1265-1269.

Li, Y. and Feng, Z. C. (2004), Bifurcation and chaos in friction-induced vibration, *Communications in Nonlinear Science and Numerical Simulation,* **9**, pp.633-647.

Lu, C. (2007), Existence of slip and stick periodic motions in a non-smooth dynamical system, *Chaos, Solitons and Fractals,* **35**, pp.949-959.

Luo, A. C. J. (2005a), A theory for non-smooth dynamical systems on connectable domains, *Communication in Nonlinear Science and Numerical Simulation,* **10**, pp.1-55.

Luo, A. C. J. (2005b), Imaginary, sink and source flows in the vicinity of the separatrix of non-smooth dynamic system, *Journal of Sound and Vibration,* **285**, pp.443-456.

Luo, A. C. J.(2006), *Singularity and Dynamics on Discontinuous Vector Fields*, Amsterdam: Elsevier.

Luo, A. C. J. (2007a), Differential geometry of flows in nonlinear dynamical systems, *Proceedings of IDECT'07*, ASME International Design Engineering Technical Conferences, September 4-7, 2007, Las Vegas, Nevada, USA. DETC2007-84754.

Luo, A. C. J. (2007b), On flow switching bifurcations in discontinuous dynamical system, *Communications in Nonlinear Science and Numerical Simulation,* **12**, pp.100-116.

Luo, A. C. J. (2008a), A theory for flow swtichability in discontinuous dynamical systems, *Nonlinear Analysis: Hybrid Systems,* **2**, pp.1030-1061.

Luo, A. C. J. (2008b), *Global Transversality, Resonance and Chaotic dynamics*, Singapore: World Scientific.

Luo, A. C. J., and Chen, L. D. (2005), Periodic motion and grazing in a harmonically forced, piecewise linear, oscillator with impacts, *Chaos, Solitons and Fractals,* **24**, pp.567-578.

Luo, A. C. J. and Chen, L. D. (2006), The grazing mechanism of the strange attractor fragmentation of a harmonically forced, piecewise, linear oscillator with impacts, *IMechE Part K: Journal of Multi-body Dynamics*, **220**, pp.35-51.

Luo, A. C. J. and Chen, L. D. (2007), Arbitrary periodic motions and grazing switching of a forced piecewise-linear, impacting oscillator, *ASME Journal of Vibration and Acoustics*, **129**, pp.276-284.

Luo, A. C. J. and Gegg, B. C. (2005a), On the mechanism of stick and non-stick periodic motion in a forced oscillator including dry-friction, *ASME Journal of Vibration and Acoustics*, **128**, pp.97-105.

Luo, A. C. J. and Gegg, B. C. (2005b), Stick and non-stick periodic motions in a periodically forced, linear oscillator with dry friction, *Journal of Sound and Vibration*, **291**, pp.132-168.

Luo, A. C. J. and Gegg, B. C. (2006a), Periodic motions in a periodically forced oscillator moving on an oscillating belt with dry friction, *ASME Journal of Computational and Nonlinear Dynamics*, **1**, pp.212-220.

Luo, A. C. J. and Gegg, B. C. (2006b), Dynamics of a periodically excited oscillator with dry friction on a sinusoidally time-varying, traveling surface, *International Journal of Bifurcation and Chaos,* **16**, pp.3539-3566.

Luo, A. C. J. and O'Connor, D. (2007a), Nonlinear dynamics of a gear transmission system, Part I: mechanism of impacting chatter with stick, *Proceedings of IDETC'07*, 2007 ASME International Design Engineering Conferences and Exposition, September 4-7, 2007, Las Vegas, Nevada. IDETC2007-34881.

Luo, A. C. J. and O'Connor, D. (2007b), Nonlinear dynamics of a gear transmission system, Part II: periodic impacting chatter and stick, *Proceedings of IMECE'07*, 2007 ASME International Mechanical Engineering Congress and Exposition, November 10-16, 2007, Seattle, Washington. IMECE2007-43192.

Luo, A. C. J. and Thapa, S. (2007), On nonlinear dynamics of simplified brake dynamical systems, *Proceedings of IMECE2007*, 2007 ASME International Mechanical Engineering Congress and Exposition, November 5-10, 2007, Seattle, Washington, USA. IMECE2007-42349.

Natsiavas, S. (1998), Stability of piecewise linear oscillators with viscous and dry friction damping, *Journal of Sound and Vibration*, **217**, pp.507-522.

Natsiavas, S. and Verros, G. (1999), Dynamics of oscillators with strongly nonlinear asymmetric damping, *Nonlinear Dynamics*, **20**, pp.221-246.

Nordmark, A. B. (1991), Non-periodic motion caused by grazing incidence in an impact oscillator, *Journal of Sound and Vibration*, **145**, pp.279-297.

Nusse, H. E. and Yorke J. A. (1992), Border-collision bifurcations including "period two to period three" for piecewise smooth systems, *Physica D*, 1992, **57**, pp.39-57.

Ozguven, H. N. and Houser, D. R. (1988), Mathematical models used in gear dynamics—a review, *Journal of Sound and Vibration*, **121**, pp.383-411.

Pfeiffer, F. (1984), Mechanische Systems mit unstetigen Ubergangen, *Ingeniuer-Archiv*, **54**, pp.232-240.

Pilipchuk, V. N. and Tan, C. A. (2004), Creep-slip capture as a possible source of squeal during decelerating sliding, *Nonlinear Dynamics* **35**, pp.258-285.

Shaw, S. W. (1986), On the dynamic response of a system with dry-friction, *Journal of Sound and Vibration*, **108**, pp.305-325.

Theodossiades, S. and Natsiavas, S. (2000), Non-linear dynamics of gear-pair systems with periodic stiffness and backlash, *Journal of Sound and Vibration*, **229**, pp.287-310.

Thomsen, J. J. and Fidlin, A. (2003), Analytical approximations for stick-slip vibration amplitudes, *International Journal of Non-Linear Mechanics*, **38**, pp.389-403.

Zhusubaliyev, Z. and Mosekilde, E. (2003), *Bifurcations and Chaos in Piecewise-Smooth Dynamical Systems*, Singapore: World Scientific.

Chapter 2
Flow Switchability

In this chapter, from Luo (2008a,b), a theory of flow switchability to the boundary in discontinuous dynamical systems will be presented. The G-functions for discontinuous dynamical systems will be introduced to investigate singularity in discontinuous dynamical systems. Based on the G-functions, the full and half sink and source, non-passable flows to the separation boundary in the systems will be presented, and the switchability of a flow from a domain to an adjacent one will also be discussed. Therefore, the switching bifurcations between the passable and non-passable flows will be presented. The basic theory will be applied to determine the complexity in the systems with time-varying domains in Chapters 4~6.

2.1 Discontinuous dynamic systems

As considered by Luo (2005a, 2006), a dynamic system consists of N sub-dynamic systems in a universal domain $\mho \subset \mathbb{R}^n$. The accessible domain in phase space means that a continuous dynamical system can be defined on such a domain. The inaccessible domain in phase space means that none dynamical system can be defined on such a domain. A universal domain in phase space is divided into N accessible sub-domains Ω_i plus the inaccessible domain Ω_0. The union of all the accessible sub-domains is $\cup_{i=1}^{N}\Omega_i$ and the universal domain is $\mho = \cup_{i=1}^{N}\Omega_i \cup \Omega_0$, which can be expressed through an n_1-dimensional sub-vector $\mathbf{x}_{n_1}$ and an $(n-n_1)$-dimensional sub-vector $\mathbf{x}_{n-n_1}$. Ω_0 is the union of the inaccessible domains. $\Omega_0 = \mho \setminus \cup_{i=1}^{N}\Omega_i$ is the complement of the union of the accessible sub-domains. If all the accessible domains are connected, the universal domain in phase space is called the connectable domain. If the accessible domains are separated by the inaccessible domain, the universal domain is called the separable domains, as shown in Fig.2.1. To investigate the relation between two disconnected domains without any common boundary, the specific transport laws should be inserted. Such an issue can be referred to Luo (2006). Herein, the flow switchability in discontinuous dynamical system focuses on dynamics in the two connected domains with a common boundary. For example, the boundary between the two domains Ω_i and Ω_j is $\partial\Omega_{ij} = \bar{\Omega}_i \cap \bar{\Omega}_j$, as sketched in Fig.2.2. This boundary is formed by the intersection of the closed sub-domains.

On the ith open sub-domain Ω_i, there is a C^{r_α}-continuous system ($r_i \geq 1$) in a form of

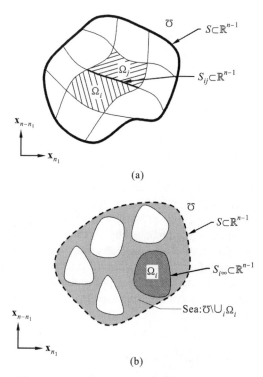

Fig. 2.1 Phase space: (a) connectable and (b) separable domains

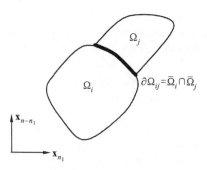

Fig. 2.2 Two adjacent sub-domains Ω_i and Ω_j, the corresponding boundary $\partial\Omega_{ij}$

2.2 G-functions

$$\dot{\mathbf{x}}^{(i)} \equiv \mathbf{F}^{(i)}\left(\mathbf{x}^{(i)}, t, \mathbf{p}_i\right) \in \mathbb{R}^n, \mathbf{x}^{(i)} = \left(x_1^{(i)}, x_2^{(i)}, \ldots, x_n^{(i)}\right)^{\mathrm{T}} \in \Omega_i. \quad (2.1)$$

The time is t and $\dot{\mathbf{x}}^{(i)} = d\mathbf{x}^{(i)}/dt$. In an accessible sub-domain Ω_i, the vector field $\mathbf{F}^{(i)}(\mathbf{x}^{(i)}, t, \mathbf{p}_i)$ with parameter vectors $\mathbf{p} = (p_i^{(1)}, p_i^{(2)}, \ldots, p_i^{(l)})^{\mathrm{T}} \in \mathbb{R}^l$ is C^{r_i}-continuous ($r_i \geqslant 1$) in a state vector $\mathbf{x}^{(i)}$ and for all time t; and the continuous flow in Eq.(2.1) $\mathbf{x}^{(i)}(t) = \mathbf{\Phi}^{(i)}(\mathbf{x}^{(i)}(t_0), t, \mathbf{p}_i)$ with $\mathbf{x}^{(i)}(t_0) = \mathbf{\Phi}^{(i)}(\mathbf{x}^{(i)}(t_0), t_0, \mathbf{p}_i)$ is C^{r+1}-continuous for time t. The discontinuous dynamics theory presented herein holds for the following hypothesis.

H1: The switching between two adjacent sub-systems possesses time-continuity.
H2: For an unbounded, accessible sub-domain Ω_i, there is a bounded domain $D_i \subset \Omega_i$ and the corresponding vector field and its flow are bounded, i.e.,

$$||\mathbf{F}^{(i)}|| \leqslant K_1 \text{ (const) and } ||\mathbf{\Phi}^{(i)}|| \leqslant K_2 \text{(const) on } D_i, \quad \text{for } t \in [0, \infty). \quad (2.2)$$

H3: For a bounded, accessible domain Ω_i, there is a bounded domain $D_i \subset \Omega_i$ and the corresponding vector field is bounded, but the flow may be unbounded, i.e.,

$$||\mathbf{F}^{(i)}|| \leqslant K_1 \text{(const) and } ||\mathbf{\Phi}^{(i)}|| < \infty \text{ on } D_i, \quad \text{for } t \in [0, \infty). \quad (2.3)$$

Because dynamical systems on the different accessible sub-domains are different, a relation between two flows in the two sub-domains should be developed for continuation. For a sub-domain Ω_i, there are k_i-piece boundaries ($k_i \leqslant n-1$). Consider a boundary set of any two adjacent sub-domains.

Definition 2.1. The boundary in an n-D phase space is defined as

$$S_{ij} \equiv \partial \Omega_{ij} = \bar{\Omega}_i \cap \bar{\Omega}_j$$
$$= \{\mathbf{x} \mid \varphi_{ij}(\mathbf{x}, t, \lambda) = 0, \varphi_{ij} \text{ is } C^r\text{-continuous } (r \geqslant 1)\} \subset \mathbb{R}^{n-1}. \quad (2.4)$$

Based on the boundary definition, we have $\partial \Omega_{ij} = \partial \Omega_{ji}$. On the separation boundary $\partial \Omega_{ij}$ with $\varphi_{ij}(\mathbf{x}, t, \lambda) = 0$, there is a dynamical system as

$$\dot{\mathbf{x}}^{(0)} = \mathbf{F}^{(0)}(\mathbf{x}^{(0)}, t, \lambda) \quad (2.5)$$

where $\mathbf{x}^{(0)} = (x_1^{(0)}, x_2^{(0)}, \ldots, x_n^{(0)})^{\mathrm{T}}$. The flow of $\mathbf{x}^{(0)}(t) = \mathbf{\Phi}^{(0)}(\mathbf{x}^{(0)}(t_0), t, \lambda)$ with $\mathbf{x}^{(0)}(t_0) = \mathbf{\Phi}^{(0)}(\mathbf{x}^{(0)}(t_0), t_0, \lambda)$ is C^{r+1}-continuous for time t.

2.2 G-functions

Consider two infinitesimal time intervals $[t_m - \varepsilon, t_m)$ and $(t_m, t_m + \varepsilon]$. There are two flows in domain $\Omega_\alpha (\alpha = i, j)$ and on the boundary $\partial \Omega_{ij}$ in Eqs.(2.1) and (2.5). As in Luo (2008a,b), the vector difference between the two flows for three time instants are given by $\mathbf{x}_{t_m-\varepsilon}^{(\alpha)} - \mathbf{x}_{t_m-\varepsilon}^{(0)}, \mathbf{x}_{t_m}^{(\alpha)} - \mathbf{x}_{t_m}^{(0)}$ and $\mathbf{x}_{t_m+\varepsilon}^{(\alpha)} - \mathbf{x}_{t_m+\varepsilon}^{(0)}$. The normal vector of the boundary relative to the corresponding flow $\mathbf{x}^{(0)}(t)$ are expressed by ${}^{t_m-\varepsilon}\mathbf{n}_{\partial\Omega_{ij}}$, ${}^{t_m}\mathbf{n}_{\partial\Omega_{ij}}$ and ${}^{t_m+\varepsilon}\mathbf{n}_{\partial\Omega_{ij}}$, respectively, and the corresponding tangential vectors of the flow $\mathbf{x}^{(0)}(t)$ on the boundary are the tangential vectors expressed as $\mathbf{t}_{\mathbf{x}^{(0)}_{t_m-\varepsilon}}$, $\mathbf{t}_{\mathbf{x}^{(0)}_{t_m}}$, and $\mathbf{t}_{\mathbf{x}^{(0)}_{t_m+\varepsilon}}$, From the normal vectors of the boundary $\partial\Omega_{ij}$, the normal component of the state vector difference between two flows in domain and on the boundary is defined as

$$\left.\begin{aligned} d^{(\alpha)}_{t_m-\varepsilon} &= {}^{t_m-\varepsilon}\mathbf{n}^T_{\partial\Omega_{ij}} \cdot (\mathbf{x}^{(\alpha)}_{t_m-\varepsilon} - \mathbf{x}^{(0)}_{t_m-\varepsilon}), \\ d^{(\alpha)}_{t_m} &= {}^{t_m}\mathbf{n}^T_{\partial\Omega_{ij}} \cdot (\mathbf{x}^{(\alpha)}_{t_m} - \mathbf{x}^{(0)}_{t_m}), \\ d^{(\alpha)}_{t_m+\varepsilon} &= {}^{t_m+\varepsilon}\mathbf{n}^T_{\partial\Omega_{ij}} \cdot (\mathbf{x}^{(\alpha)}_{t_m+\varepsilon} - \mathbf{x}^{(0)}_{t_m+\varepsilon}), \end{aligned}\right\} \qquad (2.6)$$

where the normal vector of the boundary surface $\partial\Omega_{ij}$ at point $\mathbf{x}^{(0)}(t)$ is given by

$${}^t\mathbf{n}_{\partial\Omega_{ij}}(\mathbf{x}^{(0)},t,\lambda) = \nabla\varphi_{ij}(\mathbf{x}^{(0)},t,\lambda) = \left(\frac{\partial\varphi_{ij}}{\partial x^{(0)}_1}, \frac{\partial\varphi_{ij}}{\partial x^{(0)}_2}, \ldots, \frac{\partial\varphi_{ij}}{\partial x^{(0)}_n}\right)^T_{(t,\mathbf{x}^{(0)})} \qquad (2.7)$$

for time t. The normal component is the distance of the two points of two flows in the normal direction of the boundary surface.

Definition 2.2. Consider a dynamical system in Eq.(2.1) in domain Ω_α ($\alpha \in \{i,j\}$) which has the flow $\mathbf{x}^{(\alpha)}_t = \mathbf{\Phi}(t_0,\mathbf{x}^{(\alpha)}_0,\mathbf{p},t)$ with the initial condition $(t_0,\mathbf{x}^{(\alpha)}_0)$, and on the boundary $\partial\Omega_{ij}$, there is a flow $\mathbf{x}^{(0)}_t = \mathbf{\Phi}(t_0,\mathbf{x}^{(0)}_0,\lambda,t)$ with the initial condition $(t_0,\mathbf{x}^{(0)}_0)$. For an arbitrarily small $\varepsilon > 0$, there are two time intervals $[t-\varepsilon,t)$ or $(t,t+\varepsilon]$ for flow $\mathbf{x}^{(\alpha)}_t$ ($\alpha \in \{i,j\}$). The G-functions ($G^{(\alpha)}_{\partial\Omega_{ij}}$) of the flow $\mathbf{x}^{(\alpha)}_t$ to the flow $\mathbf{x}^{(0)}_t$ on the boundary in the normal direction of the boundary $\partial\Omega_{ij}$ are defined as

$$\begin{aligned} &G^{(\alpha)}_{\partial\Omega_{ij}}(\mathbf{x}^{(0)}_t,t_-,\mathbf{x}^{(\alpha)}_{t_-},\mathbf{p}_\alpha,\lambda) \\ &= \lim_{\varepsilon\to 0}\frac{1}{\varepsilon}\left[{}^t\mathbf{n}^T_{\partial\Omega_{ij}}\cdot(\mathbf{x}^{(\alpha)}_{t_-}-\mathbf{x}^{(0)}_t) - {}^{t-\varepsilon}\mathbf{n}^T_{\partial\Omega_{ij}}\cdot(\mathbf{x}^{(\alpha)}_{t-\varepsilon}-\mathbf{x}^{(0)}_{t-\varepsilon})\right], \\ &G^{(\alpha)}_{\partial\Omega_{ij}}(\mathbf{x}^{(0)}_t,t_+,\mathbf{x}^{(\alpha)}_{t_+},\mathbf{p}_\alpha,\lambda) \\ &= \lim_{\varepsilon\to 0}\frac{1}{\varepsilon}\left[{}^{t+\varepsilon}\mathbf{n}^T_{\partial\Omega_{ij}}\cdot(\mathbf{x}^{(\alpha)}_{t+\varepsilon}-\mathbf{x}^{(0)}_{t+\varepsilon}) - {}^t\mathbf{n}^T_{\partial\Omega_{ij}}\cdot(\mathbf{x}^{(\alpha)}_{t_+}-\mathbf{x}^{(0)}_t)\right]. \end{aligned} \qquad (2.8)$$

$t_\pm = t \pm 0$, which means the flow in the domain instead of on the boundary. From Eq.(2.8), since $\mathbf{x}^{(\alpha)}_t$ and $\mathbf{x}^{(0)}_t$ are the solutions of Eqs.(2.1) and (2.5), their derivatives exist. Further, by use of the Taylor series expansion, equation(2.8) gives

$$\begin{aligned} &G^{(\alpha)}_{\partial\Omega_{ij}}(\mathbf{x}^{(0)}_t,t_\pm,\mathbf{x}^{(\alpha)}_{t_\pm},\mathbf{p}_\alpha,\lambda) \\ &= D_{\mathbf{x}^{(0)}_t}{}^t\mathbf{n}^T_{\partial\Omega_{ij}}\cdot(\mathbf{x}^{(\alpha)}_{t_\pm}-\mathbf{x}^{(0)}_t) + {}^t\mathbf{n}^T_{\partial\Omega_{ij}}\cdot(\dot{\mathbf{x}}^{(\alpha)}_{t_\pm}-\dot{\mathbf{x}}^{(0)}_t), \end{aligned} \qquad (2.9)$$

2.2 G-functions

where the total derivative $D_{\mathbf{x}_t^{(0)}}(\cdot) = \frac{\partial(\cdot)}{\partial \mathbf{x}_t^{(0)}} \dot{\mathbf{x}}_t^{(0)} + \frac{\partial(\cdot)}{\partial t}$. Using Eqs.(2.1) and (2.5), the G-function in Eq.(2.9) becomes

$$G_{\partial \Omega_{ij}}^{(\alpha)}(\mathbf{x}_t^{(0)}, t_\pm, \mathbf{x}_{t_\pm}^{(\alpha)}, \mathbf{p}_\alpha, \boldsymbol{\lambda})$$
$$= \left[\frac{\partial^t \mathbf{n}_{\partial \Omega_{ij}}^T}{\partial \mathbf{x}_t^{(0)}} \dot{\mathbf{x}}_t^{(0)} + \frac{\partial^t \mathbf{n}_{\partial \Omega_{ij}}^T}{\partial t} \right] \cdot (\mathbf{x}_{t_\pm}^{(\alpha)} - \mathbf{x}_t^{(0)})$$
$$+ {}^t\mathbf{n}_{\partial \Omega_{ij}}^T \cdot (\mathbf{F}^{(\alpha)}(\mathbf{x}^{(\alpha)}, t_\pm, \mathbf{p}) - \mathbf{F}^{(0)}(\mathbf{x}^{(0)}, t, \boldsymbol{\lambda})). \quad (2.10)$$

If a flow $\mathbf{x}^{(\alpha)}(t)$ approaches the separation boundary with the zero-order contact at t_m (i.e., $\mathbf{x}^{(\alpha)}(t_{m-}) = \mathbf{x}_m = \mathbf{x}^{(0)}(t_m)$), the G-function of the zero-order is defined as

$$G_{\partial \Omega_{ij}}^{(\alpha)}(\mathbf{x}_m, t_{m\pm}, \mathbf{p}_\alpha, \boldsymbol{\lambda})$$
$$\equiv \mathbf{n}_{\partial \Omega_{ij}}^T(\mathbf{x}^{(0)}, t, \boldsymbol{\lambda}) \cdot [\dot{\mathbf{x}}^{(\alpha)}(t) - \dot{\mathbf{x}}^{(0)}(t)]\Big|_{(\mathbf{x}_m^{(0)}, \mathbf{x}_m^{(\alpha)}, t_{m\pm})}$$
$$= \mathbf{n}_{\partial \Omega_{ij}}^T(\mathbf{x}^{(0)}, t, \boldsymbol{\lambda}) \cdot \dot{\mathbf{x}}^{(\alpha)}(t) + \frac{\partial \varphi_{ij}(\mathbf{x}^{(0)}, t, \boldsymbol{\lambda})}{\partial t}\Big|_{(\mathbf{x}_m^{(0)}, \mathbf{x}_m^{(\alpha)}, t_{m\pm})}$$
$$= \nabla \varphi_{ij}(\mathbf{x}^{(0)}, t, \boldsymbol{\lambda}) \cdot \dot{\mathbf{x}}^{(\alpha)}(t) + \frac{\partial \varphi_{ij}(\mathbf{x}^{(0)}, t, \boldsymbol{\lambda})}{\partial t}\Big|_{(\mathbf{x}_m^{(0)}, \mathbf{x}_m^{(\alpha)}, t_{m\pm})}. \quad (2.11)$$

With Eqs.(2.1) and (2.5), equation(2.11) can be rewritten as

$$G_{\partial \Omega_{ij}}^{(\alpha)}(\mathbf{x}_m, t_{m\pm}, \mathbf{p}_\alpha, \boldsymbol{\lambda})$$
$$\equiv \mathbf{n}_{\partial \Omega_{ij}}^T(\mathbf{x}^{(0)}, t, \boldsymbol{\lambda}) \cdot [\mathbf{F}(\mathbf{x}^{(\alpha)}, t, \mathbf{p}_\alpha) - \mathbf{F}^{(0)}(\mathbf{x}^{(0)}, t, \boldsymbol{\lambda})]\Big|_{(\mathbf{x}_m^{(0)}, \mathbf{x}_m^{(\alpha)}, t_{m\pm})}$$
$$= \mathbf{n}_{\partial \Omega_{ij}}^T(\mathbf{x}^{(0)}, t, \boldsymbol{\lambda}) \cdot \mathbf{F}(\mathbf{x}^{(\alpha)}, t, \mathbf{p}_\alpha) + \frac{\partial \varphi_{ij}(\mathbf{x}^{(0)}, t, \boldsymbol{\lambda})}{\partial t}\Big|_{(\mathbf{x}_m^{(0)}, \mathbf{x}_m^{(\alpha)}, t_{m\pm})}$$
$$= \nabla \varphi_{ij}(\mathbf{x}^{(0)}, t, \boldsymbol{\lambda}) \cdot \mathbf{F}(\mathbf{x}^{(\alpha)}, t, \mathbf{p}_\alpha) + \frac{\partial \varphi_{ij}(\mathbf{x}^{(0)}, t, \boldsymbol{\lambda})}{\partial t}\Big|_{(\mathbf{x}_m^{(0)}, \mathbf{x}_m^{(\alpha)}, t_{m\pm})}. \quad (2.12)$$

Here, $t_{m\pm} = t_m \pm 0$ reflects the responses in domains rather than on the boundary.

Definition 2.3. Consider a dynamical system in Eq.(2.1) in domain Ω_α ($\alpha \in \{i, j\}$) which has the flow $\mathbf{x}_t^{(\alpha)} = \Phi(t_0, \mathbf{x}_0^{(\alpha)}, \mathbf{p}, t)$ with the initial condition $(t_0, \mathbf{x}_0^{(\alpha)})$ and on the boundary $\partial \Omega_{ij}$, there is a flow $\mathbf{x}_t^{(0)} = \Phi(t_0, \mathbf{x}_0^{(0)}, \boldsymbol{\lambda}, t)$ with the initial condition $(t_0, \mathbf{x}_0^{(0)})$. For an arbitrarily small $\varepsilon > 0$, there are two time intervals $[t - \varepsilon, t)$ for flow $\mathbf{x}_t^{(\alpha)}$ ($\alpha \in \{i, j\}$) and $(t, t + \varepsilon]$ for flow $\mathbf{x}_t^{(\beta)}$ ($\beta \in \{i, j\}$). The vector fields $\mathbf{F}^{(\alpha)}(\mathbf{x}^{(\alpha)}, t, \mathbf{p}_\alpha)$ and $\mathbf{F}^{(0)}(\mathbf{x}^{(0)}, t, \boldsymbol{\lambda})$ are $C_{[t-\varepsilon, t+\varepsilon]}^r$-continuous ($r_\alpha \geq k$) for time t. The flow $\mathbf{x}_t^{(\alpha)}$ ($\alpha \in \{i, j\}$) and $\mathbf{x}_t^{(0)}$ are $C_{[t-\varepsilon, t)}^{r_\alpha}$ or $C_{(t, t+\varepsilon]}^{r_\alpha}$-continuous ($r_\alpha \geq k+1$) for time

t, $||d^{r_\alpha+1}\mathbf{x}_t^{(\alpha)}/dt^{r_\alpha+1}|| < \infty$ and $||d^{r_\alpha+1}\mathbf{x}_t^{(0)}/dt^{r_\alpha+1}|| < \infty$. The G-functions of kth-order for a flow $\mathbf{x}_t$ to a boundary flow $\mathbf{x}_t^{(0)}$ in the normal direction of the boundary $\partial\Omega_{ij}$ are defined as

$$\left. \begin{aligned} & G^{(k,\alpha)}_{\partial\Omega_{ij}}(\mathbf{x}_t^{(0)}, t_-, \mathbf{x}_{t_-}^{(\alpha)}, \mathbf{p}_\alpha, \boldsymbol{\lambda}) \\ &= \lim_{\varepsilon \to 0} \frac{(-1)^{k+2}}{\varepsilon^{k+1}} \left[{}^t\mathbf{n}^{\text{T}}_{\partial\Omega_{ij}} \cdot (\mathbf{x}_{t_-}^{(\alpha)} - \mathbf{x}_t^{(0)}) - {}^{t-\varepsilon}\mathbf{n}^{\text{T}}_{\partial\Omega_{ij}} \cdot (\mathbf{x}_{t-\varepsilon}^{(\alpha)} - \mathbf{x}_{t-\varepsilon}^{(0)}) \right. \\ &\quad + \sum_{s=0}^{k-1} G^{(s,\alpha)}_{\partial\Omega_{ij}}(\mathbf{x}_t^{(0)}, t_-, \mathbf{x}_{t_-}^{(\alpha)}, \mathbf{p}_\alpha, \boldsymbol{\lambda})(-\varepsilon)^{s+1} \bigg], \text{ or} \\ & G^{(k,\alpha)}_{\partial\Omega_{ij}}(\mathbf{x}_t^{(0)}, t_+, \mathbf{x}_{t_+}^{(\alpha)}, \mathbf{p}_\alpha, \boldsymbol{\lambda}) \\ &= \lim_{\varepsilon \to 0} \frac{1}{\varepsilon^{k+1}} \left[{}^{t+\varepsilon}\mathbf{n}^{\text{T}}_{\partial\Omega_{ij}} \cdot (\mathbf{x}_{t+\varepsilon}^{(\alpha)} - \mathbf{x}_{t+\varepsilon}^{(0)}) - {}^t\mathbf{n}^{\text{T}}_{\partial\Omega_{ij}} \cdot (\mathbf{x}_{t_+}^{(\alpha)} - \mathbf{x}_t^{(0)}) \right. \\ &\quad - \sum_{s=0}^{k-1} G^{(s,\alpha)}_{\partial\Omega_{ij}}(\mathbf{x}_t^{(0)}, t_+, \mathbf{x}_{t_+}^{(\alpha)}, \mathbf{p}_\alpha, \boldsymbol{\lambda})\varepsilon^{s+1} \bigg]. \end{aligned} \right\} \quad (2.13)$$

Again, the Taylor series expansion applying to Eq.(2.13) yields

$$G^{(k,\alpha)}_{\partial\Omega_{ij}}(\mathbf{x}_t^{(0)}, t_\pm, \mathbf{x}_{t_\pm}^{(\alpha)}, \mathbf{p}_\alpha, \boldsymbol{\lambda})$$
$$= \sum_{s=0}^{k+1} C^s_{k+1} D^{k+1-s}_{\mathbf{x}_t^{(0)}} {}^t\mathbf{n}^{\text{T}}_{\partial\Omega_{ij}} \cdot \left(\frac{d^s \mathbf{x}_t^{(\alpha)}}{dt^s} - \frac{d^s \mathbf{x}_t^{(0)}}{dt^s} \right) \bigg|_{(\mathbf{x}_t^{(0)}, t, \mathbf{x}_{t_\pm}^{(\alpha)})} \quad (2.14)$$

Using Eqs.(2.1) and (2.5), the kth-order G-function becomes

$$G^{(k,\alpha)}_{\partial\Omega_{ij}}(\mathbf{x}_t^{(0)}, t_\pm, \mathbf{x}_{t_\pm}^{(\alpha)}, \mathbf{p}_\alpha, \boldsymbol{\lambda}) = \sum_{s=1}^{k+1} C^s_{k+1} D^{k+1-s}_{\mathbf{x}_t^{(0)}} {}^t\mathbf{n}^{\text{T}}_{\partial\Omega_{ij}} \cdot (D^{s-1}_{\mathbf{x}_t^{(\alpha)}} \mathbf{F}(\mathbf{x}_t^{(\alpha)}, t, \mathbf{p})$$
$$- D^{s-1}_{\mathbf{x}_t^{(0)}} \mathbf{F}^{(0)}(\mathbf{x}_t^{(0)}, t, \boldsymbol{\mu})) \bigg|_{(\mathbf{x}_t^{(0)}, t_\pm, \mathbf{x}_{t_\pm}^{(\alpha)})} + D^{k+1}_{\mathbf{x}_t^{(0)}} {}^t\mathbf{n}^{\text{T}}_{\partial\Omega_{ij}} \cdot (\mathbf{x}_{t_\pm}^{(\alpha)} - \mathbf{x}_t^{(0)}), \quad (2.15)$$

with $C^s_{k+1} = \dfrac{(k+1)k(k-1)\cdots(k+2-s)}{s!}$ and $C^0_{k+1} = 1$ with $s! = 1 \times 2 \times \cdots \times s$. The G-function $G^{(k,\alpha)}_{\partial\Omega_{ij}}$ is the time-rate of $G^{(k-1,\alpha)}_{\partial\Omega_{ij}}$. If the flow contacting with the boundary $\partial\Omega_{ij}$ at time t_m (i.e., $\mathbf{x}_{t_m}^{(\alpha)} = \mathbf{x}_{t_m}^{(0)}$) and ${}^t\mathbf{n}^{\text{T}}_{\partial\Omega_{ij}} \equiv \mathbf{n}^{\text{T}}_{\partial\Omega_{ij}}$, the kth-order G-function is computed by

$$G^{(k,\alpha)}_{\partial\Omega_{ij}}(\mathbf{x}_m, t_{m\pm}, \mathbf{p}_\alpha, \boldsymbol{\lambda})$$
$$= \sum_{r=1}^{k+1} C^r_{k+1} D^{k+1-s}_{\mathbf{x}^{(0)}} \mathbf{n}^{\text{T}}_{\partial\Omega_{ij}} \cdot \left(\frac{d^s \mathbf{x}}{dt^s} - \frac{d^s \mathbf{x}^{(0)}}{dt^r} \right) \bigg|_{(\mathbf{x}_m^{(0)}, \mathbf{x}_{m\pm}^{(\alpha)}, t_{m\pm})}$$

2.3 Passable flows

$$= \sum_{s=1}^{k+1} C_{k+1}^s D_{\mathbf{x}^{(0)}}^{k+1-s} \mathbf{n}_{\partial \Omega_{ij}}^T \cdot \left[D_{\mathbf{x}}^{s-1} \mathbf{F}(\mathbf{x},t,\mathbf{p}_\alpha) - D_{\mathbf{x}^{(0)}}^{s-1} \mathbf{F}^{(0)}(\mathbf{x}^{(0)},t,\lambda) \right] \Big|_{(\mathbf{x}_m^{(0)}, \mathbf{x}_{m\pm}^{(\alpha)}, t_{m\pm})}. \tag{2.16}$$

For $k = 0$, we have $G_{\partial \Omega_{ij}}^{(\alpha,k)}(\mathbf{x}_m, t_{m\pm}, \mathbf{p}_\alpha, \lambda) = G_{\partial \Omega_{ij}}^{(\alpha)}(\mathbf{x}_m, t_{m\pm}, \mathbf{p}_\alpha, \lambda)$.

2.3 Passable flows

Before we discuss the flow passability in discontinuous dynamical systems, as Luo (2005a,b, 2006), the imaginary flow is introduced first.

Definition 2.4. The $C^{r_j+1}(r_j \geq 1)$-continuous flow $\mathbf{x}_i^{(j)}(t)$ is termed the jth-imaginary flow in the ith open sub-domain Ω_i if a flow $\mathbf{x}_i^{(j)}(t)$ is determined by application of the C^{r_j}-continuous vector field in the jth open sub-domain Ω_j, to the ith open sub-domain Ω_i, i.e.,

$$\dot{\mathbf{x}}_i^{(j)} \equiv \mathbf{F}^{(j)}\left(\mathbf{x}_i^{(j)}, t, \boldsymbol{\mu}_j\right) \in \mathbb{R}^n, \mathbf{x}_i^{(j)} = \left(x_{1i}^{(j)}, x_{2i}^{(j)}, \ldots, x_{ni}^{(j)}\right)^T \in \Omega_i, \tag{2.17}$$

with the initial condition

$$\mathbf{x}_i^{(j)}(t_0) = \boldsymbol{\Phi}^{(j)}(\mathbf{x}_i^{(j)}(t_0), t_0, \mathbf{p}_j). \tag{2.18}$$

For discontinuous dynamical systems, a passable flow to the separation boundary is discussed first, as sketched in Fig.2.3. The *real* flows $\mathbf{x}^{(i)}(t)$ and $\mathbf{x}^{(j)}(t)$ in Ω_i and Ω_j are depicted by thin solid curves. The *imaginary* flows $\mathbf{x}_i^{(j)}(t)$ and $\mathbf{x}_j^{(i)}(t)$ in Ω_i and Ω_j are controlled respectively by the vector fields in Ω_j and Ω_i, which are depicted by dashed curves. The hollow and shaded circles are switching points and starting points, respectively. The detail discussion of the real and imaginary flows can be found in Luo (2005b, 2006). The flow on the boundary is described by the thick curve. The normal and tangential vectors $\mathbf{n}_{\partial \Omega_{ij}}$ and $\mathbf{t}_{\partial \Omega_{ij}}$ on the boundary are depicted as well. The definition of the passable flow to the boundary is given as follows.

Definition 2.5. For a discontinuous dynamical system in Eq.(2.1), $\mathbf{x}(t_m) \equiv \mathbf{x}_m \in \partial \Omega_{ij}$ at time t_m. For an arbitrarily small $\varepsilon > 0$, there are two time intervals $[t_{m-\varepsilon}, t_m)$ and $(t_m, t_{m+\varepsilon}]$. Suppose $\mathbf{x}^{(i)}(t_{m-}) = \mathbf{x}_m = \mathbf{x}^{(j)}(t_{m+})$. The flow $\mathbf{x}^{(i)}(t)$ and $\mathbf{x}^{(j)}(t)$ to $\partial \Omega_{ij}$ is *semi-passable* from domain Ω_i to Ω_j (or the semi-possible flow) if

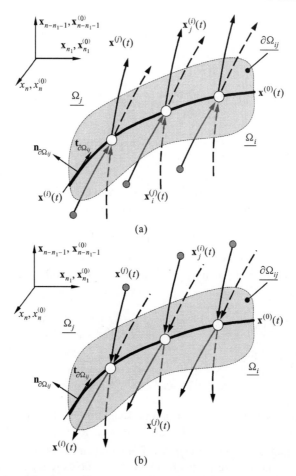

Fig. 2.3 Passable flows: (a) from Ω_i to Ω_j with $(2k_i : 2k_j)$-order and (b) from Ω_j to Ω_i with $(2k_j : 2k_i)$-order. $\mathbf{x}^{(i)}(t)$ and $\mathbf{x}^{(j)}(t)$ represent the *real* flows in domains Ω_i and Ω_j, respectively, which are depicted by the thin solid curves. $\mathbf{x}_i^{(j)}(t)$ and $\mathbf{x}_j^{(i)}(t)$ represent the *imaginary* flows in domains Ω_i and Ω_j, respectively controlled by the vector fields in Ω_j and Ω_i, which are depicted by the dashed curves. The flow on the boundary is described by $\mathbf{x}^{(0)}(t)$. The normal and tangential vectors $\mathbf{n}_{\partial\Omega_{ij}}$ and $\mathbf{t}_{\partial\Omega_{ij}}$ on the boundary are depicted. The hollow circles are the switching points and the shaded circles are the starting points

$$\left.\begin{array}{l}\text{either }\left.\begin{array}{l}\mathbf{n}_{\partial\Omega_{ij}}^{\mathrm{T}}(\mathbf{x}_{m-\varepsilon}^{(0)})\cdot\left[\mathbf{x}_{m-\varepsilon}^{(0)}-\mathbf{x}_{m-\varepsilon}^{(i)}\right]>0\\ \mathbf{n}_{\partial\Omega_{ij}}^{\mathrm{T}}(\mathbf{x}_{m+\varepsilon}^{(0)})\cdot\left[\mathbf{x}_{m+\varepsilon}^{(j)}-\mathbf{x}_{m+\varepsilon}^{(0)}\right]>0\end{array}\right\}\text{ for }\mathbf{n}_{\partial\Omega_{ij}}\to\Omega_j\\ \text{or }\left.\begin{array}{l}\mathbf{n}_{\partial\Omega_{ij}}^{\mathrm{T}}(\mathbf{x}_{m-\varepsilon}^{(0)})\cdot\left[\mathbf{x}_{m-\varepsilon}^{(0)}-\mathbf{x}_{m-\varepsilon}^{(i)}\right]<0\\ \mathbf{n}_{\partial\Omega_{ij}}^{\mathrm{T}}(\mathbf{x}_{m+\varepsilon}^{(0)})\cdot\left[\mathbf{x}_{m+\varepsilon}^{(j)}-\mathbf{x}_{m+\varepsilon}^{(0)}\right]<0\end{array}\right\}\text{ for }\mathbf{n}_{\partial\Omega_{ij}}\to\Omega_i.\end{array}\right\} \quad (2.19)$$

2.3 Passable flows

Because flow properties in domains Ω_i and Ω_j are different, so at point $(t_m, \mathbf{x}_m)$ we have $G^{(i)}_{\partial \Omega_{ij}} \neq G^{(j)}_{\partial \Omega_{ij}}$ on the boundary $\partial \Omega_{ij}$. The necessary and sufficient conditions for such a passable flow are given as follows.

Theorem 2.1. *For a discontinuous dynamical system in Eq.(2.1)*, $\mathbf{x}(t_m) = \mathbf{x}_m \in \partial \Omega_{ij}$ *at time* t_m. *For an arbitrarily small* $\varepsilon > 0$, *there are two time intervals* $[t_{m-\varepsilon}, t_m)$ *and* $(t_m, t_{m+\varepsilon}]$. *Suppose* $\mathbf{x}^{(i)}(t_{m-}) = \mathbf{x}_m = \mathbf{x}^{(j)}(t_{m+})$. *Both flows* $\mathbf{x}^{(i)}(t)$ *and* $\mathbf{x}^{(j)}(t)$ *are* $C^r_{[t_{m-\varepsilon}, t_m)}$ *and* $C^r_{(t_m, t_{m+\varepsilon}]}$-*continuous* $(r \geqslant 1)$ *for time* t, *respectively, and* $\|d^{r+1}\mathbf{x}^{(\alpha)}/dt^{r+1}\| < \infty (\alpha \in \{i, j\})$. *The flow* $\mathbf{x}^{(i)}(t)$ *and* $\mathbf{x}^{(j)}(t)$ *to the boundary* $\partial \Omega_{ij}$ *is semi-passable from* Ω_i *to* Ω_j *iff*

$$\begin{aligned}\text{either} &\left.\begin{array}{l} G^{(i)}_{\partial \Omega_{ij}}(\mathbf{x}_m, t_{m-}, \mathbf{p}_i, \lambda) > 0 \\ G^{(j)}_{\partial \Omega_{ij}}(\mathbf{x}_m, t_{m+}, \mathbf{p}_j, \lambda) > 0 \end{array}\right\} \text{for } \mathbf{n}_{\partial \Omega_{ij}} \to \Omega_j \\ \text{or} &\left.\begin{array}{l} G^{(i)}_{\partial \Omega_{ij}}(\mathbf{x}_m, t_{m-}, \mathbf{p}_i, \lambda) < 0 \\ G^{(j)}_{\partial \Omega_{ij}}(\mathbf{x}_m, t_{m+}, \mathbf{p}_j, \lambda) < 0 \end{array}\right\} \text{for } \mathbf{n}_{\partial \Omega_{ij}} \to \Omega_i.\end{aligned} \quad (2.20)$$

Proof. See Luo (2008a,b).

Definition 2.6. For a discontinuous dynamical system in Eq.(2.1), $\mathbf{x}(t_m) \equiv \mathbf{x}_m \in \partial \Omega_{ij}$ at time t_m. Suppose $\mathbf{x}^{(i)}(t_{m-}) = \mathbf{x}_m = \mathbf{x}^{(j)}(t_{m+})$. For an arbitrarily small $\varepsilon > 0$, there are two time intervals $[t_{m-\varepsilon}, t_m)$ and $(t_m, t_{m+\varepsilon}]$. The flow $\mathbf{x}^{(i)}(t)$ is $C^{r_i}_{[t_{m-\varepsilon}, t_m)}$-continuous $(r_i \geqslant 2k_i + 1)$ for time t and $\|d^{r_i+1}\mathbf{x}^{(i)}/dt^{r_i+1}\| < \infty$. The flow $\mathbf{x}^{(j)}(t)$ is $C^{r_j}_{(t_m, t_{m+\varepsilon}]}$-continuous for time t and $\|d^{r_j+1}\mathbf{x}^{(j)}/dt^{r_j+1}\| < \infty$ $(r_j \geqslant m_j + 1)$. The flow $\mathbf{x}^{(i)}(t)$ of the $(2k_i)^{\text{th}}$-order singularity and $\mathbf{x}^{(j)}(t)$ of the $(m_j)^{\text{th}}$-order singularity to the boundary $\partial \Omega_{ij}$ is $(2k_i : m_j)$-*semi-passable* from domain Ω_i to Ω_j if

$$\left.\begin{array}{l} G^{(s,i)}_{\partial \Omega_{ij}}(\mathbf{x}_m, t_{m-}, \mathbf{p}_i, \lambda) = 0, \text{ for } s = 0, 1, \ldots, 2k_i - 1, \\ G^{(2k_i, i)}_{\partial \Omega_{ij}}(\mathbf{x}_m, t_{m-}, \mathbf{p}_i, \lambda) \neq 0, \end{array}\right\} \quad (2.21)$$

$$\left.\begin{array}{l} G^{(s,j)}_{\partial \Omega_{ij}}(\mathbf{x}_m, t_{m+}, \mathbf{p}_j, \lambda) = 0, \text{ for } s = 0, 1, \ldots, m_j - 1, \\ G^{(m_j, j)}_{\partial \Omega_{ij}}(\mathbf{x}_m, t_{m+}, \mathbf{p}_j, \lambda) \neq 0, \end{array}\right\} \quad (2.22)$$

$$\begin{aligned}\text{either} &\left.\begin{array}{l} \mathbf{n}^{\text{T}}_{\partial \Omega_{ij}}(\mathbf{x}^{(0)}_{m-\varepsilon}) \cdot [\mathbf{x}^{(0)}_{m-\varepsilon} - \mathbf{x}^{(i)}_{m-\varepsilon}] > 0 \\ \mathbf{n}^{\text{T}}_{\partial \Omega_{ij}}(\mathbf{x}^{(0)}_{m+\varepsilon}) \cdot [\mathbf{x}^{(j)}_{m+\varepsilon} - \mathbf{x}^{(0)}_{m+\varepsilon}] > 0 \end{array}\right\} \text{for } \mathbf{n}_{\partial \Omega_{ij}} \to \Omega_j \\ \text{or} &\left.\begin{array}{l} \mathbf{n}^{\text{T}}_{\partial \Omega_{ij}}(\mathbf{x}^{(0)}_{m-\varepsilon}) \cdot [\mathbf{x}^{(0)}_{m-\varepsilon} - \mathbf{x}^{(i)}_{m-\varepsilon}] < 0 \\ \mathbf{n}^{\text{T}}_{\partial \Omega_{ij}}(\mathbf{x}^{(0)}_{m+\varepsilon}) \cdot [\mathbf{x}^{(j)}_{m+\varepsilon} - \mathbf{x}^{(0)}_{m+\varepsilon}] < 0 \end{array}\right\} \text{for } \mathbf{n}_{\partial \Omega_{ij}} \to \Omega_i.\end{aligned} \quad (2.23)$$

If $m_j = 2k_j$, the $(2k_i : 2k_j)$-passable flow can be shown in Fig.2.3. However, for $m_j = 2k_j - 1$, the $(2k_i : 2k_j - 1)$-passable flow from Ω_i to Ω_j is sketched in Fig.2.4(a). The tangential flow of the $(2k_j - 1)^{\text{th}}$-order exists in Ω_j. The dotted curves represent tangential flow for $t \in [t_{m-\varepsilon}, t_m)$. The starting point of a flow is $(t_{m-\varepsilon}, \mathbf{x}_{m-\varepsilon}^{(i)})$ in Ω_i. Once a flow arrives to a point $(t_m, \mathbf{x}_m)$ of $\partial \Omega_{ij}$, such a flow will follow the tangential flow in Ω_j. The $(2k_j : 2k_i - 1)$-passable flow from Ω_j to Ω_i is presented in Fig.2.4(b) with the same behavior as in Fig.2.4(a). The necessary and sufficient conditions for the $(2k_i : m_j)$-passable flow are given as follows.

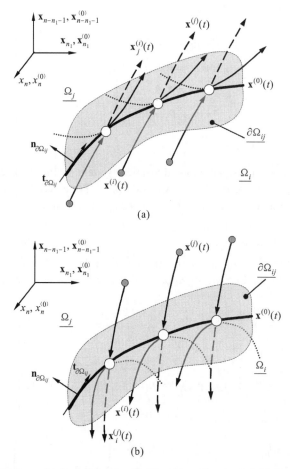

Fig. 2.4 Passable flows: (a) from Ω_i to Ω_j with $(2k_i : 2k_j - 1)$-order and (b) from Ω_j to Ω_i with $(2k_j : 2k_i - 1)$-order. $\mathbf{x}^{(i)}(t)$ and $\mathbf{x}^{(j)}(t)$ represent the *real* flows in domains Ω_i and Ω_j, respectively, which are depicted by the thin solid curves. $\mathbf{x}_i^{(j)}(t)$ and $\mathbf{x}_j^{(i)}(t)$ represent the *imaginary* flows in domains Ω_i and Ω_j, respectively controlled by the vector fields in Ω_j and Ω_i, which are depicted by the dashed curves. The flow on the boundary is described by $\mathbf{x}^{(0)}(t)$. The normal and tangential vectors $\mathbf{n}_{\partial \Omega_{ij}}$ and $\mathbf{t}_{\partial \Omega_{ij}}$ on the boundary are depicted. The hollow circles are the switching points and the shaded circles are the starting points

2.4 Non-passable flows

Theorem 2.2. *For a discontinuous dynamical system in Eq.(2.1), $\mathbf{x}(t_m) \equiv \mathbf{x}_m \in \partial\Omega_{ij}$ at time t_m. For an arbitrarily small $\varepsilon > 0$, there are two time intervals $[t_{m-\varepsilon}, t_m)$ and $(t_m, t_{m+\varepsilon}]$. Suppose $\mathbf{x}^{(i)}(t_{m-}) = \mathbf{x}_m = \mathbf{x}^{(j)}(t_{m+})$. The flow $\mathbf{x}^{(i)}(t)$ is $C^{r_i}_{[t_{m-\varepsilon}, t_m)}$-continuous for time t and $\|d^{r_i+1}\mathbf{x}^{(i)}/dt^{r_i+1}\| < \infty$ ($r_i \geq 2k_i + 1$). The flow $\mathbf{x}^{(j)}(t)$ is $C^{r_j}_{(t_m, t_{m+\varepsilon}]}$-continuous ($r_j \geq m_j + 1$) for time t and $\|d^{r_j+1}\mathbf{x}^{(j)}/dt^{r_j+1}\| < \infty$. The flow $\mathbf{x}^{(i)}(t)$ of the $(2k_i)^{\text{th}}$-order singularity and $\mathbf{x}^{(j)}(t)$ of the $(m_j)^{\text{th}}$-order singularity to the boundary $\partial\Omega_{ij}$ is $(2k_i : m_j)$-semi-passable from domain Ω_i to Ω_j iff*

$$G^{(s,i)}_{\partial\Omega_{ij}}(\mathbf{x}_m, t_{m-}, \mathbf{p}_i, \boldsymbol{\lambda}) = 0, \text{ for } s = 0, 1, \ldots, 2k_i - 1, \tag{2.24}$$

$$G^{(s,j)}_{\partial\Omega_{ij}}(\mathbf{x}_m, t_{m+}, \mathbf{p}_j, \boldsymbol{\lambda}) = 0, \text{ for } s = 0, 1, \ldots, m_j - 1, \tag{2.25}$$

$$\left. \begin{array}{l} \text{either } \left. \begin{array}{l} G^{(2k_i,i)}_{\partial\Omega_{ij}}(\mathbf{x}_m, t_{m-}, \mathbf{p}_i, \boldsymbol{\lambda}) > 0 \\ G^{(m_j,j)}_{\partial\Omega_{ij}}(\mathbf{x}_m, t_{m+}, \mathbf{p}_j, \boldsymbol{\lambda}) > 0 \end{array} \right\} \text{ for } \mathbf{n}_{\partial\Omega_{ij}} \to \Omega_j \\ \text{or } \left. \begin{array}{l} G^{(2k_i,i)}_{\partial\Omega_{ij}}(\mathbf{x}_m, t_{m-}, \mathbf{p}_i, \boldsymbol{\lambda}) < 0 \\ G^{(m_j,j)}_{\partial\Omega_{ij}}(\mathbf{x}_m, t_{m+}, \mathbf{p}_j, \boldsymbol{\lambda}) < 0 \end{array} \right\} \text{ for } \mathbf{n}_{\partial\Omega_{ij}} \to \Omega_i. \end{array} \right\} \tag{2.26}$$

Proof. See Luo(2008a,b).

2.4 Non-passable flows

In this section, the non-passable flows to the boundary in discontinuous systems will be discussed, which can be referred to Luo (2008a,b) for details. The full non-passable flow of the first kind (sink flows) and of the second kind (source flows) are sketched in Fig.2.5(a) and (b), respectively. The sink flow is defined first.

Definition 2.7. *For a discontinuous dynamical system in Eq.(2.1), $\mathbf{x}(t_m) \equiv \mathbf{x}_m \in \partial\Omega_{ij}$ at time t_m. For an arbitrarily small $\varepsilon > 0$, there is a time interval $[t_{m-\varepsilon}, t_m)$. Suppose $\mathbf{x}^{(i)}(t_{m-}) = \mathbf{x}_m = \mathbf{x}^{(j)}(t_{m-})$. The flow $\mathbf{x}^{(i)}(t)$ and $\mathbf{x}^{(j)}(t)$ to the boundary $\partial\Omega_{ij}$ is non-passable with the first kind (or called the sink flow) if*

$$\left. \begin{array}{l} \text{either } \left. \begin{array}{l} \mathbf{n}^{\text{T}}_{\partial\Omega_{ij}}(\mathbf{x}^{(0)}_{m-\varepsilon}) \cdot \left[\mathbf{x}^{(0)}_{m-\varepsilon} - \mathbf{x}^{(i)}_{m-\varepsilon}\right] > 0 \\ \mathbf{n}^{\text{T}}_{\partial\Omega_{ij}}(\mathbf{x}^{(0)}_{m-\varepsilon}) \cdot \left[\mathbf{x}^{(0)}_{m-\varepsilon} - \mathbf{x}^{(j)}_{m-\varepsilon}\right] < 0 \end{array} \right\} \text{ for } \mathbf{n}_{\partial\Omega_{ij}} \to \Omega_j \\ \text{or } \left. \begin{array}{l} \mathbf{n}^{\text{T}}_{\partial\Omega_{ij}}(\mathbf{x}^{(0)}_{m-\varepsilon}) \cdot \left[\mathbf{x}^{(0)}_{m-\varepsilon} - \mathbf{x}^{(i)}_{m-\varepsilon}\right] < 0 \\ \mathbf{n}^{\text{T}}_{\partial\Omega_{ij}}(\mathbf{x}^{(0)}_{m-\varepsilon}) \cdot \left[\mathbf{x}^{(0)}_{m-\varepsilon} - \mathbf{x}^{(j)}_{m-\varepsilon}\right] > 0 \end{array} \right\} \text{ for } \mathbf{n}_{\partial\Omega_{ij}} \to \Omega_i. \end{array} \right\} \tag{2.27}$$

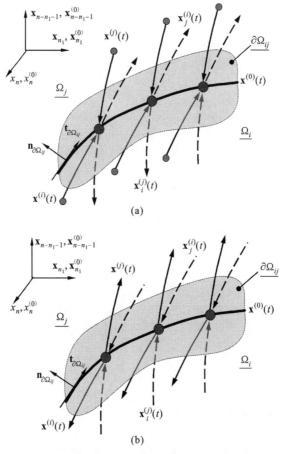

Fig. 2.5 The $(2k_i : 2k_j)$-non-passable flows of: (a) the first kind (sink flows) and (b) the second kind (source flows). $\mathbf{x}^{(i)}(t)$ and $\mathbf{x}^{(j)}(t)$ represent the *real* flows in domains Ω_i and Ω_j, respectively, which are depicted by the thin solid curves. $\mathbf{x}_i^{(j)}(t)$ and $\mathbf{x}_j^{(i)}(t)$ represent the *imaginary* flows in domains Ω_i and Ω_j, respectively controlled by the vector fields in Ω_j and Ω_i, which are depicted by the dashed curves. The flow on the boundary is described by $\mathbf{x}^{(0)}(t)$. The normal and tangential vectors $\mathbf{n}_{\partial \Omega_{ij}}$ and $\mathbf{t}_{\partial \Omega_{ij}}$ on the boundary are depicted. The black circles are sink and source points and the small shaded circles are the starting points

Theorem 2.3. *For a discontinuous dynamical system in Eq.(2.1), $\mathbf{x}(t_m) \equiv \mathbf{x}_m \in \partial \Omega_{ij}$ at time t_m. For an arbitrarily small $\varepsilon > 0$, there is a time interval $[t_{m-\varepsilon}, t_m)$. Suppose $\mathbf{x}^{(i)}(t_{m-}) = \mathbf{x}_m = \mathbf{x}^{(j)}(t_{m-})$. Both flows $\mathbf{x}^{(i)}(t)$ and $\mathbf{x}^{(j)}(t)$ are $C^r_{[t_{m-\varepsilon}, t_m)}$-continuous for time t and $||d^{r+1}\mathbf{x}^{(\alpha)}/dt^{r+1}|| < \infty$ $(\alpha \in \{i,j\}, (r \geqslant 1))$. The flow $\mathbf{x}^{(i)}(t)$ and $\mathbf{x}^{(j)}(t)$ to the boundary $\partial \Omega_{ij}$ is non-passable with the first kind iff*

2.4 Non-passable flows

$$\left.\begin{array}{l} \text{either} \quad \begin{array}{l} G^{(i)}_{\partial\Omega_{ij}}(\mathbf{x}_m, t_{m-}, \mathbf{p}_i, \lambda) > 0 \\ G^{(j)}_{\partial\Omega_{ij}}(\mathbf{x}_m, t_{m-}, \mathbf{p}_j, \lambda) < 0 \end{array} \right\} \text{for } \mathbf{n}_{\partial\Omega_{ij}} \to \Omega_j \\ \text{or} \quad \begin{array}{l} G^{(i)}_{\partial\Omega_{ij}}(\mathbf{x}_m, t_{m-}, \mathbf{p}_i, \lambda) < 0 \\ G^{(j)}_{\partial\Omega_{ij}}(\mathbf{x}_m, t_{m-}, \mathbf{p}_j, \lambda) > 0 \end{array} \right\} \text{for } \mathbf{n}_{\partial\Omega_{ij}} \to \Omega_i. \end{array} \quad (2.28)$$

Proof. See Luo (2008a,b).

The higher-order sink flow will be given by the following definition.

Definition 2.8. For a discontinuous dynamical system in Eq.(2.1), $\mathbf{x}(t_m) \equiv \mathbf{x}_m \in \partial\Omega_{ij}$ at time t_m. For an arbitrarily small $\varepsilon > 0$, there is a time interval $[t_{m-\varepsilon}, t_m)$. Suppose $\mathbf{x}^{(i)}(t_{m-}) = \mathbf{x}_m = \mathbf{x}^{(j)}(t_{m-})$. The flow $\mathbf{x}^{(i)}(t)$ is $C^{r_i}_{[t_{m-\varepsilon}, t_m)}$-continuous for time t and $||d^{r_i+1}\mathbf{x}^{(i)}/dt^{r_i+1}|| < \infty$ ($r_i \geq 2k_i + 1$). The flow $\mathbf{x}^{(j)}(t)$ is $C^{r_j}_{[t_{m-\varepsilon}, t_m)}$-continuous for time t and $||d^{r_j+1}\mathbf{x}^{(j)}/dt^{r_j+1}|| < \infty$ ($r_j \geq 2k_j + 1$). The flow $\mathbf{x}^{(i)}(t)$ of the $(2k_i)^{\text{th}}$-order singularity and $\mathbf{x}^{(j)}(t)$ of the $(2k_j)^{\text{th}}$-order singularity to the boundary $\partial\Omega_{ij}$ is $(2k_i : 2k_j)$-non-passable with the first kind (or the $(2k_i : 2k_j)$-sink flow) if

$$\left.\begin{array}{l} G^{(s,i)}_{\partial\Omega_{ij}}(\mathbf{x}_m, t_{m-}, \mathbf{p}_i, \lambda) = 0, \text{ for } s = 0, 1, \ldots, 2k_i - 1, \\ G^{(2k_i,i)}_{\partial\Omega_{ij}}(\mathbf{x}_m, t_{m-}, \mathbf{p}_i, \lambda) \neq 0, \end{array}\right\} \quad (2.29)$$

$$\left.\begin{array}{l} G^{(s,j)}_{\partial\Omega_{ij}}(\mathbf{x}_m, t_{m-}, \mathbf{p}_j, \lambda) = 0, \text{ for } s = 0, 1, \ldots, 2k_j - 1, \\ G^{(2k_j,j)}_{\partial\Omega_{ij}}(\mathbf{x}_m, t_{m-}, \mathbf{p}_j, \lambda) \neq 0, \end{array}\right\} \quad (2.30)$$

$$\left.\begin{array}{l} \text{either} \quad \begin{array}{l} \mathbf{n}^{\text{T}}_{\partial\Omega_{ij}}(\mathbf{x}^{(0)}_{m-\varepsilon}) \cdot [\mathbf{x}^{(0)}_{m-\varepsilon} - \mathbf{x}^{(i)}_{m-\varepsilon}] > 0 \\ \mathbf{n}^{\text{T}}_{\partial\Omega_{ij}}(\mathbf{x}^{(0)}_{m-\varepsilon}) \cdot [\mathbf{x}^{(0)}_{m-\varepsilon} - \mathbf{x}^{(j)}_{m-\varepsilon}] < 0 \end{array} \right\} \text{for } \mathbf{n}_{\partial\Omega_{ij}} \to \Omega_j \\ \text{or} \quad \begin{array}{l} \mathbf{n}^{\text{T}}_{\partial\Omega_{ij}}(\mathbf{x}^{(0)}_{m-\varepsilon}) \cdot [\mathbf{x}^{(0)}_{m-\varepsilon} - \mathbf{x}^{(i)}_{m-\varepsilon}] < 0 \\ \mathbf{n}^{\text{T}}_{\partial\Omega_{ij}}(\mathbf{x}^{(0)}_{m-\varepsilon}) \cdot [\mathbf{x}^{(0)}_{m-\varepsilon} - \mathbf{x}^{(j)}_{m-\varepsilon}] > 0 \end{array} \right\} \text{for } \mathbf{n}_{\partial\Omega_{ij}} \to \Omega_i. \end{array} \quad (2.31)$$

Theorem 2.4. *For a discontinuous dynamical system in Eq.(2.1), $\mathbf{x}(t_m) \equiv \mathbf{x}_m \in \partial\Omega_{ij}$ at time t_m. For an arbitrarily small $\varepsilon > 0$, there is a time interval $[t_{m-\varepsilon}, t_m)$. Suppose $\mathbf{x}^{(i)}(t_{m-}) = \mathbf{x}_m = \mathbf{x}^{(j)}(t_{m-})$. The flow $\mathbf{x}^{(i)}(t)$ is $C^{r_i}_{[t_{m-\varepsilon}, t_m)}$-continuous for time t and $||d^{r_i+1}\mathbf{x}^{(i)}/dt^{r_i+1}|| < \infty$ ($r_i \geq 2k_i + 1$). The flow $\mathbf{x}^{(j)}(t)$ is and $C^{r_j}_{[t_{m-\varepsilon}, t_m)}$-continuous for time t and $||d^{r_j+1}\mathbf{x}^{(j)}/dt^{r_j+1}|| < \infty$ ($r_j \geq 2k_j + 1$). The flow $\mathbf{x}^{(i)}(t)$ of the $(2k_i)^{\text{th}}$-order singularity and $\mathbf{x}^{(j)}(t)$ of the $(2k_j)^{\text{th}}$-order singularity to $\partial\Omega_{ij}$ is $(2k_i : 2k_j)$-non-passable with the first kind (or the $(2k_i : 2k_j)$-sink flow) iff*

$$G^{(s,i)}_{\partial\Omega_{ij}}(\mathbf{x}_m, t_{m-}, \mathbf{p}_i, \lambda) = 0, \text{ for } s = 0, 1, \ldots, 2k_i - 1, \quad (2.32)$$

$$G^{(s,j)}_{\partial\Omega_{ij}}(\mathbf{x}_m,t_{m-},\mathbf{p}_j,\boldsymbol{\lambda})=0,\ for\ s=0,1,\ldots,2k_j-1, \qquad (2.33)$$

$$\text{either}\ \left.\begin{array}{l}G^{(2k_i,i)}_{\partial\Omega_{ij}}(\mathbf{x}_m,t_{m-},\mathbf{p}_i,\boldsymbol{\lambda})>0\\ G^{(2k_j,j)}_{\partial\Omega_{ij}}(\mathbf{x}_m,t_{m-},\mathbf{p}_j,\boldsymbol{\lambda})<0\end{array}\right\}\ for\ \mathbf{n}_{\partial\Omega_{ij}}\to\Omega_j$$

$$\text{or}\ \left.\begin{array}{l}G^{(2k_i,i)}_{\partial\Omega_{ij}}(\mathbf{x}_m,t_{m-},\mathbf{p}_i,\boldsymbol{\lambda})<0\\ G^{(2k_j,j)}_{\partial\Omega_{ij}}(\mathbf{x}_m,t_{m-},\mathbf{p}_j,\boldsymbol{\lambda})>0\end{array}\right\}\ for\ \mathbf{n}_{\partial\Omega_{ij}}\to\Omega_i. \qquad (2.34)$$

Proof. See Luo (2008a,b).

Definition 2.9. For a discontinuous dynamical system in Eq.(2.1), $\mathbf{x}(t_m)\equiv\mathbf{x}_m\in\partial\Omega_{ij}$ at time t_m. For an arbitrarily small $\varepsilon>0$, there is a time interval $(t_m,t_{m+\varepsilon}]$. Suppose $\mathbf{x}^{(i)}(t_{m+})=\mathbf{x}_m=\mathbf{x}^{(j)}(t_{m+})$. The flow $\mathbf{x}^{(i)}(t)$ and $\mathbf{x}^{(j)}(t)$ to the boundary $\partial\Omega_{ij}$ is *non-passable* with the second kind (or called the source flow) if

$$\text{either}\ \left.\begin{array}{l}\mathbf{n}^{T}_{\partial\Omega_{ij}}(\mathbf{x}^{(0)}_{m+\varepsilon})\cdot\left[\mathbf{x}^{(i)}_{m+\varepsilon}-\mathbf{x}^{(0)}_{m+\varepsilon}\right]<0\\ \mathbf{n}^{T}_{\partial\Omega_{ij}}(\mathbf{x}^{(0)}_{m+\varepsilon})\cdot\left[\mathbf{x}^{(j)}_{m+\varepsilon}-\mathbf{x}^{(0)}_{m+\varepsilon}\right]>0\end{array}\right\}\ for\ \mathbf{n}_{\partial\Omega_{ij}}\to\Omega_j$$

$$\text{or}\ \left.\begin{array}{l}\mathbf{n}^{T}_{\partial\Omega_{ij}}(\mathbf{x}^{(0)}_{m+\varepsilon})\cdot\left[\mathbf{x}^{(i)}_{m+\varepsilon}-\mathbf{x}^{(0)}_{m+\varepsilon}\right]>0\\ \mathbf{n}^{T}_{\partial\Omega_{ij}}(\mathbf{x}^{(0)}_{m+\varepsilon})\cdot\left[\mathbf{x}^{(j)}_{m+\varepsilon}-\mathbf{x}^{(0)}_{m+\varepsilon}\right]<0\end{array}\right\}\ for\ \mathbf{n}_{\partial\Omega_{ij}}\to\Omega_i. \qquad (2.35)$$

Theorem 2.5. *For a discontinuous dynamical system in Eq.(2.1), $\mathbf{x}(t_m)\equiv\mathbf{x}_m\in\partial\Omega_{ij}$ at time t_m. For an arbitrarily small $\varepsilon>0$, there is a time interval $(t_m,t_{m+\varepsilon}]$. Suppose $\mathbf{x}^{(i)}(t_{m+})=\mathbf{x}_m=\mathbf{x}^{(j)}(t_{m+})$. The flow $\mathbf{x}^{(i)}(t)$ is $C^r_{(t_m,t_{m+\varepsilon}]}$-continuous ($r\geqslant 1$) for time t and $||d^{r+1}\mathbf{x}^{(i)}/dt^{r+1}||<\infty$. The flow $\mathbf{x}^j(t)$ is and $C^r_{(t_m,t_{m+\varepsilon}]}$-continuous for time t and $||d^{r+1}\mathbf{x}^{(j)}/dt^{r+1}||<\infty$ ($r\geqslant 1$). The flow $\mathbf{x}^{(i)}(t)$ and $\mathbf{x}^{(j)}(t)$ to the boundary $\partial\Omega_{ij}$ is non-passable with the second kind (or source flow) iff*

$$\text{either}\ \left.\begin{array}{l}G^{(i)}_{\partial\Omega_{ij}}(\mathbf{x}_m,t_{m+},\mathbf{p}_i,\boldsymbol{\lambda})<0\\ G^{(j)}_{\partial\Omega_{ij}}(\mathbf{x}_m,t_{m+},\mathbf{p}_j,\boldsymbol{\lambda})>0\end{array}\right\}\ for\ \mathbf{n}_{\partial\Omega_{ij}}\to\Omega_j$$

$$\text{or}\ \left.\begin{array}{l}G^{(i)}_{\partial\Omega_{ij}}(\mathbf{x}_m,t_{m+},\mathbf{p}_i,\boldsymbol{\lambda})>0\\ G^{(j)}_{\partial\Omega_{ij}}(\mathbf{x}_m,t_{m+},\mathbf{p}_j,\boldsymbol{\lambda})<0\end{array}\right\}\ for\ \mathbf{n}_{\partial\Omega_{ij}}\to\Omega_i. \qquad (2.36)$$

Proof. See Luo (2008a.b).

Definition 2.10. For a discontinuous dynamical system in Eq.(2.1), $\mathbf{x}(t_m)\equiv\mathbf{x}_m\in\partial\Omega_{ij}$ at time t_m. For an arbitrarily small $\varepsilon>0$, there is a time interval $(t_m,t_{m+\varepsilon}]$.

2.4 Non-passable flows

Suppose $\mathbf{x}^{(i)}(t_{m+}) = \mathbf{x}_m = \mathbf{x}^{(j)}(t_{m+})$. The flow $\mathbf{x}^{(i)}(t)$ is $C^{r_i}_{(t_m,t_{m+\varepsilon}]}$-continuous for time t and $||d^{r_i+1}\mathbf{x}^{(i)}/dt^{r_i+1}|| < \infty$ ($r_i \geq 2k_i + 1$). The flow $\mathbf{x}^{(j)}(t)$ is $C^{r_j}_{(t_m,t_{m+\varepsilon}]}$-continuous for time t and $||d^{r_j+1}\mathbf{x}^{(j)}/dt^{r_j+1}|| < \infty$ ($r_j \geq 2k_j + 1$). The flow $\mathbf{x}^{(i)}(t)$ of the $(2k_i)^{\text{th}}$-order singularity and $\mathbf{x}^{(j)}(t)$ of the $(2k_j)^{\text{th}}$-order singularity to the boundary $\partial\Omega_{ij}$ is $(2k_i : 2k_j)$-*non-passable* with the second kind (or $(2k_i : 2k_j)$-*source flow*) if

$$\left. \begin{aligned} G^{(s,i)}_{\partial\Omega_{ij}}(\mathbf{x}_m, t_{m+}, \mathbf{p}_i, \boldsymbol{\lambda}) &= 0, \text{ for } s = 0, 1, \ldots, 2k_i - 1, \\ G^{(2k_i,i)}_{\partial\Omega_{ij}}(\mathbf{x}_m, t_{m+}, \mathbf{p}_i, \boldsymbol{\lambda}) &\neq 0, \end{aligned} \right\} \quad (2.37)$$

$$\left. \begin{aligned} G^{(s,j)}_{\partial\Omega_{ij}}(\mathbf{x}_m, t_{m+}, \mathbf{p}_j, \boldsymbol{\lambda}) &= 0, \text{ for } s = 0, 1, \ldots, 2k_j - 1, \\ G^{(2k_j,j)}_{\partial\Omega_{ij}}(\mathbf{x}_m, t_{m+}, \mathbf{p}_j, \boldsymbol{\lambda}) &\neq 0, \end{aligned} \right\} \quad (2.38)$$

either
$$\left. \begin{aligned} \mathbf{n}^{\text{T}}_{\partial\Omega_{ij}}(\mathbf{x}^{(0)}_{m+\varepsilon}) \cdot \left[\mathbf{x}^{(i)}_{m+\varepsilon} - \mathbf{x}^{(0)}_{m+\varepsilon}\right] &< 0 \\ \mathbf{n}^{\text{T}}_{\partial\Omega_{ij}}(\mathbf{x}^{(0)}_{m+\varepsilon}) \cdot \left[\mathbf{x}^{(j)}_{m+\varepsilon} - \mathbf{x}^{(0)}_{m+\varepsilon}\right] &> 0 \end{aligned} \right\} \text{ for } \mathbf{n}_{\partial\Omega_{ij}} \to \Omega_j \quad (2.39\text{a})$$

or
$$\left. \begin{aligned} \mathbf{n}^{\text{T}}_{\partial\Omega_{ij}}(\mathbf{x}^{(0)}_{m+\varepsilon}) \cdot \left[\mathbf{x}^{(i)}_{m+\varepsilon} - \mathbf{x}^{(0)}_{m+\varepsilon}\right] &> 0 \\ \mathbf{n}^{\text{T}}_{\partial\Omega_{ij}}(\mathbf{x}^{(0)}_{m+\varepsilon}) \cdot \left[\mathbf{x}^{(j)}_{m+\varepsilon} - \mathbf{x}^{(0)}_{m+\varepsilon}\right] &< 0 \end{aligned} \right\} \text{ for } \mathbf{n}_{\partial\Omega_{ij}} \to \Omega_i. \quad (2.39\text{b})$$

Theorem 2.6. *For a discontinuous dynamical system in Eq.(2.1), $\mathbf{x}(t_m) \equiv \mathbf{x}_m \in \partial\Omega_{ij}$ at time t_m. For an arbitrarily small $\varepsilon > 0$, there is a time interval $(t_m, t_{m+\varepsilon}]$. Suppose $\mathbf{x}^{(i)}(t_{m+}) = \mathbf{x}_m = \mathbf{x}^{(j)}(t_{m+})$. The flow $\mathbf{x}^{(i)}(t)$ is $C^{r_i}_{(t_m,t_{m+\varepsilon}]}$-continuous for time t and $||d^{r_i+1}\mathbf{x}^{(i)}/dt^{r_i+1}|| < \infty$ ($r_i \geq 2k_i + 1$). The flow $\mathbf{x}^{(j)}(t)$ is $C^{r_j}_{(t_m,t_{m+\varepsilon}]}$-continuous for time t and $||d^{r_j+1}\mathbf{x}^{(j)}/dt^{r_j+1}|| < \infty$ ($r_j \geq 2k_j + 1$). The flow $\mathbf{x}^{(i)}(t)$ of the $(2k_i)^{\text{th}}$-order singularity and $\mathbf{x}^{(j)}(t)$ of the $(2k_j)^{\text{th}}$-order singularity to the boundary $\partial\Omega_{ij}$ is $(2k_i : 2k_j)$-non-passable with the second kind (or $(2k_i : 2k_j)$-source flow) iff*

$$G^{(s,i)}_{\partial\Omega_{ij}}(\mathbf{x}_m, t_{m+}, \mathbf{p}_i, \boldsymbol{\lambda}) = 0, for \; s = 0, 1, \ldots, 2k_i - 1, \quad (2.40)$$

$$G^{(s,j)}_{\partial\Omega_{ij}}(\mathbf{x}_m, t_{m+}, \mathbf{p}_j, \boldsymbol{\lambda}) = 0, for \; s = 0, 1, \ldots, 2k_j - 1, \quad (2.41)$$

either
$$\left. \begin{aligned} G^{(2k_i,i)}_{\partial\Omega_{ij}}(\mathbf{x}_m, t_{m+}, \mathbf{p}_i, \boldsymbol{\lambda}) &< 0 \\ G^{(2k_j,j)}_{\partial\Omega_{ij}}(\mathbf{x}_m, t_{m+}, \mathbf{p}_j, \boldsymbol{\lambda}) &> 0 \end{aligned} \right\} \text{ for } \mathbf{n}_{\partial\Omega_{ij}} \to \Omega_j$$

or
$$\left. \begin{aligned} G^{(2k_i,i)}_{\partial\Omega_{ij}}(\mathbf{x}_m, t_{m+}, \mathbf{p}_i, \boldsymbol{\lambda}) &> 0 \\ G^{(2k_j,j)}_{\partial\Omega_{ij}}(\mathbf{x}_m, t_{m+}, \mathbf{p}_j, \boldsymbol{\lambda}) &< 0 \end{aligned} \right\} \text{ for } \mathbf{n}_{\partial\Omega_{ij}} \to \Omega_i.$$
$$(2.42)$$

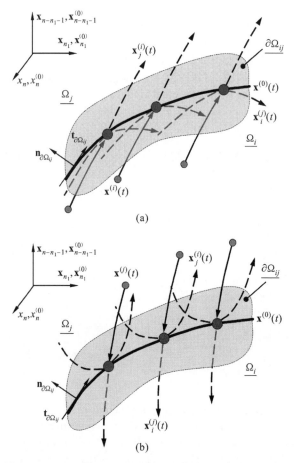

Fig. 2.6 The half-sink flows: (a) in Ω_i with $(2k_i : 2k_j - 1)$-order and (b) in Ω_j with $(2k_j : 2k_i - 1)$-order. $\mathbf{x}^{(i)}(t)$ and $\mathbf{x}^{(j)}(t)$ represent the *real* flows in domains Ω_i and Ω_j, respectively, which are depicted by the thin solid curves. $\mathbf{x}_{(i)}^{(j)}(t)$ and $\mathbf{x}_{(j)}^{(i)}(t)$ represent the *imaginary* flows in domains Ω_i and Ω_j, respectively controlled by the vector fields in Ω_j and Ω_i, which are depicted by the dashed curves. The flow on the boundary is described by $\mathbf{x}^{(0)}(t)$. The normal and tangential vectors $\mathbf{n}_{\partial\Omega_{ij}}$ and $\mathbf{t}_{\partial\Omega_{ij}}$ on the boundary are depicted. The black circles are half-sink points and the small shaded circles are the starting points

Proof. See Luo(2008a,b).

The full sink and source flows to the separation boundary have been discussed. Next, the half-non-passable flows to the boundary will be discussed. The half-sink flow to the separation boundary is sketched in Fig.2.6. The half-non-passable flow of the first kind is termed *the half-sink flow*. A half-sink flow in Ω_i is shown in Fig.2.6(a). Only $\mathbf{x}^{(i)}(t)$ for time $t \in [t_{m-\varepsilon}, t_m)$ is a real flow, and the other imaginary flows $\mathbf{x}_j^{(i)}(t)$ for time $t \in [t_{m-\varepsilon}, t_{m+\varepsilon}]$ and $\mathbf{x}_i^{(j)}(t)$ for $t \in (t_m, t_{m+\varepsilon}]$ are represented by dashed curves. To the same boundary $\partial\Omega_{ij}$, a half sink flow in Ω_j is sketched in

2.4 Non-passable flows

Fig.2.6(b). The input flow $\mathbf{x}^{(j)}(t)$ for time $t \in [t_{m-\varepsilon}, \varepsilon)$ is only a real flow. The strict mathematical description is given as follows.

Definition 2.11. For a discontinuous dynamical system in Eq.(2.1), $\mathbf{x}(t_m) \equiv \mathbf{x}_m \in \partial \Omega_{ij}$ at time t_m. For an arbitrarily small $\varepsilon > 0$, there are two time intervals $[t_{m-\varepsilon}, t_m)$ and $[t_{m-\varepsilon}, t_{m+\varepsilon}]$. Suppose $\mathbf{x}^{(i)}(t_{m-}) = \mathbf{x}_m = \mathbf{x}_i^{(j)}(t_{m\pm})$. The flow $\mathbf{x}^{(i)}(t)$ is $C^{r_i}_{[t_{m-\varepsilon}, t_m)}$-continuous for time t and $||d^{r_i+1}\mathbf{x}^{(i)}/dt^{r_i+1}|| < \infty$ $(r_i \geqslant 2k_i+1)$. The imaginary flow $\mathbf{x}_i^{(j)}(t)$ is $C^{r_j}_{[t_{m-\varepsilon}, t_{m+\varepsilon}]}$-continuous for time t and $||d^{r_j+1}\mathbf{x}^{(j)}/dt^{r_j+1}|| < \infty$ $(r_j \geqslant 2k_j)$. The flow $\mathbf{x}^{(i)}(t)$ of the $(2k_i)^{\text{th}}$-order singularity and $\mathbf{x}_i^{(j)}(t)$ of the $(2k_j-1)^{\text{th}}$-order singularity to the boundary $\partial \Omega_{ij}$ is $(2k_i : 2k_j - 1)$-*half-non-passable* with the first kind in domain Ω_i (or $(2k_i : 2k_j - 1)$-half-sink flow) if

$$\left. \begin{aligned} G^{(s,i)}_{\partial \Omega_{ij}}(\mathbf{x}_m, t_{m-}, \mathbf{p}_i, \boldsymbol{\lambda}) &= 0, \text{ for } s = 0, 1, \ldots, 2k_i - 1, \\ G^{(2k_i,i)}_{\partial \Omega_{ij}}(\mathbf{x}_m, t_{m-}, \mathbf{p}_i, \boldsymbol{\lambda}) &\neq 0, \end{aligned} \right\} \quad (2.43)$$

$$\left. \begin{aligned} G^{(s,j)}_{\partial \Omega_{ij}}(\mathbf{x}_m, t_{m\pm}, \mathbf{p}_j, \boldsymbol{\lambda}) &= 0, \text{ for } s = 0, 1, \ldots, 2k_j - 2, \\ G^{(2k_j-1,j)}_{\partial \Omega_{ij}}(\mathbf{x}_m, t_{m\pm}, \mathbf{p}_j, \boldsymbol{\lambda}) &\neq 0, \end{aligned} \right\} \quad (2.44)$$

$$\left. \begin{aligned} \text{either } \mathbf{n}^{\text{T}}_{\partial \Omega_{ij}}(\mathbf{x}^{(0)}_{m-\varepsilon}) \cdot \left[\mathbf{x}^{(0)}_{m-\varepsilon} - \mathbf{x}^{(i)}_{m-\varepsilon} \right] &> 0, \text{ for } \mathbf{n}_{\partial \Omega_{ij}} \to \Omega_j \\ \text{or } \mathbf{n}^{\text{T}}_{\partial \Omega_{ij}}(\mathbf{x}^{(0)}_{m-\varepsilon}) \cdot \left[\mathbf{x}^{(0)}_{m-\varepsilon} - \mathbf{x}^{(i)}_{m-\varepsilon} \right] &< 0, \text{ for } \mathbf{n}_{\partial \Omega_{ij}} \to \Omega_i, \end{aligned} \right\} \quad (2.45)$$

$$\left. \begin{aligned} \text{either } & \left. \begin{aligned} \mathbf{n}^{\text{T}}_{\partial \Omega_{ij}}(\mathbf{x}^{(0)}_{m-\varepsilon}) \cdot \left[\mathbf{x}^{(0)}_{m-\varepsilon} - \mathbf{x}^{(j)}_{i(m-\varepsilon)} \right] &> 0 \\ \mathbf{n}^{\text{T}}_{\partial \Omega_{ij}}(\mathbf{x}^{(0)}_{m+\varepsilon}) \cdot \left[\mathbf{x}^{(j)}_{i(m+\varepsilon)} - \mathbf{x}^{(0)}_{m+\varepsilon} \right] &< 0 \end{aligned} \right\} \text{ for } \mathbf{n}_{\partial \Omega_{ij}} \to \Omega_j \\ \text{or } & \left. \begin{aligned} \mathbf{n}^{\text{T}}_{\partial \Omega_{ij}}(\mathbf{x}^{(0)}_{m-\varepsilon}) \cdot \left[\mathbf{x}^{(0)}_{m-\varepsilon} - \mathbf{x}^{(j)}_{i(m-\varepsilon)} \right] &< 0 \\ \mathbf{n}^{\text{T}}_{\partial \Omega_{ij}}(\mathbf{x}^{(0)}_{m+\varepsilon}) \cdot \left[\mathbf{x}^{(j)}_{i(m+\varepsilon)} - \mathbf{x}^{(0)}_{m+\varepsilon} \right] &> 0 \end{aligned} \right\} \text{ for } \mathbf{n}_{\partial \Omega_{ij}} \to \Omega_i. \end{aligned} \right\} \quad (2.46)$$

From the above definition, the corresponding theorem gives the necessary and sufficient conditions for such a half-non-passable flow of the first kind (or half-sink flow).

Theorem 2.7. *For a discontinuous dynamical system in Eq.(2.1), $\mathbf{x}(t_m) \equiv \mathbf{x}_m \in \partial \Omega_{ij}$ at time t_m. Suppose $\mathbf{x}^{(i)}(t_{m-}) = \mathbf{x}_m = \mathbf{x}_i^{(j)}(t_{m\pm})$. For an arbitrarily small $\varepsilon > 0$, there are two time intervals $[t_{m-\varepsilon}, t_m)$ and $[t_{m-\varepsilon}, t_{m+\varepsilon}]$. The flow $\mathbf{x}^{(i)}(t)$ is $C^{r_i}_{[t_{m-\varepsilon}, t_m)}$-continuous for time t and $||d^{r_i+1}\mathbf{x}^{(i)}/dt^{r_i+1}|| < \infty (r_i \geqslant 2k_i+1)$. The imaginary flow $\mathbf{x}_i^{(j)}(t)$ is $C^{r_i}_{[t_{m-\varepsilon}, t_{m+\varepsilon}]}$-continuous for time t and $||d^{r_j+1}\mathbf{x}_i^{(j)}/dt^{r_j+1}|| < \infty$ $(r_j \geqslant 2k_j)$. The flow $\mathbf{x}^{(i)}(t)$ of the $(2k_i)^{\text{th}}$–order singularity and $\mathbf{x}_i^{(j)}(t)$ of the $(2k_j-1)^{\text{th}}$-order singularity to the boundary $\partial \Omega_{ij}$ is $(2k_i : 2k_j - 1)$-half-non-passable with the first kind in domain Ω_i (or $(2k_i : 2k_j - 1)$-half sink flow) iff*

$$G_{\partial \Omega_{ij}}^{(s,i)}(\mathbf{x}_m, t_{m-}, \mathbf{p}_i, \boldsymbol{\lambda}) = 0, \text{ for } s = 0, 1, \ldots, 2k_i - 1, \tag{2.47}$$

$$G_{\partial \Omega_{ij}}^{(s,j)}(\mathbf{x}_m, t_{m\pm}, \mathbf{p}_j, \boldsymbol{\lambda}) = 0, \text{ for } s = 0, 1, \ldots, 2k_j - 2, \tag{2.48}$$

$$\left. \begin{array}{l} \text{either} \left. \begin{array}{l} G_{\partial \Omega_{ij}}^{(2k_i,i)}(\mathbf{x}_m, t_{m-}, \mathbf{p}_i, \boldsymbol{\lambda}) > 0 \\ G_{\partial \Omega_{ij}}^{(2k_j-1,j)}(\mathbf{x}_m, t_{m\pm}, \mathbf{p}_j, \boldsymbol{\lambda}) < 0 \end{array} \right\} \text{ for } \mathbf{n}_{\partial \Omega_{ij}} \to \Omega_j \\ \text{or} \left. \begin{array}{l} G_{\partial \Omega_{ij}}^{(2k_i,i)}(\mathbf{x}_m, t_{m-}, \mathbf{p}_i, \boldsymbol{\lambda}) < 0 \\ G_{\partial \Omega_{ij}}^{(2k_j-1,j)}(\mathbf{x}_m, t_{m\pm}, \mathbf{p}_j, \boldsymbol{\lambda}) > 0 \end{array} \right\} \text{ for } \mathbf{n}_{\partial \Omega_{ij}} \to \Omega_i. \end{array} \right\} \tag{2.49}$$

Proof. See Luo (2008a,b).

After the description of a non-passable flow of the first kind, the half-non-passable flow of the second kind will be discussed. Before the mathematical description is given, the intuitive illustration of such a half-non-passable flow is shown in Fig.2.7. The half-non-passable flow of the second kind is termed *the half-source flow*. A half-source flow in Ω_i is sketched in Fig.2.7(a). Only $\mathbf{x}^{(i)}(t)$ is a real flow for $t \in (t_m, t_{m+\varepsilon}]$. The imaginary flows $\mathbf{x}_j^{(i)}(t)$ for $t \in [t_{m-\varepsilon}, t_{m+\varepsilon}]$ and $\mathbf{x}_i^{(j)}(t)$ for $t \in [t_{m-\varepsilon}, t_m)$ are depicted by dashed curves. To the same boundary $\partial \Omega_{ij}$, a half-source flow in Ω_j is shown in Fig.2.7(b). The flow $\mathbf{x}^{(j)}(t)$ for $t \in (t_m, t_{m+\varepsilon}]$ is only a real flow.

Definition 2.12. For a discontinuous dynamical system in Eq.(2.1), $\mathbf{x}(t_m) \equiv \mathbf{x}_m \in \partial \Omega_{ij}$ at time t_m. For an arbitrarily small $\varepsilon > 0$, there are two time intervals $[t_{m-\varepsilon}, t_m)$ and $[t_{m-\varepsilon}, t_{m+\varepsilon}]$. Suppose $\mathbf{x}^{(i)}(t_{m+}) = \mathbf{x}_m = \mathbf{x}_i^{(j)}(t_{m\pm})$. The flow $\mathbf{x}^{(i)}(t)$ is $C_{[t_{m-\varepsilon}, t_m)}^{r_i}$-continuous for time t and $||d^{r_i+1}\mathbf{x}^{(i)}/dt^{r_i+1}|| < \infty$ ($r_i \geq 2k_i + 1$). The imaginary flow $\mathbf{x}_i^{(j)}(t)$ is $C_{[t_{m-\varepsilon}, t_{m+\varepsilon}]}^{r_j}$-continuous for time t and $||d^{r_j+1}\mathbf{x}^{(j)}/dt^{r_j+1}|| < \infty$ ($r_j \geq 2k_j$). The flow $\mathbf{x}^{(i)}(t)$ of the $(2k_i)^{\text{th}}$-order singularity and $\mathbf{x}_i^{(j)}(t)$ of the $(2k_j - 1)^{\text{th}}$-order singularity to the boundary $\partial \Omega_{ij}$ is $(2k_i : 2k_j - 1)$-*half-non-passable* with the second kind in domain Ω_i (or $(2k_i : 2k_j - 1)$-half source flow) if

$$\left. \begin{array}{l} G_{\partial \Omega_{ij}}^{(s,i)}(\mathbf{x}_m, t_{m+}, \mathbf{p}_i, \boldsymbol{\lambda}) = 0, \text{ for } s = 0, 1, \ldots, 2k_i - 1, \\ G_{\partial \Omega_{ij}}^{(2k_i,i)}(\mathbf{x}_m, t_{m+}, \mathbf{p}_i, \boldsymbol{\lambda}) \neq 0, \end{array} \right\} \tag{2.50}$$

$$\left. \begin{array}{l} G_{\partial \Omega_{ij}}^{(s,j)}(\mathbf{x}_m, t_{m\pm}, \mathbf{p}_j, \boldsymbol{\lambda}) = 0, \text{ for } s = 0, 1, \ldots, 2k_j - 2, \\ G_{\partial \Omega_{ij}}^{(2k_j-1,j)}(\mathbf{x}_m, t_{m\pm}, \mathbf{p}_j, \boldsymbol{\lambda}) \neq 0, \end{array} \right\} \tag{2.51}$$

$$\left. \begin{array}{l} \text{either} \quad \mathbf{n}_{\partial \Omega_{ij}}^{\text{T}}(\mathbf{x}_{m+\varepsilon}^{(0)}) \cdot \left[\mathbf{x}_{m+\varepsilon}^{(i)} - \mathbf{x}_{m+\varepsilon}^{(0)}\right] < 0, \text{ for } \mathbf{n}_{\partial \Omega_{ij}} \to \Omega_j \\ \text{or} \quad \mathbf{n}_{\partial \Omega_{ij}}^{\text{T}}(\mathbf{x}_{m+\varepsilon}^{(0)}) \cdot \left[\mathbf{x}_{m+\varepsilon}^{(i)} - \mathbf{x}_{m+\varepsilon}^{(0)}\right] > 0, \text{ for } \mathbf{n}_{\partial \Omega_{ij}} \to \Omega_i, \end{array} \right\} \tag{2.52}$$

2.4 Non-passable flows

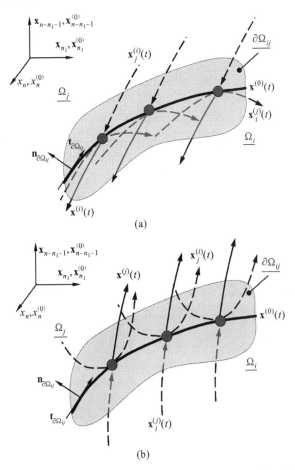

Fig. 2.7 Half-source flows: (a) in Ω_i with $(2k_i : 2k_j - 1)$-order and (b) in Ω_j with $(2k_i : 2k_j - 1)$-order. $\mathbf{x}^{(i)}(t)$ and $\mathbf{x}^{(j)}(t)$ represent the *real* flows in domains Ω_i and Ω_j, respectively, which are depicted by the thin solid curves. $\mathbf{x}_i^{(j)}(t)$ and $\mathbf{x}_j^{(i)}(t)$ represent the *imaginary* flows in domains Ω_i and Ω_j, respectively controlled by the vector fields in Ω_j and Ω_i, which are depicted by the dashed curves. The flow on the boundary is described by $\mathbf{x}^{(0)}(t)$. The normal and tangential vectors $\mathbf{n}_{\partial\Omega_{ij}}$ and $\mathbf{t}_{\partial\Omega_{ij}}$ on the boundary are depicted. The black circles are half-source points.

$$\left.\begin{array}{l}\text{either }\left.\begin{array}{l}\mathbf{n}_{\partial\Omega_{ij}}^{\mathrm{T}}(\mathbf{x}_{m-\varepsilon}^{(0)})\cdot\left[\mathbf{x}_{m-\varepsilon}^{(0)}-\mathbf{x}_{i(m-\varepsilon)}^{(j)}\right]>0\\ \mathbf{n}_{\partial\Omega_{ij}}^{\mathrm{T}}(\mathbf{x}_{m+\varepsilon}^{(0)})\cdot\left[\mathbf{x}_{i(m+\varepsilon)}^{(j)}-\mathbf{x}_{m+\varepsilon}^{(0)}\right]<0\end{array}\right\}\text{ for }\mathbf{n}_{\partial\Omega_{ij}}\to\Omega_j\\ \text{or }\left.\begin{array}{l}\mathbf{n}_{\partial\Omega_{ij}}^{\mathrm{T}}(\mathbf{x}_{m-\varepsilon}^{(0)})\cdot\left[\mathbf{x}_{m-\varepsilon}^{(0)}-\mathbf{x}_{i(m-\varepsilon)}^{(j)}\right]<0\\ \mathbf{n}_{\partial\Omega_{ij}}^{\mathrm{T}}(\mathbf{x}_{m+\varepsilon}^{(0)})\cdot\left[\mathbf{x}_{i(m+\varepsilon)}^{(j)}-\mathbf{x}_{m+\varepsilon}^{(0)}\right]>0\end{array}\right\}\text{ for }\mathbf{n}_{\partial\Omega_{ij}}\to\Omega_i.\end{array}\right\} \quad (2.53)$$

From the above definition, the necessary and sufficient conditions for such a $(2k_i : 2k_j - 1)$-half-non-passable flow of the second kind (or $(2k_i : 2k_j - 1)$-half-source flow) are stated in the following theorem.

Theorem 2.8. *For a discontinuous dynamical system in Eq.(2.1), $\mathbf{x}(t_m) \equiv \mathbf{x}_m \in \partial \Omega_{ij}$ at time t_m. For an arbitrarily small $\varepsilon > 0$, there are two time intervals $[t_{m-\varepsilon}, t_m)$ and $[t_{m-\varepsilon}, t_{m+\varepsilon}]$. Suppose $\mathbf{x}^{(i)}(t_{m+}) = \mathbf{x}_m = \mathbf{x}_i^{(j)}(t_{m\pm})$. The flow $\mathbf{x}^{(i)}(t)$ is $C_{[t_{m-\varepsilon},t_m]}^{r_i}$-continuous for time t and $\|d^{r_j+1}\mathbf{x}^{(i)}/dt^{r_i+1}\| < \infty$ $(r_i \geqslant 2k_i + 1)$. The imaginary flow $\mathbf{x}_i^{(j)}(t)$ is $C_{[t_{m-\varepsilon},t_{m+\varepsilon}]}^{r_j}$-continuous for time t and $\|d^{r_j+1}\mathbf{x}^{(j)}/dt^{r_j+1}\| < \infty$ $(r_j \geqslant 2k_j)$. The flow $\mathbf{x}^{(i)}(t)$ of the $(2k_i)^{\text{th}}$-order singularity and $\mathbf{x}_i^{(j)}(t)$ of the $(2k_j - 1)^{\text{th}}$-order singularity to the boundary $\partial \Omega_{ij}$ is $(2k_i : 2k_j - 1)$-half-non-passable with the second kind in domain Ω_i ($(2k_i : 2k_j - 1)$-source flow) iff*

$$G_{\partial \Omega_{ij}}^{(s,i)}(\mathbf{x}_m, t_{m+}, \mathbf{p}_i, \lambda) = 0, \text{ for } s = 0, 1, \ldots, 2k_i - 1, \quad (2.54)$$

$$G_{\partial \Omega_{ij}}^{(s,j)}(\mathbf{x}_m, t_{m\pm}, \mathbf{p}_j, \lambda) = 0, \text{ for } s = 0, 1, \ldots, 2k_j - 2, \quad (2.55)$$

$$\text{either} \left. \begin{array}{l} G_{\partial \Omega_{ij}}^{(2k_i,i)}(\mathbf{x}_m, t_{m+}, \mathbf{p}_i, \lambda) < 0 \\ G_{\partial \Omega_{ij}}^{(2k_j-1,j)}(\mathbf{x}_m, t_{m\pm}, \mathbf{p}_j, \lambda) < 0 \end{array} \right\} \text{ for } \mathbf{n}_{\partial \Omega_{ij}} \to \Omega_j \\ \text{or} \left. \begin{array}{l} G_{\partial \Omega_{ij}}^{(2k_i,i)}(\mathbf{x}_m, t_{m+}, \mathbf{p}_i, \lambda) > 0 \\ G_{\partial \Omega_{ij}}^{(2k_j-1,j)}(\mathbf{x}_m, t_{m\pm}, \mathbf{p}_j, \lambda) > 0 \end{array} \right\} \text{ for } \mathbf{n}_{\partial \Omega_{ij}} \to \Omega_i. \quad (2.56)$$

Proof. See Luo (2008a,b).

2.5 Tangential flows

The tangency of a flow to the boundary is the same as the global tangency of the flow to the separatrix in Luo(2007a,b; 2008c,d). The corresponding definition of the tangential flow is given as follows.

Definition 2.13. For a discontinuous dynamical system in Eq.(2.1), $\mathbf{x}(t_m) \equiv \mathbf{x}_m \in \partial \Omega_{ij}$ at time t_m. For an arbitrarily small $\varepsilon > 0$, there is a time interval $[t_{m-\varepsilon}, t_{m+\varepsilon}]$. Suppose $\mathbf{x}^{(\alpha)}(t_{m\pm}) = \mathbf{x}_m$ ($\alpha \in \{i, j\}$), the flow $\mathbf{x}^{(\alpha)}(t)$ is $C_{[t_{m-\varepsilon},t_{m+\varepsilon}]}^{r_\alpha}$-continuous ($r_\alpha \geqslant 2$) for time t. A flow $\mathbf{x}^{(\alpha)}(t)$ in Ω_α is *tangential* to the boundary $\partial \Omega_{ij}$ if

$$G_{\partial \Omega_{ij}}^{(\alpha)}(\mathbf{x}_m, t_m, \mathbf{p}_\alpha, \lambda) = 0 \text{ and } G_{\partial \Omega_{ij}}^{(1,\alpha)}(\mathbf{x}_m, t_m, \mathbf{p}_\alpha, \lambda) \neq 0, \quad (2.57)$$

2.5 Tangential flows

$$\text{either} \left.\begin{array}{l} \mathbf{n}^T_{\partial\Omega_{ij}}(\mathbf{x}^{(0)}_{m-\varepsilon}) \cdot \left[\mathbf{x}^{(0)}_{m-\varepsilon} - \mathbf{x}^{(\alpha)}_{m-\varepsilon}\right] > 0 \\ \mathbf{n}^T_{\partial\Omega_{ij}}(\mathbf{x}^{(0)}_{m+\varepsilon}) \cdot \left[\mathbf{x}^{(\alpha)}_{m+\varepsilon} - \mathbf{x}^{(0)}_{m+\varepsilon}\right] < 0 \end{array}\right\} \text{ for } \mathbf{n}_{\partial\Omega_{ij}} \to \Omega_\beta$$

$$\text{or} \left.\begin{array}{l} \mathbf{n}^T_{\partial\Omega_{ij}}(\mathbf{x}^{(0)}_{m-\varepsilon}) \cdot \left[\mathbf{x}^{(0)}_{m-\varepsilon} - \mathbf{x}^{(\alpha)}_{m-\varepsilon}\right] < 0 \\ \mathbf{n}^T_{\partial\Omega_{ij}}(\mathbf{x}^{(0)}_{m+\varepsilon}) \cdot \left[\mathbf{x}^{(\alpha)}_{m+\varepsilon} - \mathbf{x}^{(0)}_{m+\varepsilon}\right] > 0 \end{array}\right\} \text{ for } \mathbf{n}_{\partial\Omega_{ij}} \to \Omega_\alpha. \quad (2.58)$$

Theorem 2.9. *For a discontinuous dynamical system in Eq.(2.1), $\mathbf{x}(t_m) \equiv \mathbf{x}_m \in \partial\Omega_{ij}$ at time t_m. For an arbitrarily small $\varepsilon > 0$, there is a time interval $[t_{m-\varepsilon}, t_{m+\varepsilon}]$. Suppose $\mathbf{x}^{(\alpha)}(t_{m\pm}) = \mathbf{x}_m$. The flow $\mathbf{x}^{(\alpha)}(t)$ is $C^{r_\alpha}_{[t_{m-\varepsilon}, t_{m+\varepsilon}]}$-continuous ($r_\alpha \geqslant 2$) for time t, and $\|d^{r_\alpha+1}\mathbf{x}^{(\alpha)}/dt^{r_\alpha+1}\| < \infty$ ($\alpha \in \{i,j\}$). A flow $\mathbf{x}^{(\alpha)}(t)$ in Ω_α is tangential to the boundary $\partial\Omega_{ij}$ iff*

$$G^{(0,\alpha)}_{\partial\Omega_{ij}}(\mathbf{x}_m, t_m, \mathbf{p}_\alpha, \boldsymbol{\lambda}) = 0, \quad (2.59)$$

$$\left.\begin{array}{l} \text{either } G^{(1,\alpha)}_{\partial\Omega_{ij}}(\mathbf{x}_m, t_m, \mathbf{p}_\alpha, \boldsymbol{\lambda}) < 0, \text{ for } \mathbf{n}_{\partial\Omega_{ij}} \to \Omega_\beta \\ \text{or } \quad G^{(1,\alpha)}_{\partial\Omega_{ij}}(\mathbf{x}_m, t_m, \mathbf{p}_\alpha, \boldsymbol{\lambda}) > 0, \text{ for } \mathbf{n}_{\partial\Omega_{ij}} \to \Omega_\alpha. \end{array}\right\} \quad (2.60)$$

Proof. See Luo (2008a,b).

The higher order tangency of a flow to the boundary can be defined as follows.

Definition 2.14. For a discontinuous dynamical system in Eq.(2.1), $\mathbf{x}(t_m) \equiv \mathbf{x}_m \in \partial\Omega_{ij}$ at time t_m. For an arbitrarily small $\varepsilon > 0$, there is a time interval $[t_{m-\varepsilon}, t_{m+\varepsilon}]$. Suppose $\mathbf{x}^{(\alpha)}(t_{m\pm}) = \mathbf{x}_m$. The flow $\mathbf{x}^{(\alpha)}(t)$ is $C^r_{[t_{m-\varepsilon}, t_{m+\varepsilon}]}$-continuous ($r \geqslant k_\alpha + 1$) for time t and $\|d^{r+1}\mathbf{x}^{(\alpha)}/dt^{r+1}\| < \infty$ ($\alpha \in \{i,j\}$). A flow $\mathbf{x}^{(\alpha)}(t)$ in Ω_α is *tangential* to the boundary $\partial\Omega_{ij}$ with the $(2k_\alpha - 1)^{\text{th}}$-order if

$$\left.\begin{array}{l} G^{(s,\alpha)}_{\partial\Omega_{ij}}(\mathbf{x}_m, t_m, \mathbf{p}_\alpha, \boldsymbol{\lambda}) = 0, \text{ for } s = 0, 1, \ldots, 2k_\alpha - 2, \\ G^{(2k_\alpha - 1, \alpha)}_{\partial\Omega_{ij}}(\mathbf{x}_m, t_m, \mathbf{p}_\alpha, \boldsymbol{\lambda}) \neq 0, \end{array}\right\} \quad (2.61)$$

$$\text{either} \left.\begin{array}{l} \mathbf{n}^T_{\partial\Omega_{ij}}(\mathbf{x}^{(0)}_{m-\varepsilon}) \cdot \left[\mathbf{x}^{(0)}_{m-\varepsilon} - \mathbf{x}^{(\alpha)}_{m-\varepsilon}\right] > 0 \\ \mathbf{n}^T_{\partial\Omega_{ij}}(\mathbf{x}^{(0)}_{m+\varepsilon}) \cdot \left[\mathbf{x}^{(\alpha)}_{m+\varepsilon} - \mathbf{x}^{(0)}_{m+\varepsilon}\right] < 0 \end{array}\right\} \text{ for } \mathbf{n}_{\partial\Omega_{ij}} \to \Omega_\beta$$

$$\text{or} \left.\begin{array}{l} \mathbf{n}^T_{\partial\Omega_{ij}}(\mathbf{x}^{(0)}_{m-\varepsilon}) \cdot \left[\mathbf{x}^{(0)}_{m-\varepsilon} - \mathbf{x}^{(\alpha)}_{m-\varepsilon}\right] < 0 \\ \mathbf{n}^T_{\partial\Omega_{ij}}(\mathbf{x}^{(0)}_{m+\varepsilon}) \cdot \left[\mathbf{x}^{(\alpha)}_{m+\varepsilon} - \mathbf{x}^{(0)}_{m+\varepsilon}\right] > 0 \end{array}\right\} \text{ for } \mathbf{n}_{\partial\Omega_{ij}} \to \Omega_\alpha. \quad (2.62)$$

Theorem 2.10. *For a discontinuous dynamical system in Eq.(2.1), $\mathbf{x}(t_m) \equiv \mathbf{x}_m \in \partial\Omega_{ij}$ at time t_m. For an arbitrarily small $\varepsilon > 0$, there is a time interval $[t_{m-\varepsilon}, t_{m+\varepsilon}]$.*

Suppose $\mathbf{x}^{(\alpha)}(t_{m\pm}) = \mathbf{x}_m$, the flow $\mathbf{x}^{(\alpha)}(t)$ is $C^r_{[t_{m-\varepsilon}, t_{m+\varepsilon}]}$-continuous ($r \geq k_\alpha + 1$) for time t and $||d^{r+1}\mathbf{x}^{(\alpha)}/dt^{r+1}|| < \infty$ ($\alpha \in \{i,j\}$). A flow $\mathbf{x}^{(\alpha)}(t)$ in Ω_α is tangential to the boundary $\partial\Omega_{ij}$ with the $(2k_\alpha - 1)^{th}$-order iff

$$G^{(s,\alpha)}_{\partial\Omega_{ij}}(\mathbf{x}_m, t_m, \mathbf{p}_\alpha, \boldsymbol{\lambda}) = 0, \; for \; s = 0, 1, \ldots, 2k_\alpha - 2, \tag{2.63}$$

$$\left. \begin{aligned} either \quad & G^{(2k_\alpha-1,\alpha)}_{\partial\Omega_{ij}}(\mathbf{x}_m, t_m, \mathbf{p}_\alpha, \boldsymbol{\lambda}) < 0, \; for \; \mathbf{n}_{\partial\Omega_{ij}} \to \Omega_\beta \\ or \quad & G^{(2k_\alpha-1,\alpha)}_{\partial\Omega_{ij}}(\mathbf{x}_m, t_m, \mathbf{p}_\alpha, \boldsymbol{\lambda}) > 0, \; for \; \mathbf{n}_{\partial\Omega_{ij}} \to \Omega_\alpha. \end{aligned} \right\} \tag{2.64}$$

Proof. See Luo (2008a,b).

The conditions for grazing bifurcation in domain Ω_i can be given by Luo(2007a). The grazing bifurcation occurring at the boundary can be determined through the G-function (i.e., $G^{(2k_\alpha-1,\alpha)}_{\partial\Omega_{ij}}(\mathbf{x}_m, t_m, \mathbf{p}_\alpha, \boldsymbol{\lambda})$). The conditions for grazing on the boundary are $G^{(s,\alpha)}_{\partial\Omega_{ij}}(\mathbf{x}_m, t_m, \mathbf{p}_\alpha, \boldsymbol{\lambda}) = 0$ ($s = 0, 1, \ldots, 2k_\alpha - 2$) and $G^{(2k_\alpha-1,\alpha)}_{\partial\Omega_{ij}}(\mathbf{x}_m, t_m, \mathbf{p}_\alpha, \boldsymbol{\lambda}) < 0$ for the boundary $\partial\Omega_{ij}$ with $\mathbf{n}_{\partial\Omega_{ij}} \to \Omega_\beta$ and $G^{(2k_\alpha-1,\alpha)}_{\partial\Omega_{ij}}(\mathbf{x}_m, t_m, \mathbf{p}_\alpha, \boldsymbol{\lambda}) > 0$ for the boundary $\partial\Omega_{ij}$ with $\mathbf{n}_{\partial\Omega_{ij}} \to \Omega_\alpha$. To develop a uniform theory of tangential flows in Ω_α with half-non-passable flows in Ω_β as in the previous section, real and imaginary flows will be adopted. Therefore, the tangential flow can be re-defined as follows.

Definition 2.15. For a discontinuous dynamical system in Eq.(2.1), $\mathbf{x}(t_m) \equiv \mathbf{x}_m \in \partial\Omega_{ij}$ at time t_m. For an arbitrarily small $\varepsilon > 0$, there are two time intervals $[t_{m-\varepsilon}, t_m)$ and $[t_{m-\varepsilon}, t_{m+\varepsilon}]$. Suppose $\mathbf{x}^{(i)}(t_{m+}) = \mathbf{x}_m = \mathbf{x}^{(j)}(t_{m\pm})$. The flow $\mathbf{x}^{(i)}(t)$ is $C^{r_i}_{[t_{m-\varepsilon}, t_{m+\varepsilon}]}$-continuous ($r_i \geq 2k_i + 1$) for time t and $||d^{r_i+1}\mathbf{x}^{(i)}/dt^{r_i+1}|| < \infty$. The sink or source flow $\mathbf{x}^{(j)}(t)$ is $C^{r_j}_{[t_{m-\varepsilon}, t_m)}$ or $C^{r_j}_{(t_m, t_{m+\varepsilon}]}$-continuous ($r_j \geq 2k_j$) for time t and $||d^{r_j+1}\mathbf{x}^{(j)}/dt^{r_j+1}|| < \infty$. The flow $\mathbf{x}^{(i)}(t)$ of the $(2k_i-1)^{th}$-order singularity with $\mathbf{x}^{(j)}(t)$ of the $(2k_j)^{th}$-order singularity to the boundary $\partial\Omega_{ij}$ is a $(2k_i - 1 : 2k_j)$-tangential flow in domain Ω_i if

$$\left. \begin{aligned} & G^{(s,i)}_{\partial\Omega_{ij}}(\mathbf{x}_m, t_{m\pm}, \mathbf{p}_i, \boldsymbol{\lambda}) = 0, \; for \; s = 0, 1, \ldots, 2k_i - 2, \\ & G^{(2k_i-1,i)}_{\partial\Omega_{ij}}(\mathbf{x}_m, t_{m\pm}, \mathbf{p}_i, \boldsymbol{\lambda}) \neq 0, \end{aligned} \right\} \tag{2.65}$$

$$\left. \begin{aligned} & G^{(s,j)}_{\partial\Omega_{ij}}(\mathbf{x}_m, t_{m\pm}, \mathbf{p}_j, \boldsymbol{\lambda}) = 0, \; for \; s = 0, 1, \ldots, 2k_j - 1, \\ & G^{(2k_j,j)}_{\partial\Omega_{ij}}(\mathbf{x}_m, t_{m\pm}, \mathbf{p}_j, \boldsymbol{\lambda}) \neq 0, \end{aligned} \right\} \tag{2.66}$$

2.5 Tangential flows

$$\left.\begin{array}{l}\text{either } \mathbf{n}^T_{\partial\Omega_{ij}}(\mathbf{x}^{(0)}_{m-\varepsilon}) \cdot \left[\mathbf{x}^{(0)}_{m-\varepsilon} - \mathbf{x}^{(i)}_{m-\varepsilon}\right] > 0 \\ \quad \mathbf{n}^T_{\partial\Omega_{ij}}(\mathbf{x}^{(0)}_{m+\varepsilon}) \cdot \left[\mathbf{x}^{(i)}_{m+\varepsilon} - \mathbf{x}^{(0)}_{m+\varepsilon}\right] < 0 \end{array}\right\}, \text{ for } \mathbf{n}_{\partial\Omega_{ij}} \to \Omega_j$$

or

$$\left.\begin{array}{l}\mathbf{n}^T_{\partial\Omega_{ij}}(\mathbf{x}^{(0)}_{m-\varepsilon}) \cdot \left[\mathbf{x}^{(0)}_{m-\varepsilon} - \mathbf{x}^{(i)}_{m-\varepsilon}\right] < 0 \\ \mathbf{n}^T_{\partial\Omega_{ij}}(\mathbf{x}^{(0)}_{m+\varepsilon}) \cdot \left[\mathbf{x}^{(i)}_{m+\varepsilon} - \mathbf{x}^{(0)}_{m+\varepsilon}\right] > 0 \end{array}\right\}, \text{ for } \mathbf{n}_{\partial\Omega_{ij}} \to \Omega_i,$$

(2.67)

or

$$\left.\begin{array}{l}\text{either } \mathbf{n}^T_{\partial\Omega_{ij}}(\mathbf{x}^{(0)}_{m+\varepsilon}) \cdot \left[\mathbf{x}^{(j)}_{m+\varepsilon} - \mathbf{x}^{(0)}_{m+\varepsilon}\right] > 0, \text{ for } \mathbf{n}_{\partial\Omega_{ij}} \to \Omega_j \\ \text{or } \quad \mathbf{n}^T_{\partial\Omega_{ij}}(\mathbf{x}^{(0)}_{m+\varepsilon}) \cdot \left[\mathbf{x}^{(j)}_{m+\varepsilon} - \mathbf{x}^{(0)}_{m+\varepsilon}\right] < 0, \text{ for } \mathbf{n}_{\partial\Omega_{ij}} \to \Omega_i, \\ \\ \text{either } \mathbf{n}^T_{\partial\Omega_{ij}}(\mathbf{x}^{(0)}_{m-\varepsilon}) \cdot \left[\mathbf{x}^{(0)}_{m-\varepsilon} - \mathbf{x}^{(j)}_{m-\varepsilon}\right] < 0, \text{ for } \mathbf{n}_{\partial\Omega_{ij}} \to \Omega_j \\ \text{or } \quad \mathbf{n}^T_{\partial\Omega_{ij}}(\mathbf{x}^{(0)}_{m-\varepsilon}) \cdot \left[\mathbf{x}^{(0)}_{m-\varepsilon} - \mathbf{x}^{(j)}_{m-\varepsilon}\right] > 0, \text{ for } \mathbf{n}_{\partial\Omega_{ij}} \to \Omega_i. \end{array}\right\}$$

(2.68)

Such a $(2k_i - 1 : 2k_j)$-tangential flows in Ω_i are sketched in Fig.2.8(a) with source in Ω_j and in Fig.2.8(b) with sink in Ω_j. The $(2k_j - 1 : 2k_i)$-tangential flow in Ω_j are sketched in Fig.2.9(a) with source in Ω_i and in Fig.2.8(b) with sink in Ω_i. The half-sink and source flows are represented by the dotted curves. The tangential flows are presented by solid curves. The dashed curves denote the imaginary flows. If the starting point is on the flow $\mathbf{x}^{(j)}(t)$ (or $\mathbf{x}^{(j)}(t)$) in Fig.2.8(b) (or Fig.2.9(a)), the passable flow from Ω_j to Ω_i (or Ω_j to Ω_i) is formed.

Theorem 2.11. *For a discontinuous dynamical system in Eq.(2.1), $\mathbf{x}(t_m) \equiv \mathbf{x}_m \in \partial\Omega_{ij}$ at time t_m. For an arbitrarily small $\varepsilon > 0$, there are two time intervals $[t_{m-\varepsilon}, t_m)$ and $[t_{m-\varepsilon}, t_{m+\varepsilon}]$. Suppose $\mathbf{x}^{(i)}(t_{m\pm}) = \mathbf{x}_m = \mathbf{x}^{(j)}(t_{m\pm})$. The flow $\mathbf{x}^{(i)}(t)$ is $C^{r_i}_{[t_{m-\varepsilon}, t_{m+\varepsilon}]}$-continuous for time t and $||d^{r_i+1}\mathbf{x}^{(i)}/dt^{r_i+1}|| < \infty$ $(r_i \geq 2k_i + 1)$. The sink or source flow $\mathbf{x}^{(j)}(t)$ is $C^{r_j}_{[t_{m-\varepsilon}, t_m)}$ or $C^{r_j}_{(t_m, t_{m+\varepsilon}]}$-continuous $(r_j \geq 2k_j)$ for time t and $||d^{r_j+1}\mathbf{x}^{(j)}/dt^{r_j+1}|| < \infty$. The flow $\mathbf{x}^{(i)}(t)$ of the $(2k_i - 1)^{th}$-order singularity and $\mathbf{x}^{(j)}(t)$ of the $(2k_j)^{th}$-order singularity to the boundary $\partial\Omega_{ij}$ is $(2k_i - 1 : 2k_j)$-tangential flow in domain Ω_i iff*

$$G^{(s,i)}_{\partial\Omega_{ij}}(\mathbf{x}_m, t_{m\pm}, \mathbf{p}_i, \lambda) = 0, \; \text{for } s = 0, 1, \ldots, 2k_i - 2, \quad (2.69)$$

$$G^{(s,j)}_{\partial\Omega_{ij}}(\mathbf{x}_m, t_{m\pm}, \mathbf{p}_j, \lambda) = 0, \; \text{for } s = 0, 1, \ldots, 2k_j - 1, \quad (2.70)$$

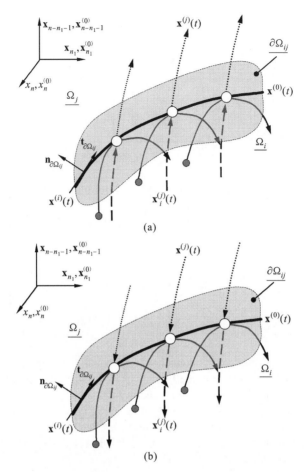

Fig. 2.8 The $(2k_i - 1 : 2k_j)$-tangential flows in Ω_i: (a) with source in Ω_j and (b) with sink in Ω_j. $\mathbf{x}^{(i)}(t)$ and $\mathbf{x}^{(j)}(t)$ represent the *real* flows in domains Ω_i and Ω_j, respectively, which are depicted by the thin solid curves. $\mathbf{x}_i^{(j)}(t)$ and $\mathbf{x}_j^{(i)}(t)$ represent the *imaginary* flows in domains Ω_i and Ω_j, respectively controlled by the vector fields in Ω_j and Ω_i, which are depicted by the dashed curves. The flow on the boundary is described by $\mathbf{x}^{(0)}(t)$. The normal and tangential vectors $\mathbf{n}_{\partial\Omega_{ij}}$ and $\mathbf{t}_{\partial\Omega_{ij}}$ on the boundary are depicted. The hollow circles are tangential (or grazing) points and the small shaded circles are the starting points

$$\left. \begin{array}{l} \left. \begin{array}{l} G_{\partial\Omega_{ij}}^{(2k_i-1,i)}(\mathbf{x}_m, t_{m\pm}, \mathbf{p}_i, \boldsymbol{\lambda}) < 0 \\ \textit{either} \quad G_{\partial\Omega_{ij}}^{(2k_j,j)}(\mathbf{x}_m, t_{m-}, \mathbf{p}_j, \boldsymbol{\lambda}) < 0 \\ \textit{or} \; G_{\partial\Omega_{ij}}^{(2k_j,j)}(\mathbf{x}_m, t_{m+}, \mathbf{p}_j, \boldsymbol{\lambda}) > 0 \end{array} \right\} \textit{for } \mathbf{n}_{\partial\Omega_{ij}} \to \Omega_j \\ \left. \begin{array}{l} G_{\partial\Omega_{ij}}^{(2k_i-1,i)}(\mathbf{x}_m, t_{m\pm}, \mathbf{p}_i, \boldsymbol{\lambda}) > 0 \\ \textit{or} \quad G_{\partial\Omega_{ij}}^{(2k_j,j)}(\mathbf{x}_m, t_{m-}, \mathbf{p}_j, \boldsymbol{\lambda}) > 0 \\ \textit{or} \; G_{\partial\Omega_{ij}}^{(2k_j,j)}(\mathbf{x}_m, t_{m+}, \mathbf{p}_j, \boldsymbol{\lambda}) < 0 \end{array} \right\} \textit{for } \mathbf{n}_{\partial\Omega_{ij}} \to \Omega_j \end{array} \right\} \quad (2.71)$$

2.5 Tangential flows

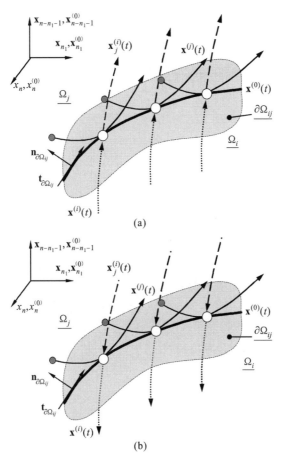

Fig. 2.9 The $(2k_j - 1 : 2k_i)$-tangential flows in Ω_j: (a) with sink in Ω_i and (b) with source in Ω_i. $\mathbf{x}^{(i)}(t)$ and $\mathbf{x}^{(j)}(t)$ represent the *real* flows in domains Ω_i and Ω_j, respectively, which are depicted by the thin solid curves. $\mathbf{x}_i^{(j)}(t)$ and $\mathbf{x}_j^{(i)}(t)$ represent the *imaginary* flows in domains Ω_i and Ω_j, respectively controlled by the vector fields in Ω_j and Ω_i, which are depicted by the dashed curves. The flow on the boundary is described by $\mathbf{x}^{(0)}(t)$. The normal and tangential vectors $\mathbf{n}_{\partial\Omega_{ij}}$ and $\mathbf{t}_{\partial\Omega_{ij}}$ on the boundary are depicted. The hollow circles are tangential (or grazing) points and the small shaded circles are the starting points

Proof. See Luo (2008a,b).

Definition 2.16. For a discontinuous dynamical system in Eq.(2.1), $\mathbf{x}(t_m) \equiv \mathbf{x}_m \in \partial\Omega_{ij}$ at time t_m. For an arbitrarily small $\varepsilon > 0$, there is a time interval $[t_{m-\varepsilon}, t_{m+\varepsilon}]$. Suppose $\mathbf{x}^{(\alpha)}(t_{m\pm}) = \mathbf{x}_m = \mathbf{x}_\alpha^{(\beta)}(t_{m\pm})$ ($\alpha, \beta \in \{i, j\}$ and $\alpha \neq \beta$). The flow $\mathbf{x}^{(\alpha)}(t)$ is $C_{[t_{m-\varepsilon}, t_{m+\varepsilon}]}^{r_\alpha}$-continuous ($r_\alpha \geq 2k_\alpha$) for time t and $\|d^{r_\alpha+1}\mathbf{x}^{(\alpha)}/dt^{r_\alpha+1}\| < \infty$. The imaginary flow $\mathbf{x}_\alpha^{(\beta)}(t)$ is $C_{[t_{m-\varepsilon}, t_{m+\varepsilon}]}^{r}$-continuous ($r_\beta \geq 2k_\beta$) for time t and $\|d^{r_\beta+1}\mathbf{x}_\alpha^{(\beta)}/dt^{r_\beta+1}\| < \infty$. The flow $\mathbf{x}^{(\alpha)}(t)$ of the $(2k_\alpha - 1)^{\text{th}}$-order singularity and

$\mathbf{x}_\alpha^{(\beta)}(t)$ of the $(2k_\beta-1)^{\text{th}}$-order singularity to $\partial\Omega_{ij}$ is $(2k_\alpha-1:2k_\beta-1)$-tangential flow in domain Ω_α if

$$\left.\begin{aligned}G^{(s,\alpha)}_{\partial\Omega_{\alpha\beta}}(\mathbf{x}_m,t_{m\pm},\mathbf{p}_\alpha,\lambda)&=0,\text{ for }s=0,1,\ldots,2k_\alpha-2,\\ G^{(2k_\alpha-1,\alpha)}_{\partial\Omega_{\alpha\beta}}(\mathbf{x}_m,t_{m\pm},\mathbf{p}_\alpha,\lambda)&\neq 0,\end{aligned}\right\} \quad (2.72)$$

$$\left.\begin{aligned}G^{(s,\beta)}_{\partial\Omega_{\alpha\beta}}(\mathbf{x}_m,t_{m\pm},\mathbf{p}_\beta,\lambda)&=0,\text{ for }s=0,1,\ldots,2k_\beta-2,\\ G^{(2k_\beta-1,\beta)}_{\partial\Omega_{\alpha\beta}}(\mathbf{x}_m,t_{m-},\mathbf{p}_\beta,\lambda)&\neq 0,\end{aligned}\right\} \quad (2.73)$$

either
$$\left.\begin{aligned}\mathbf{n}^{\text{T}}_{\partial\Omega_{ij}}(\mathbf{x}^{(0)}_{m-\varepsilon})\cdot\left[\mathbf{x}^{(0)}_{m-\varepsilon}-\mathbf{x}^{(\alpha)}_{m-\varepsilon}\right]&>0\\ \mathbf{n}^{\text{T}}_{\partial\Omega_{ij}}(\mathbf{x}^{(0)}_{m+\varepsilon})\cdot\left[\mathbf{x}^{(\alpha)}_{m+\varepsilon}-\mathbf{x}^{(0)}_{m+\varepsilon}\right]&<0\end{aligned}\right\}\text{ for }\mathbf{n}_{\partial\Omega_{\alpha\beta}}\to\Omega_\beta \quad (2.74a)$$

or
$$\left.\begin{aligned}\mathbf{n}^{\text{T}}_{\partial\Omega_{ij}}(\mathbf{x}^{(0)}_{m-\varepsilon})\cdot\left[\mathbf{x}^{(0)}_{m-\varepsilon}-\mathbf{x}^{(\alpha)}_{m-\varepsilon}\right]&<0\\ \mathbf{n}^{\text{T}}_{\partial\Omega_{ij}}(\mathbf{x}^{(0)}_{m+\varepsilon})\cdot\left[\mathbf{x}^{(\alpha)}_{m+\varepsilon}-\bar{\mathbf{x}}^{(0)}_{m+\varepsilon}\right]&>0\end{aligned}\right\}\text{ for }\mathbf{n}_{\partial\Omega_{\alpha\beta}}\to\Omega_\alpha, \quad (2.74b)$$

either
$$\mathbf{n}^{\text{T}}_{\partial\Omega_{ij}}(\mathbf{x}^{(0)}_{m-\varepsilon})\cdot\left[\mathbf{x}^{(0)}_{m-\varepsilon}-\mathbf{x}^{(\beta)}_{\alpha(m-\varepsilon)}\right]>0$$
$$\mathbf{n}^{\text{T}}_{\partial\Omega_{ij}}(\mathbf{x}^{(0)}_{m+\varepsilon})\cdot\left[\mathbf{x}^{(\beta)}_{\alpha(m+\varepsilon)}-\mathbf{x}^{(0)}_{m+\varepsilon}\right]<0 \quad\text{for }\mathbf{n}_{\partial\Omega_{\alpha\beta}}\to\Omega_\beta$$

or
$$\mathbf{n}^{\text{T}}_{\partial\Omega_{ij}}(\mathbf{x}^{(0)}_{m-\varepsilon})\cdot\left[\mathbf{x}^{(0)}_{m-\varepsilon}-\mathbf{x}^{(\beta)}_{\alpha(m-\varepsilon)}\right]<0$$
$$\mathbf{n}^{\text{T}}_{\partial\Omega_{ij}}(\mathbf{x}^{(0)}_{m+\varepsilon})\cdot\left[\mathbf{x}^{(\beta)}_{\alpha(m+\varepsilon)}-\mathbf{x}^{(0)}_{m+\varepsilon}\right]>0 \quad\text{for }\mathbf{n}_{\partial\Omega_{\alpha\beta}}\to\Omega_\alpha. \quad (2.75)$$

The $(2k_\alpha-1:2k_\beta-1)$-tangential flows in Ω_α and Ω_β ($\alpha,\beta\in\{i,j\}$ and $\alpha\neq\beta$) are sketched in Fig.2.10 with the corresponding imaginary tangential flows. The real tangential flows are presented by the solid curves. The dashed curves denote the imaginary tangential flows.

Theorem 2.12. *For a discontinuous dynamical system in Eq.(2.1), $\mathbf{x}(t_m)\equiv\mathbf{x}_m\in\partial\Omega_{ij}$ at time t_m. For an arbitrarily small $\varepsilon>0$, there is a time interval $[t_{m-\varepsilon},t_{m+\varepsilon}]$. Suppose $\mathbf{x}^{(\alpha)}(t_{m\pm})=\mathbf{x}_m=\mathbf{x}^{(\beta)}_\alpha(t_{m\pm})$ ($\alpha,\beta\in\{i,j\}$ and $\alpha\neq\beta$). The flow $\mathbf{x}^{(\alpha)}(t)$ is $C^{r_\alpha}_{[t_{m-\varepsilon},t_{m+\varepsilon}]}$-continuous ($r_\alpha\geq 2k_\alpha$) for time t and $\|d^{r_\alpha+1}\mathbf{x}^{(\alpha)}/dt^{r_\alpha+1}\|<\infty$. The imaginary flow $\mathbf{x}^{(\beta)}_\alpha(t)$ is $C^r_{[t_{m-\varepsilon},t_{m+\varepsilon}]}$-continuous ($r_\beta\geq 2k_\beta$) for time t and $\|d^{r_\beta+1}\mathbf{x}^{(\beta)}_\alpha/dt^{r_\beta+1}\|<\infty$. The flow $\mathbf{x}^{(\alpha)}(t)$ of the $(2k_\alpha-1)^{\text{th}}$-order singularity and $\mathbf{x}^{(\beta)}_\alpha(t)$ of the $(2k_\beta-1)^{\text{th}}$-order singularity to $\partial\Omega_{ij}$ is a $(2k_\alpha-1:2k_\beta-1)$-tangential flow in domain Ω_α iff*

2.5 Tangential flows

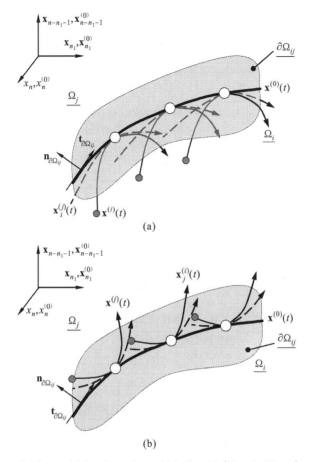

Fig. 2.10 Tangential flows with imaginary flows: (a) in Ω_i with $(2k_i - 1 : 2k_j - 1)$-order and (b) in Ω_j with $(2k_j - 1 : 2k_i - 1)$-order. $\mathbf{x}^{(i)}(t)$ and $\mathbf{x}^{(j)}(t)$ represent the *real* flows in domains Ω_i and Ω_j, respectively, which are depicted by the thin solid curves. $\mathbf{x}_i^{(j)}(t)$ and $\mathbf{x}_j^{(i)}(t)$ represent the *imaginary* flows in domains Ω_i and Ω_j, respectively controlled by the vector fields in Ω_j and Ω_i, which are depicted by the dashed curves. The flow on the boundary is described by $\mathbf{x}^{(0)}(t)$. The normal and tangential vectors $\mathbf{n}_{\partial \Omega_{ij}}$ and $\mathbf{t}_{\partial \Omega_{ij}}$ on the boundary are depicted. The hollow circles are tangential (or grazing) points and the shaded circles are the starting points

$$G_{\partial \Omega_{\alpha\beta}}^{(s,\alpha)}(\mathbf{x}_m, t_{m\pm}, \mathbf{p}_\alpha, \boldsymbol{\lambda}) = 0, \ for \ s = 0, 1, \ldots, 2k_\alpha - 2, \tag{2.76}$$

$$G_{\partial \Omega_{\alpha\beta}}^{(s,\beta)}(\mathbf{x}_m, t_{m\pm}, \mathbf{p}_\beta, \boldsymbol{\lambda}) = 0, \ for \ s = 0, 1, \ldots, 2k_\beta - 2; \tag{2.77}$$

$$either \ \left. \begin{array}{l} G_{\partial \Omega_{\alpha\beta}}^{(2k_\alpha-1,\alpha)}(\mathbf{x}_m, t_{m\pm}, \mathbf{p}_\alpha, \boldsymbol{\lambda}) < 0 \\ G_{\partial \Omega_{\alpha\beta}}^{(2k_\beta-1,\beta)}(\mathbf{x}_m, t_{m\pm}, \mathbf{p}_\beta, \boldsymbol{\lambda}) < 0 \end{array} \right\} for \ \mathbf{n}_{\partial \Omega_{\alpha\beta}} \to \Omega_\beta \tag{2.78a}$$

$$\left.\begin{array}{l} or \quad G^{(2k_\alpha-1,\alpha)}_{\partial\Omega_{\alpha\beta}}(\mathbf{x}_m,t_{m\pm},\mathbf{p}_\alpha,\boldsymbol{\lambda}) > 0 \\ G^{(2k_\beta-1,\beta)}_{\partial\Omega_{\alpha\beta}}(\mathbf{x}_m,t_{m\pm},\mathbf{p}_\beta,\boldsymbol{\lambda}) > 0 \end{array}\right\} for\ \mathbf{n}_{\partial\Omega_{\alpha\beta}} \to \Omega_\alpha. \quad (2.78b)$$

Proof. See Luo (2008a,b).

Definition 2.17. For a discontinuous dynamical system in Eq.(2.1), $\mathbf{x}(t_m) \equiv \mathbf{x}_m \in \partial\Omega_{ij}$ at time t_m. For an arbitrarily small $\varepsilon > 0$, there is a time interval $[t_{m-\varepsilon},t_{m+\varepsilon}]$. Suppose $\mathbf{x}^{(\alpha)}(t_{m\pm}) = \mathbf{x}_m = \mathbf{x}^{(\beta)}(t_{m\pm})$ ($\alpha,\beta \in \{i,j\}$ and $\alpha \neq \beta$). The flow $\mathbf{x}^{(\alpha)}(t)$ is $C^{r_\alpha}_{[t_{m-\varepsilon},t_{m+\varepsilon}]}$-continuous ($r_\alpha \geqslant 2k_\alpha$) for time t and $||d^{r_\alpha+1}\mathbf{x}^{(\alpha)}/dt^{r_\alpha+1}|| < \infty$. The flow $\mathbf{x}^{(\beta)}(t)$ is $C^{r_\beta}_{[t_{m-\varepsilon},t_{m+\varepsilon}]}$-continuous for time t and $||d^{r_\alpha+1}\mathbf{x}^{(\beta)}/dt^{r_\alpha+1}|| < \infty$ ($r_\beta \geqslant 2k_\beta$). The flow $\mathbf{x}^{(\alpha)}(t)$ of the $(2k_\alpha - 1)^{\text{th}}$-order singularity and $\mathbf{x}^{(\beta)}(t)$ of the $(2k_\beta - 1)^{\text{th}}$-order singularity to the boundary $\partial\Omega_{\alpha\beta}$ is a $(2k_\alpha - 1 : 2k_\beta - 1)$-double tangential flow if

$$\left.\begin{array}{l} G^{(s,\alpha)}_{\partial\Omega_{\alpha\beta}}(\mathbf{x}_m,t_{m\pm},\mathbf{p}_\alpha,\boldsymbol{\lambda}) = 0,\ \text{for}\ s = 0,1,\ldots,2k_\alpha - 2, \\ G^{(2k_\alpha-1,\alpha)}_{\partial\Omega_{\alpha\beta}}(\mathbf{x}_m,t_{m\pm},\mathbf{p}_\alpha,\boldsymbol{\lambda}) \neq 0, \end{array}\right\} \quad (2.79)$$

$$\left.\begin{array}{l} G^{(s,\beta)}_{\partial\Omega_{\alpha\beta}}(\mathbf{x}_m,t_{m\pm},\mathbf{p}_\beta,\boldsymbol{\lambda}) = 0,\quad \text{for}\ s = 0,1,\ldots,2k_\beta - 2, \\ G^{(2k_\beta-1,\beta)}_{\partial\Omega_{\alpha\beta}}(\mathbf{x}_m,t_{m-},\mathbf{p}_\beta,\boldsymbol{\lambda}) \neq 0, \end{array}\right\} \quad (2.80)$$

$$either \quad \left.\begin{array}{l} \mathbf{n}^{\text{T}}_{\partial\Omega_{ij}}(\mathbf{x}^{(0)}_{m-\varepsilon}) \cdot \left[\mathbf{x}^{(0)}_{m-\varepsilon} - \mathbf{x}^{(\alpha)}_{m-\varepsilon}\right] > 0 \\ \mathbf{n}^{\text{T}}_{\partial\Omega_{ij}}(\mathbf{x}^{(0)}_{m+\varepsilon}) \cdot \left[\mathbf{x}^{(\alpha)}_{m+\varepsilon} - \mathbf{x}^{(0)}_{m+\varepsilon}\right] < 0 \end{array}\right\} for\ \mathbf{n}_{\partial\Omega_{\alpha\beta}} \to \Omega_\beta$$

$$or \quad \left.\begin{array}{l} \mathbf{n}^{\text{T}}_{\partial\Omega_{ij}}(\mathbf{x}^{(0)}_{m-\varepsilon}) \cdot \left[\mathbf{x}^{(0)}_{m-\varepsilon} - \mathbf{x}^{(\alpha)}_{m-\varepsilon}\right] < 0 \\ \mathbf{n}^{\text{T}}_{\partial\Omega_{ij}}(\mathbf{x}^{(0)}_{m+\varepsilon}) \cdot \left[\mathbf{x}^{(\alpha)}_{m+\varepsilon} - \bar{\mathbf{x}}^{(0)}_{m+\varepsilon}\right] > 0 \end{array}\right\} for\ \mathbf{n}_{\partial\Omega_{\alpha\beta}} \to \Omega_\alpha, \quad (2.81)$$

$$either \quad \left.\begin{array}{l} \mathbf{n}^{\text{T}}_{\partial\Omega_{ij}}(\mathbf{x}^{(0)}_{m-\varepsilon}) \cdot \left[\mathbf{x}^{(0)}_{m-\varepsilon} - \mathbf{x}^{(\beta)}_{m-\varepsilon}\right] < 0 \\ \mathbf{n}^{\text{T}}_{\partial\Omega_{ij}}(\mathbf{x}^{(0)}_{m+\varepsilon}) \cdot \left[\mathbf{x}^{(\beta)}_{m+\varepsilon} - \mathbf{x}^{(0)}_{m+\varepsilon}\right] > 0 \end{array}\right\} for\ \mathbf{n}_{\partial\Omega_{\alpha\beta}} \to \Omega_\beta \quad (2.82a)$$

$$or \quad \left.\begin{array}{l} \mathbf{n}^{\text{T}}_{\partial\Omega_{ij}}(\mathbf{x}^{(0)}_{m-\varepsilon}) \cdot \left[\mathbf{x}^{(0)}_{m-\varepsilon} - \mathbf{x}^{(\beta)}_{m-\varepsilon}\right] > 0 \\ \mathbf{n}^{\text{T}}_{\partial\Omega_{ij}}(\mathbf{x}^{(0)}_{m+\varepsilon}) \cdot \left[\mathbf{x}^{(\beta)}_{m+\varepsilon} - \mathbf{x}^{(0)}_{m+\varepsilon}\right] < 0 \end{array}\right\} for\ \mathbf{n}_{\partial\Omega_{\alpha\beta}} \to \Omega_\alpha. \quad (2.82b)$$

The $(2k_\alpha - 1 : 2k_\beta - 1)$-double tangential flows are sketched in Fig.2.11(a) by the solid curves. The double tangential flow is formed by the two real tangential flows in both domains. The $(2k_\alpha - 1 : 2k_\beta - 1)$-double inaccessible tangential flows are

2.5 Tangential flows

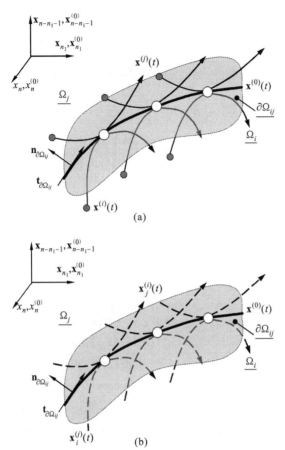

Fig. 2.11 (a) Tangential flow in both Ω_i and Ω_j and (b) imaginary tangential flow in both Ω_i and Ω_j. $\mathbf{x}^{(i)}(t)$ and $\mathbf{x}^{(j)}(t)$ represent the *real* flows in domains Ω_i and Ω_j, respectively, which are depicted by the thin solid curves. $\mathbf{x}_i^{(j)}(t)$ and $\mathbf{x}_j^{(i)}(t)$ represent the *imaginary* flows in domains Ω_i and Ω_j, respectively controlled by the vector fields in Ω_j and Ω_i, which are depicted by the dashed curves. The flow on the boundary is described by $\mathbf{x}^{(0)}(t)$. The normal and tangential vectors $\mathbf{n}_{\partial \Omega_{ij}}$ and $\mathbf{t}_{\partial \Omega_{ij}}$ on the boundary are depicted. The hollow circles are tangential (or grazing) points and the shaded circles are the starting points

sketched in Fig.2.11(b) by the dashed curves. Such a double inaccessible flow is formed by two imaginary tangential flows to the boundary. The corresponding necessary and sufficient conditions for the double-tangential flows are given as follows.

Theorem 2.13. *For a discontinuous dynamical system in Eq.(2.1), $\mathbf{x}(t_m) \equiv \mathbf{x}_m \in \partial \Omega_{ij}$ at time t_m. For an arbitrarily small $\varepsilon > 0$, there is a time interval $[t_{m-\varepsilon}, t_{m+\varepsilon}]$. Suppose $\mathbf{x}^{(\alpha)}(t_{m\pm}) = \mathbf{x}_m = \mathbf{x}^{(\beta)}(t_{m\pm})$ $(\alpha, \beta \in \{i, j\}$ and $\alpha \neq \beta)$. The flow $\mathbf{x}^{(\alpha)}(t)$ is $C^{r_\alpha}_{[t_{m-\varepsilon}, t_{m+\varepsilon}]}$-continuous $(r_\alpha \geq 2k_\alpha)$ for time t and $||d^{r_\alpha+1}\mathbf{x}^{(\alpha)}/dt^{r_\alpha+1}|| < \infty$. The flow $\mathbf{x}^{(\beta)}(t)$ is $C^{r_\beta}_{[t_{m-\varepsilon}, t_{m+\varepsilon}]}$-continuous $(r_\beta \geq 2k_\beta)$ for time t and $||d^{r_\beta+1}\mathbf{x}^{(j)}/dt^{r_\beta+1}|| < \infty$.*

The flow $\mathbf{x}^{(\alpha)}(t)$ of the $(2k_\alpha - 1)^{\text{th}}$-order singularity and $\mathbf{x}^{(\beta)}(t)$ of the $(2k_\beta - 1)^{\text{th}}$-order singularity to $\partial\Omega_{\alpha\beta}$ is a $(2k_\alpha - 1 : 2k_\beta - 1)$-double tangential flow iff

$$G^{(s,\alpha)}_{\partial\Omega_{\alpha\beta}}(\mathbf{x}_m, t_{m\pm}, \mathbf{p}_\alpha, \boldsymbol{\lambda}) = 0, \ for \ s = 0, 1, \ldots, 2k_\alpha - 2, \tag{2.83}$$

$$G^{(s,\beta)}_{\partial\Omega_{\alpha\beta}}(\mathbf{x}_m, t_{m\pm}, \mathbf{p}_\beta, \boldsymbol{\lambda}) = 0, \ for \ s = 0, 1, \ldots, 2k_\beta - 2, \tag{2.84}$$

$$\left.\begin{array}{l} either \ \begin{array}{l} G^{(2k_\alpha-1,\alpha)}_{\partial\Omega_{\alpha\beta}}(\mathbf{x}_m, t_{m\pm}, \mathbf{p}_\alpha, \boldsymbol{\lambda}) < 0 \\ G^{(2k_\beta-1,\beta)}_{\partial\Omega_{\alpha\beta}}(\mathbf{x}_m, t_{m\pm}, \mathbf{p}_\beta, \boldsymbol{\lambda}) > 0 \end{array} \right\} for \ \mathbf{n}_{\partial\Omega_{\alpha\beta}} \to \Omega_\beta \\ or \ \begin{array}{l} G^{(2k_\alpha-1,\alpha)}_{\partial\Omega_{\alpha\beta}}(\mathbf{x}_m, t_{m\pm}, \mathbf{p}_\alpha, \boldsymbol{\lambda}) > 0 \\ G^{(2k_\beta-1,\beta)}_{\partial\Omega_{\alpha\beta}}(\mathbf{x}_m, t_{m\pm}, \mathbf{p}_\beta, \boldsymbol{\lambda}) < 0 \end{array} \right\} for \ \mathbf{n}_{\partial\Omega_{\alpha\beta}} \to \Omega_\alpha. \end{array}\right\} \tag{2.85}$$

Proof. See Luo(2008a,b).

Definition 2.18. For a discontinuous dynamical system in Eq.(2.1), $\mathbf{x}(t_m) \equiv \mathbf{x}_m \in \partial\Omega_{ij}$ at time t_m. For an arbitrarily small $\varepsilon > 0$, there is a time interval $[t_{m-\varepsilon}, t_{m+\varepsilon}]$. Suppose $\mathbf{x}^{(\alpha)}_\beta(t_{m\pm}) = \mathbf{x}_m = \mathbf{x}^{(\beta)}_\alpha(t_{m\pm})$ ($\alpha, \beta \in \{i, j\}$ and $\alpha \neq \beta$). The flow $\mathbf{x}^{(\alpha)}_\beta(t)$ is $C^{r_\alpha}_{[t_{m-\varepsilon}, t_{m+\varepsilon}]}$-continuous ($r_\alpha \geq 2k_\alpha$) for time t and $\|d^{r_\alpha+1}\mathbf{x}^{(\alpha)}/dt^{r_\alpha+1}\| < \infty$. The imaginary flow $\mathbf{x}^{(\beta)}_\alpha(t)$ is $C^{r_\beta}_{[t_{m-\varepsilon}, t_{m+\varepsilon}]}$-continuous ($r_\beta \geq 2k_\beta$) for time t and $\|d^{r_\beta+1}\mathbf{x}^{(\beta)}_\alpha / dt^{r_\beta+1}\| < \infty$. The flow $\mathbf{x}^{(\alpha)}_\beta(t)$ of the $(2k_\alpha - 1)^{\text{th}}$-order singularity and $\mathbf{x}^{(\beta)}_\alpha(t)$ of the $(2k_\beta - 1)^{\text{th}}$-order singularity to $\partial\Omega_{ij}$ is a $(2k_\alpha - 1 : 2k_\beta - 1)$-double inaccessible tangential flow if

$$\left.\begin{array}{l} G^{(s,\alpha)}_{\partial\Omega_{\alpha\beta}}(\mathbf{x}_m, t_{m\pm}, \mathbf{p}_\alpha, \boldsymbol{\lambda}) = 0, \quad for \ s = 0, 1, \ldots, 2k_\alpha - 2, \\ G^{(2k_\alpha-1,\alpha)}_{\partial\Omega_{\alpha\beta}}(\mathbf{x}_m, t_{m\pm}, \mathbf{p}_\alpha, \boldsymbol{\lambda}) \neq 0, \end{array}\right\} \tag{2.86}$$

$$\left.\begin{array}{l} G^{(s,\beta)}_{\partial\Omega_{\alpha\beta}}(\mathbf{x}_m, t_{m\pm}, \mathbf{p}_\beta, \boldsymbol{\lambda}) = 0, \quad for \ s = 0, 1, \ldots, 2k_\beta - 2, \\ G^{(2k_\beta-1,\beta)}_{\partial\Omega_{\alpha\beta}}(\mathbf{x}_m, t_{m-}, \mathbf{p}_\beta, \boldsymbol{\lambda}) \neq 0, \end{array}\right\} \tag{2.87}$$

$$\left.\begin{array}{l} either \ \begin{array}{l} \mathbf{n}^T_{\partial\Omega_{ij}}(\mathbf{x}^{(0)}_{m-\varepsilon}) \cdot [\mathbf{x}^{(0)}_{m-\varepsilon} - \mathbf{x}^{(\alpha)}_{\beta(m-\varepsilon)}] < 0 \\ \mathbf{n}^T_{\partial\Omega_{ij}}(\mathbf{x}^{(0)}_{m+\varepsilon}) \cdot [\mathbf{x}^{(\alpha)}_{\beta(m+\varepsilon)} - \mathbf{x}^{(0)}_{m+\varepsilon}] > 0 \end{array} \right\} for \ \mathbf{n}_{\partial\Omega_{\alpha\beta}} \to \Omega_\beta \\ or \ \begin{array}{l} \mathbf{n}^T_{\partial\Omega_{ij}}(\mathbf{x}^{(0)}_{m-\varepsilon}) \cdot [\mathbf{x}^{(0)}_{m-\varepsilon} - \mathbf{x}^{(\alpha)}_{\beta(m-\varepsilon)}] > 0 \\ \mathbf{n}^T_{\partial\Omega_{ij}}(\mathbf{x}^{(0)}_{m+\varepsilon}) \cdot [\mathbf{x}^{(\alpha)}_{\beta(m+\varepsilon)} - \mathbf{x}^{(0)}_{m+\varepsilon}] < 0 \end{array} \right\} for \ \mathbf{n}_{\partial\Omega_{\alpha\beta}} \to \Omega_\alpha, \end{array}\right\} \tag{2.88}$$

2.6 Switching bifurcations

$$\left. \begin{array}{l} \text{either} \begin{array}{l} \mathbf{n}_{\partial\Omega_{ij}}^{\mathrm{T}}(\mathbf{x}_{m-\varepsilon}^{(0)}) \cdot [\mathbf{x}_{m-\varepsilon}^{(0)} - \mathbf{x}_{\alpha(m-\varepsilon)}^{(\beta)}] > 0 \\ \mathbf{n}_{\partial\Omega_{ij}}^{\mathrm{T}}(\mathbf{x}_{m+\varepsilon}^{(0)}) \cdot [\mathbf{x}_{\alpha(m+\varepsilon)}^{(\beta)} - \mathbf{x}_{m+\varepsilon}^{(0)}] < 0 \end{array} \right\} \text{ for } \mathbf{n}_{\partial\Omega_{\alpha\beta}} \to \Omega_{\beta} \\ \text{or} \begin{array}{l} \mathbf{n}_{\partial\Omega_{ij}}^{\mathrm{T}}(\mathbf{x}_{m-\varepsilon}^{(0)}) \cdot [\mathbf{x}_{m-\varepsilon}^{(0)} - \mathbf{x}_{\alpha(m-\varepsilon)}^{(\beta)}] < 0 \\ \mathbf{n}_{\partial\Omega_{ij}}^{\mathrm{T}}(\mathbf{x}_{m+\varepsilon}^{(0)}) \cdot [\mathbf{x}_{\alpha(m+\varepsilon)}^{(\beta)} - \mathbf{x}_{m+\varepsilon}^{(0)}] > 0 \end{array} \right\} \text{ for } \mathbf{n}_{\partial\Omega_{\alpha\beta}} \to \Omega_{\alpha}. \end{array} \right\} \quad (2.89)$$

The corresponding necessary and sufficient conditions for the tangential flows are given as follows.

Theorem 2.14. *For a discontinuous dynamical system in Eq.(2.1), $\mathbf{x}(t_m) \equiv \mathbf{x}_m \in \partial\Omega_{ij}$ at time t_m. For an arbitrarily small $\varepsilon > 0$, there is a time interval $[t_{m-\varepsilon}, t_{m+\varepsilon}]$. Suppose $\mathbf{x}_{\beta}^{(\alpha)}(t_{m\pm}) = \mathbf{x}_m = \mathbf{x}_{\alpha}^{(\beta)}(t_{m\pm})$ ($\alpha, \beta \in \{i, j\}$ and $\alpha \neq \beta$). The flow $\mathbf{x}_{\beta}^{(\alpha)}(t)$ is $C_{[t_{m-\varepsilon}, t_{m+\varepsilon}]}^{r_\alpha}$-continuous ($r_\alpha \geq 2k_\alpha$) for time t and $\|d^{r_\alpha+1}\mathbf{x}^{(\alpha)}/dt^{r_\alpha+1}\| < \infty$. The imaginary flow $\mathbf{x}_{\alpha}^{(\beta)}(t)$ is $C_{[t_{m-\varepsilon}, t_{m+\varepsilon}]}^{r_\beta}$-continuous ($r_\beta \geq 2k_\beta$) for time t and $\|d^{r_\beta+1}\mathbf{x}_{\alpha}^{(\beta)}/dt^{r_\beta+1}\| < \infty$. The flow $\mathbf{x}_{\beta}^{(\alpha)}(t)$ of the $(2k_\alpha - 1)^{\mathrm{th}}$-order singularity and $\mathbf{x}_{\alpha}^{(\beta)}(t)$ of the $(2k_\beta - 1)^{\mathrm{th}}$-order singularity to the boundary $\partial\Omega_{ij}$ is a $(2k_\alpha - 1 : 2k_\beta - 1)$-double inaccessible tangential flow iff*

$$G_{\partial\Omega_{\alpha\beta}}^{(s,\alpha)}(\mathbf{x}_m, t_{m\pm}, \mathbf{p}_\alpha, \boldsymbol{\lambda}) = 0, \text{ for } s = 0, 1, \ldots, 2k_\alpha - 2, \quad (2.90)$$

$$G_{\partial\Omega_{\alpha\beta}}^{(s,\beta)}(\mathbf{x}_m, t_{m\pm}, \mathbf{p}_\beta, \boldsymbol{\lambda}) = 0, \text{ for } s = 0, 1, \ldots, 2k_\beta - 2, \quad (2.91)$$

$$\left. \begin{array}{l} \text{either} \begin{array}{l} G_{\partial\Omega_{\alpha\beta}}^{(2k_\alpha-1,\alpha)}(\mathbf{x}_m, t_{m\pm}, \mathbf{p}_\alpha, \boldsymbol{\lambda}) > 0 \\ G_{\partial\Omega_{\alpha\beta}}^{(2k_\beta-1,\beta)}(\mathbf{x}_m, t_{m\pm}, \mathbf{p}_\beta, \boldsymbol{\lambda}) < 0 \end{array} \right\} \text{ for } \mathbf{n}_{\partial\Omega_{\alpha\beta}} \to \Omega_{\beta} \\ \text{or} \begin{array}{l} G_{\partial\Omega_{\alpha\beta}}^{(2k_\alpha-1,\alpha)}(\mathbf{x}_m, t_{m\pm}, \mathbf{p}_\alpha, \boldsymbol{\lambda}) < 0 \\ G_{\partial\Omega_{\alpha\beta}}^{(2k_\beta-1,\beta)}(\mathbf{x}_m, t_{m\pm}, \mathbf{p}_\beta, \boldsymbol{\lambda}) > 0 \end{array} \right\} \text{ for } \mathbf{n}_{\partial\Omega_{\alpha\beta}} \to \Omega_{\alpha}. \end{array} \right\} \quad (2.92)$$

Proof. See Luo (2008a,b).

2.6 Switching bifurcations

In this section, the flow switching bifurcations from the passable to non-passable flow and the sliding fragmentation bifurcation from the non-passable to passable flow will be discussed. This section will extend the idea presented in Luo (2006,

2007b). Before discussion of switching bifurcations, the product of the normal vector fields of the $(m_i : m_j)$-order on the boundary $\partial \Omega_{ij}$ is introduced as follows.

Definition 2.19. For a discontinuous dynamical system in Eq.(2.1), $\mathbf{x}(t_m) = \mathbf{x}_m \in [\mathbf{x}_{m_1}, \mathbf{x}_{m_2}] \subset \partial \Omega_{ij}$ for time t_m and $\mathbf{x}^{(\alpha)}(t_{m\pm}) = \mathbf{x}_m$, $\alpha \in \{i, j\}$. The flow $\mathbf{x}^{(i)}(t)$ is $C^{r_i}_{[t_m-\varepsilon, t_m+\varepsilon]}$-continuous for t and $||d^{r_i+1}\mathbf{x}^{(i)}/dt^{r_i+1}|| < \infty$ ($r_i \geq m_i + 1$). The flow $\mathbf{x}^{(j)}(t)$ is $C^{r_j}_{[t_m-\varepsilon, t_m+\varepsilon]}$-continuous ($r_j \geq m_j + 1$) for time t and $||d^{r_j+1}\mathbf{x}^{(j)}/dt^{r_j+1}|| < \infty$. The $(m_i : m_j)$-product of G-functions on the boundary $\partial \Omega_{ij}$ is defined as

$$L_{ij}^{(m_i:m_j)}(t_m) \equiv L_{ij}^{(m_i:m_j)}(\mathbf{x}_m, t_m, \mathbf{p}_i, \mathbf{p}_j, \lambda)$$
$$= G_{\partial \Omega_{ij}}^{(m_i,i)}(\mathbf{x}_m, t_{m-}, \mathbf{p}_i, \lambda) \times G_{\partial \Omega_{ij}}^{(m_j,j)}(\mathbf{x}_m, t_{m+}, \mathbf{p}_j, \lambda) \quad (2.93)$$

and for $m_i = m_j = 0$, we have $L_{ij}^{(0:0)} = L_{ij}$,

$$L_{ij}(t_m) \equiv L_{ij}(\mathbf{x}_m, t_m, \mathbf{p}_i, \mathbf{p}_j, \lambda)$$
$$= G_{\partial \Omega_{ij}}^{(i)}(\mathbf{x}_m, t_{m-}, \mathbf{p}_i, \lambda) \times G_{\partial \Omega_{ij}}^{(j)}(\mathbf{x}_m, t_{m+}, \mathbf{p}_j, \lambda). \quad (2.94)$$

From the foregoing definition, the normal vector field products for the full passable, sink and source on the boundary $\partial \Omega_{\alpha\beta}$ are

$$\left. \begin{array}{l} L_{\alpha\beta}^{(2k_\alpha:2k_\beta)}(t_m) > 0, \text{on } \overrightarrow{\partial \Omega}_{\alpha\beta}, \\ L_{\alpha\beta}^{(2k_\alpha:2k_\beta)}(t_m) < 0, \text{on } \overline{\partial \Omega}_{\alpha\beta} = \widetilde{\partial \Omega}_{\alpha\beta} \cup \widehat{\partial \Omega}_{\alpha\beta}. \end{array} \right\} \quad (2.95)$$

With an arrow on the boundary, it means the boundary for passable flow and with bar on the boundary, it means the boundary for non-passable flow. The $\widetilde{\partial \Omega}_{\alpha\beta}$ and $\widehat{\partial \Omega}_{\alpha\beta}$ represent the boundary for the sink and source flow. The switching bifurcation of a flow at $(t_m, \mathbf{x}_m)$ on $\partial \Omega_{\alpha\beta}$ requires

$$L_{\alpha\beta}^{(2k_\alpha:2k_\beta)}(t_m) = 0. \quad (2.96)$$

For a passable flow at $\mathbf{x}(t_m) \equiv \mathbf{x}_m \in [\mathbf{x}_{m_1}, \mathbf{x}_{m_2}] \subset \overrightarrow{\partial \Omega}_{ij}$, consider a time interval $[t_{m_1}, t_{m_2}]$ for $[\mathbf{x}_{m_1}, \mathbf{x}_{m_2}]$ on the boundary $\partial \Omega_{ij}$, and the normal vector field product for $t_m \in [t_{m_1}, t_{m_2}]$ and $\mathbf{x}_m \in [\mathbf{x}_{m_1}, \mathbf{x}_{m_2}]$ is positive, i.e., $L_{ij}^{(2k_i:2k_j)}(t_m) > 0$. To determine the switching bifurcation, the global minimum of such a normal vector field product should be determined. Because $\mathbf{x}_m$ is a function of t_m, the two total derivatives of $L_{ij}^{(2k_i:2k_j)}(t_m)$ are introduced by

2.6 Switching bifurcations

$$DL_{ij}^{(2k_i:2k_j)}(t_m)$$
$$= \nabla L_{ij}^{(2k_i:2k_j)}(\mathbf{x}_m,t_m,\mathbf{p}_i,\mathbf{p}_j,\lambda) \cdot \mathbf{F}_{ij}^{(0)}(\mathbf{x}_m,t_m) + \frac{\partial L_{ij}^{(2k_i:2k_j)}(\mathbf{x}_m,t_m,\mathbf{p}_i,\mathbf{p}_j,\lambda)}{\partial t_m}, \quad (2.97)$$

$$D^r L_{ij}^{(2k_i:2k_j)} = D^{r-1}\left\{ DL_{ij}^{(2k_i:2k_j)}(\mathbf{x}_m,t_m,\mathbf{p}_i,\mathbf{p}_j,\lambda) \right\}, \quad (2.98)$$

for $r = 1,2,\ldots$. Thus, the local minimum of $L_{ij}^{(2k_i:2k_j)}(t_m)$ is determined by

$$D^r L_{ij}^{(2k_i:2k_j)}(t_m) = 0, \text{ for } r = 1,2,\ldots,2l-1, \quad (2.99)$$

$$D^{2l} L_{ij}^{(2k_i:2k_j)}(t_m) > 0. \quad (2.100)$$

Definition 2.20. For a discontinuous dynamical system in Eq.(2.1), $\mathbf{x}(t_m) = \mathbf{x}_m \in [\mathbf{x}_{m_1},\mathbf{x}_{m_2}] \subset \overrightarrow{\partial \Omega}_{ij}$ for time t_m. For a small $\varepsilon > 0$, there are two time intervals (i.e., $[t_{m-\varepsilon},t_m)$, $(t_m,t_{m+\varepsilon}]$) and $\mathbf{x}^{(i)}(t_{m\pm}) = \mathbf{x}_m = \mathbf{x}^{(j)}(t_{m\mp})$. The flow $\mathbf{x}^{(i)}(t)$ is $C^{r_i}_{[t_{m-\varepsilon},t_m)}$-continuous for time t and $||d^{r_i+1}\mathbf{x}^{(i)}/dt^{r_i+1}|| < \infty$ ($r_i \geq 2k_i+1$). The flow $\mathbf{x}^{(j)}(t)$ is $C^{r_j}_{(t_m,t_{m+\varepsilon}]}$-continuous ($r_j \geq 2k_j+1$) for time t and $||d^{r_j+1}\mathbf{x}^{(j)}/dt^{r_j+1}|| < \infty$. The local minimum value set of the $(2k_i:2k_j)$-product of G-functions (i.e., $L_{ij}^{(2k_i:2k_j)}(t_m)$) is defined by

$$\min L_{ij}^{(2k_i:2k_j)}(t_m)$$
$$= \left\{ L_{ij}^{(2k_i:2k_j)}(t_m) \left| \begin{array}{l} \text{for } t_m \in [t_{m_1},t_{m_2}] \text{ and } \mathbf{x}_m \in [\mathbf{x}_{m_1},\mathbf{x}_{m_2}], \\ D^r L_{ij}^{(2k_i:2k_j)} = 0 \text{ for } r = 1,2,\ldots,2l-1, \\ \text{and } D^{2l} L_{ij}^{(2k_i:2k_j)} > 0. \end{array}\right. \right\} \quad (2.101)$$

From the local minimum set of $L_{ij}^{(2k_i:2k_j)}(t_m)$, the global minimum values of $L_{ij}^{(2k_i:2k_j)}(t_m)$ is defined as follows.

Definition 2.21. For a discontinuous dynamical system in Eq.(2.1), $\mathbf{x}(t_m) = \mathbf{x}_m \in [\mathbf{x}_{m_1},\mathbf{x}_{m_2}] \subset \overrightarrow{\partial \Omega}_{ij}$ for time t_m. $\mathbf{x}^{(i)}(t_{m\pm}) = \mathbf{x}_m = \mathbf{x}^{(j)}(t_{m\mp})$. For a small $\varepsilon > 0$, there are two time intervals (i.e., $[t_{m-\varepsilon},t_m),(t_m,t_{m+\varepsilon}]$). The flow $\mathbf{x}^{(i)}(t)$ is $C^{r_i}_{[t_{m-\varepsilon},t_m)}$ or $C^{r_i}_{(t_m,t_{m+\varepsilon}]}$-continuous ($r_i \geq 2k_i+1$) for time t and $||d^{r_i+1}\mathbf{x}^{(i)}/dt^{r_i+1}|| < \infty$. The flow $\mathbf{x}^{(j)}(t)$ is $C^{r_j}_{(t_m,t_{m+\varepsilon}]}$ or $C^{r_j}_{[t_{m-\varepsilon},t_m)}$-continuous ($r_j \geq 2k_j+1$) for time t and $||d^{r_j+1}\mathbf{x}^{(j)}/dt^{r_j+1}|| < \infty$. The global minimum value of the $(2k_i:2k_j)$-product of G-functions is defined by

$$G\min L_{ij}^{(2k_i:2k_j)}(t_m)$$
$$= \min_{t_m \in [t_{m_1},t_{m_2}]} \left\{ \min L_{ij}^{(2k_i:2k_j)}(t_m), L_{ij}^{(2k_i:2k_j)}(t_{m_1}), L_{ij}^{(2k_i:2k_j)}(t_{m_2}) \right\}. \quad (2.102)$$

To look into the switching bifurcation to parameters $\mathbf{q} \in \{\mathbf{p}_i, \mathbf{p}_j, \lambda\}$, $D^r L_{ij}^{(2k_i:2k_j)}$ in Eq.(2.97) is replaced by $\dfrac{d^r}{d\mathbf{q}^r} L_{ij}^{(2k_i:2k_j)}$. Similarly, the maximum set of the $(2k_i : 2k_j)$-product of G-functions (i.e., $L_{ij}^{(2k_i:2k_j)}(t_m)$) can be developed.

Definition 2.22. For a discontinuous dynamical system in Eq.(2.1), $\mathbf{x}(t_m) = \mathbf{x}_m \in [\mathbf{x}_{m_1}, \mathbf{x}_{m_2}] \subset \overline{\partial \Omega_{ij}}$ for time t_m. For a small $\varepsilon > 0$, there are a time intervals $([t_{m-\varepsilon}, t_m)$ or $(t_m, t_{m+\varepsilon}])$ and $\mathbf{x}^{(\alpha)}(t_{m\pm}) = \mathbf{x}_m$ $(\alpha \in \{i,j\})$. The flow $\mathbf{x}^{(i)}(t)$ is $C^{r_i}_{[t_{m-\varepsilon},t_m)}$ or $C^{r_i}_{(t_m,t_{m+\varepsilon}]}$-continuous $(r_i \geqslant 2k_i + 1)$ for time t and $||d^{r_i+1}\mathbf{x}^{(i)}/dt^{r_i+1}|| < \infty$. The flow $\mathbf{x}^{(j)}(t)$ is $C^{r_j}_{[t_{m-\varepsilon},t_m]}$ or $C^{r_j}_{(t_m,t_{m+\varepsilon}]}$-continuous for time t and $||d^{r_j+1}\mathbf{x}^{(j)}/dt^{r_j+1}|| < \infty$ $(r_j \geqslant 2k_j + 1)$. *The local maximum set* of the $(2k_i : 2k_j)$-product of G-functions (i.e., $L_{ij}^{(2k_i:2k_j)}(\mathbf{x}_m, t_m, \mathbf{p}_i, \mathbf{p}_j, \lambda)$) is defined by

$$\max L_{ij}^{(2k_i:2k_j)}(t_m) = \left\{ L_{ij}^{(2k_i:2k_j)}(t_m) \,\middle|\, \begin{array}{l} \text{for } t_m \in [t_{m_1}, t_{m_2}] \text{ and } \mathbf{x}_m \in [\mathbf{x}_{m_1}, \mathbf{x}_{m_2}], \\ D^s L_{ij}^{(2k_i:2k_j)} = 0 \text{ for } s = 1, 2, \dots, 2l, \\ \text{and } D^{2l+1} L_{ij}^{(2k_i:2k_j)} < 0. \end{array} \right\} \quad (2.103)$$

Definition 2.23. For a discontinuous dynamical system in Eq.(2.1), there is a point $\mathbf{x}(t_m) = \mathbf{x}_m \in [\mathbf{x}_{m_1}, \mathbf{x}_{m_2}] \subset \overline{\partial \Omega_{ij}}$ for time t_m. For a small $\varepsilon > 0$, there is a time interval $([t_{m-\varepsilon}, t_m)$ or $(t_m, t_{m+\varepsilon}])$ and $\mathbf{x}^{(\alpha)}(t_{m\pm}) = \mathbf{x}_m$ $(\alpha \in \{i,j\})$. The flow $\mathbf{x}^{(i)}(t)$ is $C^{r_i}_{[t_{m-\varepsilon},t_m)}$ or $C^{r_i}_{(t_m,t_{m+\varepsilon}]}$-continuous for time t and $||d^{r_i+1}\mathbf{x}^{(i)}/dt^{r_i+1}|| < \infty$ $(r_i \geqslant 2k_i + 1)$. The flow $\mathbf{x}^{(j)}(t)$ is $C^{r_j}_{[t_{m-\varepsilon},t_m]}$ or $C^{r_j}_{(t_m,t_{m+\varepsilon}]}$-continuous for time t and $||d^{r_j+1}\mathbf{x}^{(j)}/dt^{r_j+1}|| < \infty$ $(r_j \geqslant 2k_j + 1)$. The global maximum of *the $(2k_i : 2k_j)$-product of G-functions* (i.e., $L_{ij}^{(2k_i:2k_j)}(\mathbf{x}_m, t_m, \mathbf{p}_i, \mathbf{p}_j, \lambda)$) is defined by

$$G \max L_{ij}^{(2k_i:2k_j)}(t_m) = \max_{t_m \in [t_{m_1}, t_{m_2}]} \left\{ \max L_{ij}^{(2k_i:2k_j)}(t_m), L_{ij}^{(2k_i:2k_j)}(t_{m_1}), L_{ij}^{(2k_i:2k_j)}(t_{m_2}) \right\} \quad (2.104)$$

Definition 2.24. For a discontinuous dynamical system in Eq.(2.1), there is a point for time t_m. For a small $\varepsilon > 0$, there are two time intervals $[t_{m-\varepsilon}, t_m)$ and $(t_m, t_{m+\varepsilon}]$. Suppose $\mathbf{x}^{(i)}(t_{m-}) = \mathbf{x}_m = \mathbf{x}^{(j)}(t_{m\pm})$. Both flows $\mathbf{x}^{(i)}(t)$ and $\mathbf{x}^{(j)}(t)$ are $C^{r_i}_{[t_{m-\varepsilon},t_m)}$ and $C^{r_j}_{[t_{m-\varepsilon},t_{m+\varepsilon}]}$-continuous for time t, respectively, and $||d^{r_\alpha+1}\mathbf{x}^{(\alpha)}/dt^{r_\alpha+1}|| < \infty$ $(r_\alpha \geqslant 2, \alpha \in \{i,j\})$. The tangential bifurcation of the flow $\mathbf{x}^{(j)}(t)$ at $\mathbf{x}_m$ on the boundary $\overrightarrow{\partial \Omega_{ij}}$ is termed *the switching bifurcation of the first kind of the non-passable flow* (or called *the sliding bifurcation*) if

$$\begin{aligned} &G^{(j)}_{\partial \Omega_{ij}}(\mathbf{x}_m, t_{m\pm}, \mathbf{p}_j, \lambda) = 0 \text{ and } G^{(i)}_{\partial \Omega_{ij}}(\mathbf{x}_m, t_{m-}, \mathbf{p}_i, \lambda) \neq 0 \\ &G^{(1,j)}_{\partial \Omega_{ij}}(\mathbf{x}_m, t_{m\pm}, \mathbf{p}_j, \lambda) \neq 0 \end{aligned} \quad (2.105)$$

2.6 Switching bifurcations

$$\left.\begin{array}{l} \mathbf{n}_{\partial\Omega_{ij}}^{\mathrm{T}}(\mathbf{x}_{m-\varepsilon}^{(0)}) \cdot [\mathbf{x}_{m-\varepsilon}^{(0)} - \mathbf{x}_{m-\varepsilon}^{(j)}] < 0 \\ \mathbf{n}_{\partial\Omega_{ij}}^{\mathrm{T}}(\mathbf{x}_{m+\varepsilon}^{(0)}) \cdot [\mathbf{x}_{m+\varepsilon}^{(j)} - \mathbf{x}_{m+\varepsilon}^{(0)}] > 0 \\ \mathbf{n}_{\partial\Omega_{ij}}^{\mathrm{T}}(\mathbf{x}_{m-\varepsilon}^{(0)}) \cdot [\mathbf{x}_{m-\varepsilon}^{(0)} - \mathbf{x}_{m-\varepsilon}^{(i)}] > 0 \end{array}\right\} \text{ for } \mathbf{n}_{\partial\Omega_{ij}} \to \Omega_j, \quad (2.106a)$$

$$\left.\begin{array}{l} \mathbf{n}_{\partial\Omega_{ij}}^{\mathrm{T}}(\mathbf{x}_{m-\varepsilon}^{(0)}) \cdot [\mathbf{x}_{m-\varepsilon}^{(0)} - \mathbf{x}_{m-\varepsilon}^{(j)}] > 0 \\ \mathbf{n}_{\partial\Omega_{ij}}^{\mathrm{T}}(\mathbf{x}_{m+\varepsilon}^{(0)}) \cdot [\mathbf{x}_{m+\varepsilon}^{(j)} - \mathbf{x}_{m+\varepsilon}^{(0)}] < 0 \\ \mathbf{n}_{\partial\Omega_{ij}}^{\mathrm{T}}(\mathbf{x}_{m-\varepsilon}^{(0)}) \cdot [\mathbf{x}_{m-\varepsilon}^{(0)} - \mathbf{x}_{m-\varepsilon}^{(i)}] < 0 \end{array}\right\} \text{ for } \mathbf{n}_{\partial\Omega_{ij}} \to \Omega_i. \quad (2.106b)$$

Theorem 2.15. *For a discontinuous dynamical system in Eq.(2.1), there is a point $\mathbf{x}(t_m) = \mathbf{x}_m \in [\mathbf{x}_{m_1}, \mathbf{x}_{m_2}] \subset \overrightarrow{\partial\Omega}_{ij}$ for time t_m. For an arbitrarily small $\varepsilon > 0$, there are two time intervals $[t_{m-\varepsilon}, t_m)$ and $(t_m, t_{m+\varepsilon}]$. Suppose $\mathbf{x}^{(i)}(t_{m-}) = \mathbf{x}_m = \mathbf{x}^{(j)}(t_{m\pm})$. Both flows $\mathbf{x}^{(i)}(t)$ and $\mathbf{x}^{(j)}(t)$ are $C^{r_i}_{[t_{m-\varepsilon}, t_m)}$ and $C^{r_j}_{[t_{m-\varepsilon}, t_{m+\varepsilon}]}$-continuous for time t, respectively, and $\|d^{r_\alpha+1}\mathbf{x}^{(\alpha)}/dt^{r_\alpha+1}\| < \infty$ ($r_\alpha \geq 2, \alpha \in \{i, j\}$). The bifurcation of the passable flow of $\mathbf{x}^{(i)}(t)$ and $\mathbf{x}^{(j)}(t)$ at $\mathbf{x}_m$ switching to the non-passable flow of the first kind on the boundary $\overrightarrow{\partial\Omega}_{ij}$ occurs iff*

$$\left.\begin{array}{l} G^{(j)}_{\partial\Omega_{ij}}(\mathbf{x}_m, t_{m\pm}, \mathbf{p}_j, \boldsymbol{\lambda}) = 0, \\ G^{(i)}_{\partial\Omega_{ij}}(\mathbf{x}_m, t_{m-}, \mathbf{p}_i, \boldsymbol{\lambda}) > 0, \, for \, \mathbf{n}_{\partial\Omega_{ij}} \to \Omega_j, \\ G^{(i)}_{\partial\Omega_{ij}}(\mathbf{x}_m, t_{m-}, \mathbf{p}_i, \boldsymbol{\lambda}) < 0, \, for \, \mathbf{n}_{\partial\Omega_{ij}} \to \Omega_i, \end{array}\right\} \quad (2.107a)$$

or $\quad L_{ij}(\mathbf{x}_{m_2}, t_{m_2}, \mathbf{p}_i, \mathbf{p}_j, \boldsymbol{\lambda}) = 0$ and $G^{(i)}_{\partial\Omega_{ij}}(\mathbf{x}_m, t_{m-}, \mathbf{p}_i, \boldsymbol{\lambda}) \neq 0, \quad (2.107b)$

or $\quad G\min L_{ij}(t_m) = 0$ and $G^{(i)}_{\partial\Omega_{ij}}(\mathbf{x}_m, t_{m-}, \mathbf{p}_i, \boldsymbol{\lambda}) \neq 0, \quad (2.107c)$

$$\left.\begin{array}{l} G^{(1,j)}_{\partial\Omega_{ij}}(\mathbf{x}_m, t_{m\pm}, \mathbf{p}_j, \boldsymbol{\lambda}) > 0, \, for \, \mathbf{n}_{\partial\Omega_{ij}} \to \Omega_j, \\ G^{(1,j)}_{\partial\Omega_{ij}}(\mathbf{x}_m, t_{m\pm}, \mathbf{p}_j, \boldsymbol{\lambda}) < 0, \, for \, \mathbf{n}_{\partial\Omega_{ij}} \to \Omega_i. \end{array}\right\} \quad (2.108)$$

Proof. See Luo(2008a,b).

Definition 2.25. For a discontinuous dynamical system in Eq.(2.1), there is a point $\mathbf{x}(t_m) = \mathbf{x}_m \in [\mathbf{x}_{m_1}, \mathbf{x}_{m_2}] \subset \overrightarrow{\partial\Omega}_{ij}$ for time t_m. For an arbitrarily small $\varepsilon > 0$, there are two time intervals (i.e., $[t_{m-\varepsilon}, t_m)$ and $(t_m, t_{m+\varepsilon}]$). Suppose $\mathbf{x}^{(i)}(t_{m-}) = \mathbf{x}_m = \mathbf{x}^{(j)}(t_{m\pm})$. The flow $\mathbf{x}^{(i)}(t)$ is $C^{r_i}_{[t_{m-\varepsilon}, t_m)}$-continuous for time t and $\|d^{r_i+1}\mathbf{x}^{(i)}/dt^{r_i+1}\| < \infty (r_i \geq 2k_i + 1)$. The flow $\mathbf{x}^{(j)}(t)$ is $C^{r_j}_{[t_{m-\varepsilon}, t_{m+\varepsilon}]}$-continuous for time t and $\|d^{r_j+1}\mathbf{x}^{(j)}/dt^{r_j+1}\| < \infty$ ($r_j \geq 2k_j + 1$). The tangential bifurcation of the

$(2k_i : 2k_j)$-passable flow of $\mathbf{x}^{(i)}(t)$ and $\mathbf{x}^{(j)}(t)$ at $\mathbf{x}_m$ on the boundary $\overrightarrow{\partial\Omega}_{ij}$ is termed *the switching bifurcation of the $(2k_i : 2k_j)$-non-passable flow of the first kind* (or called *the $(2k_i : 2k_j)$-sliding bifurcation*) if

$$\left.\begin{aligned}
G^{(s,j)}_{\partial\Omega_{ij}}(\mathbf{x}_m, t_{m\pm}, \mathbf{p}_j, \boldsymbol{\lambda}) &= 0, \text{ for } s = 0, 1, \ldots, 2k_j, \\
G^{(s,i)}_{\partial\Omega_{ij}}(\mathbf{x}_m, t_{m-}, \mathbf{p}_i, \boldsymbol{\lambda}) &= 0, \text{ for } s = 0, 1, \ldots, 2k_i - 1, \\
G^{(2k_i,i)}_{\partial\Omega_{ij}}(\mathbf{x}_m, t_{m-}, \mathbf{p}_i, \boldsymbol{\lambda}) &\neq 0, \text{ and } G^{(2k_j+1,j)}_{\partial\Omega_{ij}}(\mathbf{x}_m, t_{m\pm}, \mathbf{p}_j, \boldsymbol{\lambda}) \neq 0,
\end{aligned}\right\} \quad (2.109)$$

$$\left.\begin{aligned}
\mathbf{n}^T_{\partial\Omega_{ij}}(\mathbf{x}^{(0)}_{m-\varepsilon}) \cdot \left[\mathbf{x}^{(0)}_{m-\varepsilon} - \mathbf{x}^{(j)}_{m-\varepsilon}\right] &< 0 \\
\mathbf{n}^T_{\partial\Omega_{ij}}(\mathbf{x}^{(0)}_{m+\varepsilon}) \cdot \left[\mathbf{x}^{(j)}_{m+\varepsilon} - \mathbf{x}^{(0)}_{m+\varepsilon}\right] &> 0 \\
\mathbf{n}^T_{\partial\Omega_{ij}}(\mathbf{x}^{(0)}_{m-\varepsilon}) \cdot \left[\mathbf{x}^{(0)}_{m-\varepsilon} - \mathbf{x}^{(i)}_{m-\varepsilon}\right] &> 0
\end{aligned}\right\} \text{ for } \mathbf{n}_{\partial\Omega_{ij}} \to \Omega_j,$$

$$\left.\begin{aligned}
\mathbf{n}^T_{\partial\Omega_{ij}}(\mathbf{x}^{(0)}_{m-\varepsilon}) \cdot \left[\mathbf{x}^{(0)}_{m-\varepsilon} - \mathbf{x}^{(j)}_{m-\varepsilon}\right] &> 0 \\
\mathbf{n}^T_{\partial\Omega_{ij}}(\mathbf{x}^{(0)}_{m+\varepsilon}) \cdot \left[\mathbf{x}^{(j)}_{m+\varepsilon} - \mathbf{x}^{(0)}_{m+\varepsilon}\right] &< 0 \\
\mathbf{n}^T_{\partial\Omega_{ij}}(\mathbf{x}^{(0)}_{m-\varepsilon}) \cdot \left[\mathbf{x}^{(0)}_{m-\varepsilon} - \mathbf{x}^{(i)}_{m-\varepsilon}\right] &< 0
\end{aligned}\right\} \text{ for } \mathbf{n}_{\partial\Omega_{ij}} \to \Omega_i. \quad (2.110)$$

Theorem 2.16. *For a discontinuous dynamical system in Eq.(2.1), there is a point $\mathbf{x}(t_m) = \mathbf{x}_m \in [\mathbf{x}_{m_1}, \mathbf{x}_{m_2}] \subset \overrightarrow{\partial\Omega}_{ij}$ for time t_m. For an arbitrarily small $\varepsilon > 0$, there are two time intervals $[t_{m-\varepsilon}, t_m)$ and $(t_m, t_{m+\varepsilon}]$. Suppose $\mathbf{x}^{(i)}(t_{m-}) = \mathbf{x}_m = \mathbf{x}^{(j)}(t_{m\pm})$. The flow $\mathbf{x}^{(i)}(t)$ is $C^{r_i}_{[t_{m-\varepsilon},t_m)}$-continuous for time t and $||d^{r_i+1}\mathbf{x}^{(i)}/dt^{r_i+1}|| < \infty$ $(r_i \geq 2k_i + 1)$. The flow $\mathbf{x}^{(j)}(t)$ is $C^{r_j}_{[t_{m-\varepsilon},t_{m+\varepsilon}]}$-continuous for time t and $||d^{r_j+1}\mathbf{x}^{(j)}/dt^{r_j+1}|| < \infty$ $(r_j \geq 2k_j+2)$. The bifurcation of the $(2k_i : 2k_j)$-passable flow of $\mathbf{x}^{(i)}(t)$ and $\mathbf{x}^{(j)}(t)$ at $\mathbf{x}_m$ switching to the $(2k_i : 2k_j)$-non-passable flow of the first kind on the boundary $\overrightarrow{\partial\Omega}_{ij}$ occurs iff*

$$\left.\begin{aligned}
G^{(s,j)}_{\partial\Omega_{ij}}(\mathbf{x}_m, t_{m\pm}, \mathbf{p}_j, \boldsymbol{\lambda}) &= 0, \text{ for } s = 0, 1, \ldots, 2k_j - 1, \\
G^{(s,i)}_{\partial\Omega_{ij}}(\mathbf{x}_m, t_{m-}, \mathbf{p}_i, \boldsymbol{\lambda}) &= 0, \text{ for } s = 0, 1, \ldots, 2k_i - 1,
\end{aligned}\right\} \quad (2.111)$$

$$\left.\begin{aligned}
G^{(2k_i,i)}_{\partial\Omega_{ij}}(\mathbf{x}_m, t_{m-}, \mathbf{p}_i, \boldsymbol{\lambda}) &> 0, \text{ for } \mathbf{n}_{\partial\Omega_{ij}} \to \Omega_j, \\
G^{(2k_i,i)}_{\partial\Omega_{ij}}(\mathbf{x}_m, t_{m-}, \mathbf{p}_i, \boldsymbol{\lambda}) &< 0, \text{ for } \mathbf{n}_{\partial\Omega_{ij}} \to \Omega_i,
\end{aligned}\right\} \quad (2.112a)$$

$$G^{(2k_j,j)}_{\partial\Omega_{ij}}(\mathbf{x}_m, t_{m\pm}, \mathbf{p}_j, \boldsymbol{\lambda}) = 0,$$

or $\quad L^{(2k_i:2k_j)}_{ij}(\mathbf{x}_m, t_m, \mathbf{p}_i, \mathbf{p}_j, \boldsymbol{\lambda}) = 0, G^{(2k_i,i)}_{\partial\Omega_{ij}}(\mathbf{x}_m, t_{m-}, \mathbf{p}_i, \boldsymbol{\lambda}) \neq 0, \quad (2.112b)$

or $\quad G\min L^{(2k_i:2k_j)}_{ij}(t_m) = 0 \text{ and } G^{(2k_i,i)}_{\partial\Omega_{ij}}(\mathbf{x}_m, t_{m-}, \mathbf{p}_i, \boldsymbol{\lambda}) \neq 0, \quad (2.112c)$

2.6 Switching bifurcations

$$\left.\begin{aligned}G^{(2k_j+1,j)}_{\partial\Omega_{ij}}(\mathbf{x}_m,t_{m\pm},\mathbf{p}_j,\boldsymbol{\lambda})>0\ for\ \mathbf{n}_{\partial\Omega_{ij}}\to\Omega_j,\\G^{(2k_j+1,j)}_{\partial\Omega_{ij}}(\mathbf{x}_m,t_{m\pm},\mathbf{p}_j,\boldsymbol{\lambda})<0\ for\ \mathbf{n}_{\partial\Omega_{ij}}\to\Omega_i.\end{aligned}\right\} \quad (2.113)$$

Proof. See Luo (2008a,b).

Definition 2.26. For a discontinuous dynamical system in Eq.(2.1), there is a point $\mathbf{x}(t_m) = \mathbf{x}_m \in [\mathbf{x}_{m_1}, \mathbf{x}_{m_2}] \subset \overrightarrow{\partial\Omega_{ij}}$ for time t_m. For an arbitrarily small $\varepsilon > 0$, there are two time intervals $[t_{m-\varepsilon}, t_m)$ and $(t_m, t_{m+\varepsilon}]$. Suppose $\mathbf{x}^{(i)}(t_{m\pm}) = \mathbf{x}_m = \mathbf{x}^{(j)}(t_{m+})$. Both flows $\mathbf{x}^{(i)}(t)$ and $\mathbf{x}^{(j)}(t)$ are $C^{r_i}_{[t_{m-\varepsilon},t_{m+\varepsilon}]}$ and $C^{r_j}_{[t_{m-\varepsilon},t_m)}$-continuous ($r_\alpha \geq 2$) for time t, respectively, and $||d^{r_\alpha+1}\mathbf{x}^{(\alpha)}/dt^{r_\alpha+1}|| < \infty$ ($r_\alpha \geq 2, \alpha \in \{i,j\}$). The tangential bifurcation of the flow $\mathbf{x}^{(i)}(t)$ at $\mathbf{x}_m$ on the boundary $\overrightarrow{\partial\Omega_{ij}}$ is termed *the switching bifurcation of the non-passable flow of the second kind* (or called *the source flow bifurcation*) if

$$\left.\begin{aligned}G^{(j)}_{\partial\Omega_{ij}}(\mathbf{x}_m,t_{m+},\mathbf{p}_j,\boldsymbol{\lambda})\neq 0,\\G^{(i)}_{\partial\Omega_{ij}}(\mathbf{x}_m,t_{m\pm},\mathbf{p}_i,\boldsymbol{\lambda})=0\ \text{and}\ G^{(1,i)}_{\partial\Omega_{ij}}(\mathbf{x}_m,t_{m\pm},\mathbf{p}_i,\boldsymbol{\lambda})\neq 0,\end{aligned}\right\} \quad (2.114)$$

$$\left.\begin{aligned}\mathbf{n}^T_{\partial\Omega_{ij}}(\mathbf{x}^{(0)}_{m-\varepsilon})\cdot\left[\mathbf{x}^{(0)}_{m-\varepsilon}-\mathbf{x}^{(i)}_{m-\varepsilon}\right]>0\\\mathbf{n}^T_{\partial\Omega_{ij}}(\mathbf{x}^{(0)}_{m+\varepsilon})\cdot\left[\mathbf{x}^{(i)}_{m+\varepsilon}-\mathbf{x}^{(0)}_{m+\varepsilon}\right]<0\\\mathbf{n}^T_{\partial\Omega_{ij}}(\mathbf{x}^{(0)}_{m+\varepsilon})\cdot\left[\mathbf{x}^{(j)}_{m+\varepsilon}-\mathbf{x}^{(0)}_{m+\varepsilon}\right]>0\end{aligned}\right\}\ \text{for}\ \mathbf{n}_{\partial\Omega_{ij}}\to\Omega_j, \quad (2.115a)$$

or

$$\left.\begin{aligned}\mathbf{n}^T_{\partial\Omega_{ij}}(\mathbf{x}^{(0)}_{m-\varepsilon})\cdot\left[\mathbf{x}^{(0)}_{m-\varepsilon}-\mathbf{x}^{(i)}_{m-\varepsilon}\right]<0\\\mathbf{n}^T_{\partial\Omega_{ij}}(\mathbf{x}^{(0)}_{m+\varepsilon})\cdot\left[\mathbf{x}^{(i)}_{m+\varepsilon}-\mathbf{x}^{(0)}_{m+\varepsilon}\right]>0\\\mathbf{n}^T_{\partial\Omega_{ij}}(\mathbf{x}^{(0)}_{m+\varepsilon})\cdot\left[\mathbf{x}^{(j)}_{m+\varepsilon}-\mathbf{x}^{(0)}_{m+\varepsilon}\right]<0\end{aligned}\right\}\ \text{for}\ \mathbf{n}_{\partial\Omega_{ij}}\to\Omega_i. \quad (2.115b)$$

Theorem 2.17. *For a discontinuous dynamical system in Eq.(2.1), there is a point* $\mathbf{x}(t_m) = \mathbf{x}_m \in [\mathbf{x}_{m_1}, \mathbf{x}_{m_2}] \subset \overrightarrow{\partial\Omega_{ij}}$ *for time* t_m*. For an arbitrarily small* $\varepsilon > 0$*, there are two time intervals* $[t_{m-\varepsilon}, t_m)$ *and* $(t_m, t_{m+\varepsilon}]$*. Suppose* $\mathbf{x}^{(i)}(t_{m-}) = \mathbf{x}_m = \mathbf{x}^{(j)}(t_{m+})$*. Both flows* $\mathbf{x}^{(i)}(t)$ *and* $\mathbf{x}^{(j)}(t)$ *are* $C^{r_i}_{[t_{m-\varepsilon},t_{m+\varepsilon}]}$ *and* $C^{r_j}_{[t_{m-\varepsilon},t_m)}$*-continuous* ($r_\alpha \geq 2$) *for time t, respectively, and* $||d^{r_\alpha+1}\mathbf{x}^{(\alpha)}/dt^{r_\alpha+1}|| < \infty$ ($r_\alpha \geq 2, \alpha \in \{i,j\}$)*. The bifurcation of the passable flow of* $\mathbf{x}^{(i)}(t)$ *and* $\mathbf{x}^{(j)}(t)$ *at* $\mathbf{x}_m$ *switching to the non-passable flow of the second kind on the boundary* $\overrightarrow{\partial\Omega_{ij}}$ *occurs iff*

$$\left.\begin{aligned}G^{(i)}_{\partial\Omega_{ij}}(\mathbf{x}_m,t_{m\pm},\mathbf{p}_i,\boldsymbol{\lambda})=0,\\G^{(j)}_{\partial\Omega_{ij}}(\mathbf{x}_m,t_{m+},\mathbf{p}_j,\boldsymbol{\lambda})>0,\ for\ \mathbf{n}_{\partial\Omega_{ij}}\to\Omega_j,\\G^{(j)}_{\partial\Omega_{ij}}(\mathbf{x}_m,t_{m+},\mathbf{p}_j,\boldsymbol{\lambda})<0,\ for\ \mathbf{n}_{\partial\Omega_{ij}}\to\Omega_i,\end{aligned}\right\} \quad (2.116a)$$

or $\quad L_{ij}\left(\mathbf{x}_{m_2},t_{m_2},\mathbf{p}_i,\mathbf{p}_j,\boldsymbol{\lambda}\right)=0, G^{(j)}_{\partial\Omega_{ij}}(\mathbf{x}_m,t_{m+},\mathbf{p}_i,\boldsymbol{\lambda})\neq 0,$ (2.116b)

or $\quad G_{\min}L_{ij}(t_m)=0 \text{ and } G^{(j)}_{\partial\Omega_{ij}}(\mathbf{x}_m,t_{m+},\mathbf{p}_i,\boldsymbol{\lambda})\neq 0,$ (2.116c)

$$\left.\begin{aligned}G^{(1,i)}_{\partial\Omega_{ij}}(\mathbf{x}_m,t_{m\pm},\mathbf{p}_i,\boldsymbol{\lambda}) &< 0, \text{ for } \mathbf{n}_{\partial\Omega_{ij}}\to\Omega_j,\\ G^{(1,i)}_{\partial\Omega_{ij}}(\mathbf{x}_m,t_{m\pm},\mathbf{p}_i,\boldsymbol{\lambda}) &> 0, \text{ for } \mathbf{n}_{\partial\Omega_{ij}}\to\Omega_i.\end{aligned}\right\}$$ (2.117)

Proof. See Luo (2008a,b).

Definition 2.27. For a discontinuous dynamical system in Eq.(2.1), there is a point $\mathbf{x}(t_m)=\mathbf{x}_m\in[\mathbf{x}_{m_1},\mathbf{x}_{m_2}]\subset\overrightarrow{\partial\Omega}_{ij}$ for time t_m. For an arbitrarily small $\varepsilon>0$, there are two time intervals $[t_{m-\varepsilon},t_m)$ and $(t_m,t_{m+\varepsilon}]$. The flow $\mathbf{x}^{(i)}(t)$ is $C^{r_i}_{[t_{m-\varepsilon},t_{m+\varepsilon}]}$-continuous for time t and $\|d^{r_i+1}\mathbf{x}^{(i)}/dt^{r_i+1}\|<\infty$ $(r_i\geq 2k_i+2)$. The flow $\mathbf{x}^{(j)}(t)$ is $C^{r_j}_{(t_m,t_{m+\varepsilon}]}$-continuous for time t and $\|d^{r_j+1}\mathbf{x}^{(j)}/dt^{r_j+1}\|<\infty(r_j\geq 2k_j+1)$. The tangential bifurcation of the $(2k_i:2k_j)$-passable flow of $\mathbf{x}^{(i)}(t)$ and $\mathbf{x}^{(j)}(t)$ at $\mathbf{x}_m$ on the boundary $\overrightarrow{\partial\Omega}_{ij}$ is termed *the switching bifurcation of the $(2k_i:2k_j)$, non-passable flow of the second kind* (or called *the $(2k_i:2k_j)$-source flow bifurcation*) if

$$\left.\begin{aligned}G^{(s,i)}_{\partial\Omega_{ij}}(\mathbf{x}_m,t_{m\pm},\mathbf{p}_i,\boldsymbol{\lambda}) &= 0, \text{ for } s=0,1,\ldots,2k_i,\\ G^{(s,j)}_{\partial\Omega_{ij}}(\mathbf{x}_m,t_{m+},\mathbf{p}_j,\boldsymbol{\lambda}) &= 0, \text{ for } s=0,1,\ldots,2k_j-1,\\ G^{(2k_j,j)}_{\partial\Omega_{ij}}(\mathbf{x}_m,t_{m+},\mathbf{p}_j,\boldsymbol{\lambda}) &\neq 0 \text{ and } G^{(2k_i+1,i)}_{\partial\Omega_{ij}}(\mathbf{x}_m,t_{m\pm},\mathbf{p}_i,\boldsymbol{\lambda})\neq 0,\end{aligned}\right\}$$ (2.118)

$$\left.\begin{aligned}\mathbf{n}^{\mathrm{T}}_{\partial\Omega_{ij}}(\mathbf{x}^{(0)}_{m-\varepsilon})\cdot\left[\mathbf{x}^{(0)}_{m-\varepsilon}-\mathbf{x}^{(i)}_{m-\varepsilon}\right] &> 0\\ \mathbf{n}^{\mathrm{T}}_{\partial\Omega_{ij}}(\mathbf{x}^{(0)}_{m+\varepsilon})\cdot\left[\mathbf{x}^{(i)}_{m+\varepsilon}-\mathbf{x}^{(0)}_{m+\varepsilon}\right] &< 0\\ \mathbf{n}^{\mathrm{T}}_{\partial\Omega_{ij}}(\mathbf{x}^{(0)}_{m+\varepsilon})\cdot\left[\mathbf{x}^{(j)}_{m+\varepsilon}-\mathbf{x}^{(0)}_{m+\varepsilon}\right] &> 0\end{aligned}\right\} \text{ for } \mathbf{n}_{\partial\Omega_{ij}}\to\Omega_j$$

$$\left.\begin{aligned}\mathbf{n}^{\mathrm{T}}_{\partial\Omega_{ij}}(\mathbf{x}^{(0)}_{m-\varepsilon})\cdot\left[\mathbf{x}^{(0)}_{m-\varepsilon}-\mathbf{x}^{(i)}_{m-\varepsilon}\right] &< 0\\ \mathbf{n}^{\mathrm{T}}_{\partial\Omega_{ij}}(\mathbf{x}^{(0)}_{m+\varepsilon})\cdot\left[\mathbf{x}^{(i)}_{m+\varepsilon}-\mathbf{x}^{(0)}_{m+\varepsilon}\right] &> 0\\ \mathbf{n}^{\mathrm{T}}_{\partial\Omega_{ij}}(\mathbf{x}^{(0)}_{m+\varepsilon})\cdot\left[\mathbf{x}^{(j)}_{m+\varepsilon}-\mathbf{x}^{(0)}_{m+\varepsilon}\right] &< 0\end{aligned}\right\} \text{ for } \mathbf{n}_{\partial\Omega_{ij}}\to\Omega_i.$$ (2.119)

Theorem 2.18. *For a discontinuous dynamical system in Eq.(2.1), there is a point $\mathbf{x}(t_m)=\mathbf{x}_m\in[\mathbf{x}_{m_1},\mathbf{x}_{m_2}]\subset\overrightarrow{\partial\Omega}_{ij}$ for time t_m. For an arbitrarily small $\varepsilon>0$, there are two time intervals $[t_{m-\varepsilon},t_m)$ and $(t_m,t_{m+\varepsilon}]$. Suppose $\mathbf{x}^{(i)}(t_{m-})=\mathbf{x}_m=\mathbf{x}^{(j)}(t_{m+})$. The flow $\mathbf{x}^{(i)}(t)$ is $C^{r_i}_{[t_{m-\varepsilon},t_{m+\varepsilon}]}$-continuous for time t and $\|d^{r_i+1}\mathbf{x}^{(i)}/dt^{r_i+1}\|<\infty$ $(r_i\geq 2k_i+2)$. The flow $\mathbf{x}^{(j)}(t)$ is $C^{r_j}_{(t_m,t_{m+\varepsilon}]}$-continuous for time t and $\|d^{r_j+1}\mathbf{x}^{(j)}/dt^{r_j+1}\|<\infty$ $(r_j\geq 2k_j+1)$. The bifurcation of the $(2k_i:2k_j)$-passable flow of $\mathbf{x}^{(i)}(t)$ and $\mathbf{x}^{(j)}(t)$*

2.6 Switching bifurcations

at $\mathbf{x}_m$ switching to the $(2k_i : 2k_j)$- *non-passable flow of the second kind on the boundary* $\overrightarrow{\partial \Omega}_{ij}$ *occurs iff*

$$\left. \begin{array}{l} G^{(s,j)}_{\partial \Omega_{ij}}(\mathbf{x}_m, t_{m+}, \mathbf{p}_j, \lambda) = 0, \; for \; s = 0, 1, \ldots, 2k_j - 1, \\ G^{(s,i)}_{\partial \Omega_{ij}}(\mathbf{x}_m, t_{m\pm}, \mathbf{p}_i, \lambda) = 0, \; for \; s = 0, 1, \ldots, 2k_i - 1, \end{array} \right\} \quad (2.120)$$

$$\left. \begin{array}{l} G^{(2k_j,j)}_{\partial \Omega_{ij}}(\mathbf{x}_m, t_{m+}, \mathbf{p}_j, \lambda) > 0, \; for \; \mathbf{n}_{\Omega_{ij}} \to \Omega_j, \\ G^{(2k_j,j)}_{\partial \Omega_{ij}}(\mathbf{x}_m, t_{m+}, \mathbf{p}_j, \lambda) < 0, \; for \; \mathbf{n}_{\Omega_{ij}} \to \Omega_i, \\ G^{(2k_i,i)}_{\partial \Omega_{ij}}(\mathbf{x}_m, t_{m\pm}, \mathbf{p}_i, \lambda) = 0, \end{array} \right\} \quad (2.121\mathrm{a})$$

$$or \quad L^{(2k_i:2k_j)}_{ij}(\mathbf{x}_m, t_m, \mathbf{p}_i, \mathbf{p}_j, \lambda) = 0, G^{(j,2k_j)}_{\partial \Omega_{ij}}(\mathbf{x}_m, t_{m+}, \mathbf{p}_j, \lambda) \neq 0, \quad (2.121\mathrm{b})$$

$$or \quad G\min L^{(2k_i:2k_j)}_{ij}(t_m) = 0, G^{(j,2k_j)}_{\partial \Omega_{ij}}(\mathbf{x}_m, t_{m+}, \mathbf{p}_j, \lambda) \neq 0, \quad (2.121\mathrm{c})$$

$$\left. \begin{array}{l} G^{(2k_i+1,i)}_{\partial \Omega_{ij}}(\mathbf{x}_m, t_{m\pm}, \mathbf{p}_i, \lambda) < 0, \; for \; \mathbf{n}_{\Omega_{ij}} \to \Omega_j, \\ G^{(2k_i+1,i)}_{\partial \Omega_{ij}}(\mathbf{x}_m, t_{m\pm}, \mathbf{p}_i, \lambda) > 0, \; for \; \mathbf{n}_{\Omega_{ij}} \to \Omega_i. \end{array} \right\} \quad (2.122)$$

Proof. See Luo (2008a,b).

Definition 2.28. For a discontinuous dynamical system in Eq.(2.1), there is a point $\mathbf{x}(t_m) = \mathbf{x}_m \in [\mathbf{x}_{m_1}, \mathbf{x}_{m_2}] \subset \overrightarrow{\partial \Omega}_{ij}$ for time t_m. For an small $\varepsilon > 0$, there are two time intervals $[t_{m-\varepsilon}, t_m)$ and $(t_m, t_{m+\varepsilon}]$. Suppose $\mathbf{x}^{(i)}(t_{m-}) = \mathbf{x}_m = \mathbf{x}^{(j)}(t_{m+})$. Both flows $\mathbf{x}^{(i)}(t)$ and $\mathbf{x}^{(j)}(t)$ are $C^{r_i}_{[t_{m-\varepsilon}, t_m)}$ and $C^{r_j}_{[t_{m-\varepsilon}, t_{m+\varepsilon}]}$-continuous for time t, respectively, and $\|d^{r_\alpha+1}\mathbf{x}^{(\alpha)}/dt^{r_\alpha+1}\| < \infty$ ($r_\alpha \geq 2$, $\alpha \in \{i, j\}$). The tangential bifurcation of the flow $\mathbf{x}^{(i)}(t)$ and $\mathbf{x}^{(j)}(t)$ at $\mathbf{x}_m$ on the boundary $\overrightarrow{\partial \Omega}_{ij}$ is termed *the switching bifurcation of the passable flow* from $\overrightarrow{\partial \Omega}_{ij}$ to $\overleftarrow{\partial \Omega}_{ij}$ if

$$\left. \begin{array}{l} G^{(i)}_{\partial \Omega_{ij}}(\mathbf{x}_m, t_{m-}, \mathbf{p}_i, \lambda) = 0 \; \text{and} \; G^{(j)}_{\partial \Omega_{ij}}(\mathbf{x}_m, t_{m+}, \mathbf{p}_j, \lambda) = 0, \\ G^{(1,i)}_{\partial \Omega_{ij}}(\mathbf{x}_m, t_{m-}, \mathbf{p}_i, \lambda) \neq 0 \; \text{and} \; G^{(1,j)}_{\partial \Omega_{ij}}(\mathbf{x}_m, t_{m+}, \mathbf{p}_j, \lambda) \neq 0, \end{array} \right\} \quad (2.123)$$

$$\left.\begin{array}{l}\mathbf{n}_{\partial\Omega_{ij}}^{T}(\mathbf{x}_{m-\varepsilon}^{(0)})\cdot\left[\mathbf{x}_{m-\varepsilon}^{(0)}-\mathbf{x}_{m-\varepsilon}^{(i)}\right]>0\\ \mathbf{n}_{\partial\Omega_{ij}}^{T}(\mathbf{x}_{m+\varepsilon}^{(0)})\cdot\left[\mathbf{x}_{m+\varepsilon}^{(i)}-\mathbf{x}_{m+\varepsilon}^{(0)}\right]<0\\ \mathbf{n}_{\partial\Omega_{ij}}^{T}(\mathbf{x}_{m-\varepsilon}^{(0)})\cdot\left[\mathbf{x}_{m-\varepsilon}^{(0)}-\mathbf{x}_{m-\varepsilon}^{(j)}\right]<0\\ \mathbf{n}_{\partial\Omega_{ij}}^{T}(\mathbf{x}_{m+\varepsilon}^{(0)})\cdot\left[\mathbf{x}_{m+\varepsilon}^{(j)}-\mathbf{x}_{m+\varepsilon}^{(0)}\right]>0\end{array}\right\} \text{ for } \mathbf{n}_{\partial\Omega_{ij}}\to\Omega_{j},$$
$$\left.\begin{array}{l}\mathbf{n}_{\partial\Omega_{ij}}^{T}(\mathbf{x}_{m-\varepsilon}^{(0)})\cdot\left[\mathbf{x}_{m-\varepsilon}^{(0)}-\mathbf{x}_{m-\varepsilon}^{(i)}\right]<0\\ \mathbf{n}_{\partial\Omega_{ij}}^{T}(\mathbf{x}_{m+\varepsilon}^{(0)})\cdot\left[\mathbf{x}_{m+\varepsilon}^{(i)}-\mathbf{x}_{m+\varepsilon}^{(0)}\right]>0\\ \mathbf{n}_{\partial\Omega_{ij}}^{T}(\mathbf{x}_{m-\varepsilon}^{(0)})\cdot\left[\mathbf{x}_{m-\varepsilon}^{(0)}-\mathbf{x}_{m-\varepsilon}^{(j)}\right]>0\\ \mathbf{n}_{\partial\Omega_{ij}}^{T}(\mathbf{x}_{m+\varepsilon}^{(0)})\cdot\left[\mathbf{x}_{m+\varepsilon}^{(j)}-\mathbf{x}_{m+\varepsilon}^{(0)}\right]<0\end{array}\right\} \text{ for } \mathbf{n}_{\partial\Omega_{ij}}\to\Omega_{i}.$$
(2.124)

Theorem 2.19. *For a discontinuous dynamical system in Eq.(2.1), there is a point* $\mathbf{x}(t_m)=\mathbf{x}_m\in[\mathbf{x}_{m_1},\mathbf{x}_{m_2}]\subset\overrightarrow{\partial\Omega}_{ij}$ *for time* t_m. *Suppose* $\mathbf{x}^{(i)}(t_{m\pm})=\mathbf{x}_m=\mathbf{x}^{(j)}(t_{m\pm})$. *For an arbitrarily small* $\varepsilon>0$, *there is a time interval* $[t_{m-\varepsilon},t_{m+\varepsilon}]$. *Both flows* $\mathbf{x}^{(i)}(t)$ *and* $\mathbf{x}^{(j)}(t)$ *are* $C^{r_i}_{[t_{m-\varepsilon},t_{m+\varepsilon}]}$ *and* $C^{r_j}_{[t_{m-\varepsilon},t_{m+\varepsilon}]}$ *-continuous* $(r_\alpha\geqslant 3,\alpha\in\{i,j\})$ *for time* t *and* $||d^{r_\alpha+1}\mathbf{x}^{(\alpha)}/dt^{r_\alpha+1}||<\infty$ $(\alpha\in\{i,j\})$. *The switching bifurcation of the passable flow from* $\overrightarrow{\partial\Omega}_{ij}$ *to* $\overleftarrow{\partial\Omega}_{ij}$ *occurs iff*

$$G^{(i)}_{\partial\Omega_{ij}}(\mathbf{x}_m,t_{m-},\mathbf{p}_j,\lambda)=0 \text{ and } G^{(j)}_{\partial\Omega_{ij}}(\mathbf{x}_m,t_{m+},\mathbf{p}_i,\lambda)=0, \quad (2.125a)$$

or $$L_{ij}(\mathbf{x}_m,t_m,\mathbf{p}_i,\mathbf{p}_j,\lambda)=0, \quad (2.125b)$$

or $$_G\min L_{ij}(t_m)=0, \quad (2.125c)$$

$$\left.\begin{array}{l}G^{(1,i)}_{\partial\Omega_{ij}}(\mathbf{x}_m,t_{m-},\mathbf{p}_i,\lambda)<0\\ G^{(1,j)}_{\partial\Omega_{ij}}(\mathbf{x}_m,t_m,\mathbf{p}_j,\lambda)>0\end{array}\right\} \text{ for } \mathbf{n}_{\Omega_{ij}}\to\Omega_j,$$
$$\left.\begin{array}{l}G^{(1,i)}_{\partial\Omega_{ij}}(\mathbf{x}_m,t_{m-},\mathbf{p}_i,\lambda)>0\\ G^{(1,j)}_{\partial\Omega_{ij}}(\mathbf{x}_m,t_{m-},\mathbf{p}_j,\lambda)<0\end{array}\right\} \text{ for } \mathbf{n}_{\Omega_{ij}}\to\Omega_i.$$
(2.126)

Proof. See Luo(2008a,b). ∎

Definition 2.29. For a discontinuous dynamical system in Eq.(2.1), there is a point $\mathbf{x}(t_m)=\mathbf{x}_m\in[\mathbf{x}_{m_1},\mathbf{x}_{m_2}]\subset\overrightarrow{\partial\Omega}_{ij}$ for time t_m. For an arbitrarily small $\varepsilon>0$, there is a time interval $[t_{m-\varepsilon},t_{m+\varepsilon}]$. The flow $\mathbf{x}^{(i)}(t)$ is $C^{r_i}_{[t_{m-\varepsilon},t_{m+\varepsilon}]}$-continuous ($r_i\geqslant 2k_i+1$) for time t and $||d^{r_i+1}\mathbf{x}^{(i)}/dt^{r_i+1}||<\infty$. The flow $\mathbf{x}^{(j)}(t)$ is $C^{r_j}_{[t_{m-\varepsilon},t_{m+\varepsilon}]}$-continuous ($r_j\geqslant 2k_j+1$) for time t and $||d^{r_j+1}\mathbf{x}^{(j)}/dt^{r_j+1}||<\infty$. The tangential bifurcation

2.6 Switching bifurcations

of the $(2k_i : 2k_j)$-passable flow of $\mathbf{x}^{(i)}(t)$ and $\mathbf{x}^{(j)}(t)$ at $\mathbf{x}_m$ on the boundary $\overrightarrow{\partial\Omega_{ij}}$ is termed *the switching bifurcation of the $(2k_i : 2k_j)$-passable flow* from $\overrightarrow{\partial\Omega_{ij}}$ to $\overleftarrow{\partial\Omega_{ij}}$ if

$$\left.\begin{array}{l} G_{\partial\Omega_{ij}}^{(s,i)}(\mathbf{x}_m, t_{m-}, \mathbf{p}_i, \lambda) = 0, \text{ for } s = 0, 1, \ldots, 2k_i, \\ G_{\partial\Omega_{ij}}^{(s,j)}(\mathbf{x}_m, t_{m+}, \mathbf{p}_j, \lambda) = 0, \text{ for } s = 0, 1, \ldots, 2k_j, \\ G_{\partial\Omega_{ij}}^{(2k_i+1,i)}(\mathbf{x}_m, t_{m-}, \mathbf{p}_i, \lambda) \neq 0 \text{ and } G_{\partial\Omega_{ij}}^{(2k_j+1,j)}(\mathbf{x}_m, t_{m+}, \mathbf{p}_j, \lambda) \neq 0, \end{array}\right\} \quad (2.127)$$

$$\left.\begin{array}{l} \mathbf{n}_{\partial\Omega_{ij}}^T(\mathbf{x}_{m-\varepsilon}^{(0)}) \cdot [\mathbf{x}_{m-\varepsilon}^{(0)} - \mathbf{x}_{m-\varepsilon}^{(i)}] > 0 \\ \mathbf{n}_{\partial\Omega_{ij}}^T(\mathbf{x}_{m+\varepsilon}^{(0)}) \cdot [\mathbf{x}_{m+\varepsilon}^{(i)} - \mathbf{x}_{m+\varepsilon}^{(0)}] < 0 \\ \mathbf{n}_{\partial\Omega_{ij}}^T(\mathbf{x}_{m-\varepsilon}^{(0)}) \cdot [\mathbf{x}_{m-\varepsilon}^{(0)} - \mathbf{x}_{m-\varepsilon}^{(j)}] > 0 \\ \mathbf{n}_{\partial\Omega_{ij}}^T(\mathbf{x}_{m+\varepsilon}^{(0)}) \cdot [\mathbf{x}_{m+\varepsilon}^{(j)} - \mathbf{x}_{m+\varepsilon}^{(0)}] > 0 \end{array}\right\} \text{ for } \mathbf{n}_{\partial\Omega_{ij}} \to \Omega_j, \quad (2.128a)$$

$$\left.\begin{array}{l} \mathbf{n}_{\partial\Omega_{ij}}^T(\mathbf{x}_{m-\varepsilon}^{(0)}) \cdot [\mathbf{x}_{m-\varepsilon}^{(0)} - \mathbf{x}_{m-\varepsilon}^{(i)}] < 0 \\ \mathbf{n}_{\partial\Omega_{ij}}^T(\mathbf{x}_{m+\varepsilon}^{(0)}) \cdot [\mathbf{x}_{m+\varepsilon}^{(i)} - \mathbf{x}_{m+\varepsilon}^{(0)}] > 0 \\ \mathbf{n}_{\partial\Omega_{ij}}^T(\mathbf{x}_{m-\varepsilon}^{(0)}) \cdot [\mathbf{x}_{m-\varepsilon}^{(0)} - \mathbf{x}_{m-\varepsilon}^{(j)}] < 0 \\ \mathbf{n}_{\partial\Omega_{ij}}^T(\mathbf{x}_{m+\varepsilon}^{(0)}) \cdot [\mathbf{x}_{m+\varepsilon}^{(j)} - \mathbf{x}_{m+\varepsilon}^{(0)}] < 0 \end{array}\right\} \text{ for } \mathbf{n}_{\partial\Omega_{ij}} \to \Omega_i. \quad (2.128b)$$

Theorem 2.20. *For a discontinuous dynamical system in Eq.(2.1), there is a point $\mathbf{x}(t_m) = \mathbf{x}_m \in [\mathbf{x}_{m_1}, \mathbf{x}_{m_2}] \subset \overrightarrow{\partial\Omega_{ij}}$ for time t_m. For an arbitrarily small $\varepsilon > 0$, there are two time intervals $[t_{m-\varepsilon}, t_m)$ and $(t_m, t_{m+\varepsilon}]$. Suppose $\mathbf{x}^{(i)}(t_{m-}) = \mathbf{x}_m = \mathbf{x}^{(j)}(t_{m+})$. The flow $\mathbf{x}^{(i)}(t)$ is $C_{[t_{m-\varepsilon}, t_m)}^{r_i}$-continuous for time t and $||d^{r_i+1}\mathbf{x}^{(i)}/dt^{r_i+1}|| < \infty (r_i \geq 2k_i + 1)$. The flow $\mathbf{x}^{(j)}(t)$ is $C_{(t_m, t_{m+\varepsilon}]}^{r_j}$-continuous for time t and $||d^{r_j+1}\mathbf{x}^{(j)}/dt^{r_j+1}|| < \infty (r_j \geq 2k_j + 1)$. The bifurcation of the $(2k_i : 2k_j)$-passable flow of $\mathbf{x}^{(i)}(t)$ and $\mathbf{x}^{(j)}(t)$ at $\mathbf{x}_m$ switching to the $(2k_i : 2k_j)$-passable flow on $\overrightarrow{\partial\Omega_{ij}}$ occurs iff*

$$\left.\begin{array}{l} G_{\partial\Omega_{ij}}^{(s,j)}(\mathbf{x}_m, t_{m+}, \mathbf{p}_j, \lambda) = 0, \text{ for } s = 0, 1, \ldots, 2k_j - 1, \\ G_{\partial\Omega_{ij}}^{(s,i)}(\mathbf{x}_m, t_{m-}, \mathbf{p}_i, \lambda) = 0, \text{ for } s = 0, 1, \ldots, 2k_i - 1, \end{array}\right\} \quad (2.129)$$

$$G_{\partial\Omega_{ij}}^{(2k_i,i)}(\mathbf{x}_m, t_{m\pm}, \mathbf{p}_i, \lambda) = 0, G_{\partial\Omega_{ij}}^{(2k_j,j)}(\mathbf{x}_m, t_{m\pm}, \mathbf{p}_j, \lambda) = 0, \quad (2.130a)$$

or
$$L_{ij}^{(2k_i:2k_j)}(\mathbf{x}_m, t_m, \mathbf{p}_i, \mathbf{p}_j, \lambda) = 0, \quad (2.130b)$$

or
$$G \min L_{ij}^{(2k_i:2k_j)}(t_m) = 0, \quad (2.130c)$$

$$\left.\begin{array}{l}G^{(2k_i+1,i)}_{\partial\Omega_{ij}}(\mathbf{x}_m,t_{m\pm},\mathbf{p}_i,\lambda)<0\\ G^{(2k_j+1,j)}_{\partial\Omega_{ij}}(\mathbf{x}_m,t_{m\pm},\mathbf{p}_j,\lambda)>0\end{array}\right\}\; for\; \mathbf{n}_{\partial\Omega_{ij}}\to\Omega_j,\\ \left.\begin{array}{l}G^{(2k_i+1,i)}_{\partial\Omega_{ij}}(\mathbf{x}_m,t_{m\mp},\mathbf{p}_i,\lambda)>0\\ G^{(2k_j+1,j)}_{\partial\Omega_{ij}}(\mathbf{x}_m,t_{m\mp},\mathbf{p}_j,\lambda)<0\end{array}\right\}\; for\; \mathbf{n}_{\partial\Omega_{ij}}\to\Omega_i.\right\} \quad (2.131)$$

Proof. See Luo (2008a,b).

Following the Definitions 2.22~2.28, the sliding and source fragmentation bifurcations can be similarly defined.

Definition 2.30. For a discontinuous dynamical system in Eq.(2.1), there is a point $\mathbf{x}(t_m) = \mathbf{x}_m \in [\mathbf{x}_{m_1},\mathbf{x}_{m_2}] \subset \widetilde{\partial\Omega}_{ij}$ for time t_m.
(i) The tangential bifurcation of the flow $\mathbf{x}^{(j)}(t)$ at $\mathbf{x}_m$ on the boundary $\widetilde{\partial\Omega}_{ij}$ is termed *the fragmentation bifurcation of the non-passable flow of the first kind* (or called *the sliding fragmentation bifurcation*) if Eqs.(2.105) and (2.106) hold.
(ii) The tangential bifurcation of the flow $\mathbf{x}^{(i)}(t)$ with the $(2k_i)^{\text{th}}$-order singularity and $\mathbf{x}^{(j)}(t)$ with the $(2k_j)^{\text{th}}$-order singularity at $\mathbf{x}_m$ on the boundary $\widetilde{\partial\Omega}_{ij}$ is termed *the fragmentation bifurcation of the $(2k_i:2k_j)$-non-passable flow of the first kind* (or called *the $(2k_i:2k_j)$-sliding fragmentation bifurcation*) if Eqs.(2.109) and (2.110) hold.

The necessary and sufficient conditions for the sliding fragmentation bifurcation of the non-passable flow of the first kind are given by Eqs.(2.107) and (2.108) with $_{G\max}L_{ij}(t_m)$ replacing $_{G\min}L_{ij}(t_m)$. Similarly, the necessary and sufficient conditions for the sliding fragmentation bifurcation of the $(2k_i:2k_j)$-non-passable flow of the first kind are presented by Eqs.(2.111)~(2.113) with $_{G\max}L^{(2k_i:2k_j)}_{ij}(t_m)$ replacing $_{G\min}L^{(2k_i:2k_j)}_{ij}(t_m)$.

Definition 2.31. For a discontinuous dynamical system in Eq.(2.1), there is a point $\mathbf{x}(t_m) = \mathbf{x}_m \in [\mathbf{x}_{m_1},\mathbf{x}_{m_2}] \subset \widehat{\partial\Omega}_{ij}$ for time t_m.
(i) The tangential bifurcation of the flow $\mathbf{x}^{(j)}(t)$ at $\mathbf{x}_m$ on the boundary $\widehat{\partial\Omega}_{ij}$ is termed *the fragmentation bifurcation of the non-passable flow of the second kind* (or called *the source fragmentation bifurcation*) if Eqs.(2.114) and (2.115) hold.
(ii) The tangential bifurcation of the flow $\mathbf{x}^{(i)}(t)$ with the $(2k_i)^{\text{th}}$-order singularity and $\mathbf{x}^{(j)}(t)$ with the $(2k_j)^{\text{th}}$-order singularity at $\mathbf{x}_m$ on the boundary $\widehat{\partial\Omega}_{ij}$ is termed *the fragmentation bifurcation of the $(2k_i:2k_j)$-non-passable flow of the second kind* (or called *the $(2k_i:2k_j)$-source fragmentation bifurcation*) if Eqs.(2.118) and (2.119) hold.

The necessary and sufficient conditions for the source fragmentation bifurcation of the non-passable flow of the second kind are given by Eqs.(2.116) and (2.117)

2.6 Switching bifurcations

with $_{G\max}L_{ij}(t_m)$ replacing $_{G\min}L_{ij}(t_m)$. Similarly, the necessary and sufficient conditions for the sliding fragmentation bifurcation of the $(2k_i : 2k_j)$-non-passable flow of the second kind are given by Eqs.(2.120)~(2.122) with $_{G\max}L_{ij}^{(2k_i:2k_j)}(t_m)$ replacing $_{G\min}L_{ij}^{(2k_i:2k_j)}(t_m)$.

Definition 2.32. For a discontinuous dynamical system in Eq.(2.1), there is a point $\mathbf{x}(t_m) = \mathbf{x}_m \in [\mathbf{x}_{m_1}, \mathbf{x}_{m_2}] \subset \widetilde{\partial\Omega}_{ij}$ (or $\widehat{\partial\Omega}_{ij}$) for time t_m.
(i) The tangential bifurcation of the flow $\mathbf{x}^{(i)}(t)$ and $\mathbf{x}^{(j)}(t)$ at $\mathbf{x}_m$ on the boundary $\widetilde{\partial\Omega}_{ij}$ (or $\widehat{\partial\Omega}_{ij}$) is termed *the switching bifurcation of the non-passable flow* from $\widetilde{\partial\Omega}_{ij}$ to $\widehat{\partial\Omega}_{ij}$ (or from $\widehat{\partial\Omega}_{ij}$ to $\widetilde{\partial\Omega}_{ij}$) if Eqs.(2.123) and (2.124) hold.
(ii) The tangential bifurcation of the flow $\mathbf{x}^{(i)}(t)$ with the $(2k_i)^{\text{th}}$-order singularity and $\mathbf{x}^{(j)}(t)$ with the $(2k_j)^{\text{th}}$-order singularity at $\mathbf{x}_m$ on the boundary $\widetilde{\partial\Omega}_{ij}$ (or $\widehat{\partial\Omega}_{ij}$) is termed *the switching bifurcation of the $(2k_i : 2k_j)$-non-passable flow* from $\widetilde{\partial\Omega}_{ij}$ to $\widehat{\partial\Omega}_{ij}$ (or from $\widehat{\partial\Omega}_{ij}$ to $\widetilde{\partial\Omega}_{ij}$) if Eqs.(2.127) and (2.128) hold.

The necessary and sufficient conditions for the switching bifurcation of the non-passable flow from $\widetilde{\partial\Omega}_{ij}$ to $\widehat{\partial\Omega}_{ij}$ (or from $\widehat{\partial\Omega}_{ij}$ to $\widetilde{\partial\Omega}_{ij}$) are given by Eqs.(2.125) and (2.126) with $_{G\max}L_{ij}(t_m)$ replacing $_{G\min}L_{ij}(t_m)$. Similarly, the necessary and sufficient conditions for the switching bifurcation of the $(2k_i : 2k_j)$-non-passable flow from $\widetilde{\partial\Omega}_{ij}$ to $\widehat{\partial\Omega}_{ij}$ (or from $\widehat{\partial\Omega}_{ij}$ to $\widetilde{\partial\Omega}_{ij}$) are presented by Eqs.(2.129)~(2.131) with $_{G\max}L_{ij}^{(2k_i:2k_j)}(t_m)$ replacing $_{G\min}L_{ij}^{(2k_i:2k_j)}(t_m)$. The necessary and sufficient conditions for switching bifurcations among the $(2k_\alpha : 2k_\beta)$-flows, $(2k_\alpha : 2k_\beta - 1)$-flows, $(2k_\alpha - 1 : 2k_\beta)$-flows and $(2k_\alpha - 1 : 2k_\beta)$- flows can be referred to Luo (2008a,b).

If discontinuous dynamical systems possess flow barriers on the separation boundary, the corresponding passability, non-passability and tangency to the separation boundary can be discussed similarly. The switching bifurcation can be further presented. Except the G-functions are different in two discontinuous dynamical systems, the corresponding conditions for the passable flows, non-passable flows and tangency to the separation boundary are similar to the discontinuous dynamical system without flow barriers. Therefore, a new G-function is introduced for discontinuous dynamical systems with flow barriers herein only.

From input and output flow barriers on the boundary $\partial\Omega_{\alpha\beta}$ in Luo (2006, 2007b), the input and output vector fields of the domain Ω_α on the boundary $\partial\Omega_{\alpha\beta}$ are defined as $\mathbf{F}_{in}^{(\alpha)}(\mathbf{x},t,\mathbf{p}_\alpha)$ and $\mathbf{F}_{b}^{(\alpha)}(\mathbf{x},t,\mathbf{p}_\alpha)$. The G-function for the flow barriers should be defined as

$$G_{\partial\Omega_{ij}}^{(\alpha)}(\mathbf{x}_m,t_{m-},\mathbf{p}_\alpha,\lambda)$$
$$\equiv \mathbf{n}_{\partial\Omega_{ij}}^T(\mathbf{x}^{(0)},t,\lambda)\cdot[\mathbf{F}_{in}^{(\alpha)}(\mathbf{x}^{(\alpha)},t,\mathbf{p}_\alpha) - \mathbf{F}^{(0)}(\mathbf{x}^{(0)},t,\lambda)]\Big|_{(\mathbf{x}_m^{(0)},\mathbf{x}_m^{(\alpha)},t_{m-})}$$

$$= \mathbf{n}_{\partial\Omega_{ij}}^{T}(\mathbf{x}^{(0)},t,\boldsymbol{\lambda}) \cdot \mathbf{F}_{in}^{(\alpha)}(\mathbf{x}^{(\alpha)},t,\mathbf{p}_{\alpha}) + \left.\frac{\partial \varphi_{ij}(\mathbf{x}^{(0)},t,\boldsymbol{\lambda})}{\partial t}\right|_{(\mathbf{x}_m^{(0)},\mathbf{x}_m^{(\alpha)},t_{m-})}$$

$$= \nabla \varphi_{ij}(\mathbf{x}^{(0)},t,\boldsymbol{\lambda}) \cdot \mathbf{F}_{in}^{(\alpha)}(\mathbf{x}^{(\alpha)},t,\mathbf{p}_{\alpha}) + \left.\frac{\partial \varphi_{ij}(\mathbf{x}^{(0)},t,\boldsymbol{\lambda})}{\partial t}\right|_{(\mathbf{x}_m^{(0)},\mathbf{x}_m^{(\alpha)},t_{m-})}, \quad (2.132)$$

$$G_{\partial\Omega_{ij}}^{(\alpha)}(\mathbf{x}_m,t_{m+},\mathbf{p}_{\alpha},\boldsymbol{\lambda})$$

$$\equiv \mathbf{n}_{\partial\Omega_{ij}}^{T}(\mathbf{x}^{(0)},t,\boldsymbol{\lambda}) \cdot [\mathbf{F}_{b}^{(\alpha)}(\mathbf{x}^{(\alpha)},t,\mathbf{p}_{\alpha}) - \mathbf{F}^{(0)}(\mathbf{x}^{(0)},t,\boldsymbol{\lambda})]\Big|_{(\mathbf{x}_m^{(0)},\mathbf{x}_m^{(\alpha)},t_{m+})}$$

$$= \mathbf{n}_{\partial\Omega_{ij}}^{T}(\mathbf{x}^{(0)},t,\boldsymbol{\lambda}) \cdot \mathbf{F}_{b}^{(\alpha)}(\mathbf{x}^{(\alpha)},t,\mathbf{p}_{\alpha}) + \left.\frac{\partial \varphi_{ij}(\mathbf{x}^{(0)},t,\boldsymbol{\lambda})}{\partial t}\right|_{(\mathbf{x}_m^{(0)},\mathbf{x}_m^{(\alpha)},t_{m+})}$$

$$= \nabla \varphi_{ij}(\mathbf{x}^{(0)},t,\boldsymbol{\lambda}) \cdot \mathbf{F}_{b}^{(\alpha)}(\mathbf{x}^{(\alpha)},t,\mathbf{p}_{\alpha}) + \left.\frac{\partial \varphi_{ij}(\mathbf{x}^{(0)},t,\boldsymbol{\lambda})}{\partial t}\right|_{(\mathbf{x}_m^{(0)},\mathbf{x}_m^{(\alpha)},t_{m+})}. \quad (2.133)$$

The higher-order derivatives of the *G*-function for discontinuous dynamical systems with the input and output flow barriers are defined as

$$G_{\partial\Omega_{ij}}^{(k,\alpha)}(\mathbf{x}_m,t_{m-},\mathbf{p}_{\alpha},\boldsymbol{\lambda})$$

$$= \sum_{r=1}^{k+1} C_{k+1}^{r} D_{\tilde{\mathbf{x}}}^{k+1-r} \mathbf{n}_{\partial\Omega_{ij}}^{T}(\mathbf{x}^{(0)},t,\boldsymbol{\lambda}) \cdot \left[D_{\mathbf{x}}^{r-1} \mathbf{F}_{in}^{(\alpha)}\left(\mathbf{x}^{(\alpha)},t,\mathbf{p}_{\alpha}\right) \right.$$

$$\left. - D_{\mathbf{x}^{(0)}}^{r-1} \mathbf{F}^{(0)}(\mathbf{x}^{(0)},t,\boldsymbol{\lambda})\right]\Big|_{(\mathbf{x}_m^{(0)},\mathbf{x}_m^{(\alpha)},t_{m-})}, \quad (2.134)$$

$$G_{\partial\Omega_{ij}}^{(k,\alpha)}(\mathbf{x}_m,t_{m+},\mathbf{p}_{\alpha},\boldsymbol{\lambda})$$

$$= \sum_{r=1}^{k+1} C_{k+1}^{r} D_{\tilde{\mathbf{x}}}^{k+1-r} \mathbf{n}_{\partial\Omega_{ij}}^{T}(\mathbf{x}^{(0)},t,\boldsymbol{\lambda}) \cdot \left[D_{\mathbf{x}}^{r-1} \mathbf{F}_{b}^{(\alpha)}\left(\mathbf{x}^{(\alpha)},t,\mathbf{p}_{\alpha}\right) \right.$$

$$\left. - D_{\mathbf{x}^{(0)}}^{r-1} \mathbf{F}^{(0)}(\mathbf{x}^{(0)},t,\boldsymbol{\lambda})\right]\Big|_{(\mathbf{x}_m^{(0)},\mathbf{x}_m^{(\alpha)},t_{m+})}. \quad (2.135)$$

Further the dynamics of discontinuous dynamical systems with flow barriers can be investigated. Once the afore-mentioned flow barriers exist, the conditions for the non-passable flows should be measured by the new *G*-function.

References

Luo, A. C. J. (2005a), A theory for nonsmooth dynamic systems on the connectable domains, *Communications in Nonlinear Science and Numerical Simulation*, **10**, pp.1-55.

Luo, A. C. J. (2005b), Imaginary, sink and source flows in the vicinity of separatrix of non-smooth dynamical systems, *Journal of Sound and Vibration*, **285**, pp.443-456.

Luo, A. C. J. (2006), *Singularity and Dynamics on Discontinuous Vector Fields*, Amsterdam: Elsevier.

References

Luo, A. C. J. (2007a), A theory for n-dimensional, nonlinear dynamics on continuous vector fields, *Communications in Nonlinear Science and Numerical Simulation,* **12,** pp.117-194.

Luo, A. C. J. (2007b), On flow switching bifurcations in discontinuous dynamical systems, *Communications in Nonlinear Science and Numerical Simulation,* **12,** pp.100-116.

Luo, A. C. J. (2008a), A theory for flow switchability in discontinuous dynamical systems, *Nonlinear Analysis: Hybrid Systems*, 2(4),pp.1030-1061.

Luo, A. C. J. (2008b), *Global Transversality, Resonance and Chaotic Dynamics*, Singapore: World Scientific.

Luo, A. C. J. (2008c), On the differential geometry of flows in nonlinear dynamic systems, ASME *Journal of Computational and Nonlinear Dynamics*, 021104-1∼10.

Luo, A. C. J. (2008d), Global tangency and transversality of periodic flows and chaos in a periodically forced, damped Duffing oscillator, *International Journal of Bifurcations and Chaos*, **18,** pp.1-49.

Chapter 3
Transversality and Sliding Phenomena

In this chapter, the sliding and transversality of a flow at a boundary from one domain into another in discontinuous dynamical systems will be presented through a periodically forced, discontinuous dynamical system. The inclined line boundary in phase space will be considered for the dynamical system to switch. The normal vector-field for flow switching on the separation boundary will be introduced. The transversality condition of a flow to the separation boundary will be achieved through such normal vector fields, and the sliding and grazing conditions to the separation boundary will be presented as well. Using mapping structures, the periodic motions in such a discontinuous system will be predicted analytically. With the analytical conditions of grazing and sliding motions, the parameter maps of specific motions will be developed. Illustrations of periodic and chaotic motions are given, and the normal vector fields will be presented to show the analytical conditions. The results presented in this chapter may help one better understand the sliding mode control.

3.1 A controlled system

As in Luo and Rapp (2007, 2008), consider an inclined straight line boundary in a discontinuous dynamical system through a mass-spring-damper model. The mass m is connected with a switchable spring of stiffness k_α and a switchable damper of coefficient r_α in the α-region ($\alpha = 1,2$), as shown in Fig.3.1. A periodical force exerts on the mass of the oscillator, i.e.,

$$P_\alpha = Q_0 \cos(\Omega t + \phi) + U_\alpha, \quad \alpha = \{1,2\}, \tag{3.1}$$

where Q_0 and Ω are excitation amplitude and frequency, and the constant force is represented by U_α. The coordinate system is defined by (x,t), in which t and x are time and mass displacement, respectively. The control law for this discontinuous dynamical system is given by

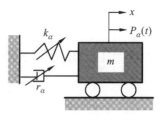

Fig. 3.1 A mechanical model for a discontinuous dynamical system with an inclined line control law

$$ax + b\dot{x} = c, \tag{3.2}$$

where $\dot{x} = dx/dt$ and a, b, c are constants. For the foregoing mass-spring-damper discontinuous system, the equation of motion is

$$\ddot{x} + 2d_\alpha \dot{x} + c_\alpha x = A_0 \cos(\Omega t + \phi) + b_\alpha, \tag{3.3}$$

where

$$c_\alpha = \frac{k_\alpha}{m}, \quad d_\alpha = \frac{r_\alpha}{2m}, \quad b_\alpha = \frac{U_\alpha}{m}, \quad A_0 = \frac{Q_0}{m}. \tag{3.4}$$

Before a comprehensive discussion is given, the history of such a class of problem can be given herein. Filippov (1964) investigated the motion in the Coulomb friction oscillator and presented a differential equation theory for dynamical systems with discontinuous right-hand sides. To determine the sliding motion along the discontinuous boundary, the differential inclusion was introduced via the set-valued analysis. The detailed discussion of such discontinuous differential equations can be referred to Filippov (1988). Since then, the Filippov's theory has been applied for control systems. Aizerman and Pyatniskii (1974a,b) extended the Filippov's concepts and presented a generalized theory for discontinuous dynamical systems. From such a generalized theory, Utkin (1976) developed methods for controlling dynamic systems through the discontinuity (i.e., sliding mode control). DeCarlo et al. (1988) gave a review on the development of the sliding mode control. Renzi and Angelis (2005) used the sliding mode control method to study the dynamics of variable stiffness structures. The Filippov's theory mainly focused on the existence and uniqueness of the solutions for non-smooth dynamical systems, and the Filippov's differential inclusion provides a set of possible candidates for motion switching on the boundary or sliding. However, the sliding mode control theory has a difficulty to achieve an efficient control because the local singularity caused by the separation boundary was discussed incompletely. Thus, the further investigation of the local singularity in the vicinity of the separation boundary should be completed.

Luo (2005a) developed a general theory for the local singularity of a flow to the boundary in non-smooth dynamical systems with connectable domains. The local singularity of non-smooth dynamical systems near the separation boundary was discussed. The imaginary, source and sink flows were introduced in Luo (2005b) to determine the sliding and source motions in non-smooth dynamical systems. The comprehensive discussion on the singularity and dynamics of discontinuous dynamical systems can be referred to Luo (2006). Based on the local singularity theory, the grazing motion to the separation boundary and the sliding motion on the separation boundary in discontinuous dynamical systems were discussed through several piecewise linear systems (e.g., Menon and Luo, 2005; Luo, 2005c; Luo and Chen, 2005) and friction-induced oscillator (e.g., Luo and Gegg, 2005, 2006a,b,c). Two classes of discontinuous systems are based on the displacement and velocity boundaries. The inclined line control law in phase space will be considered herein as a generalized case to demonstrate flows to be sliding and transversal on the discontinuous boundary. Through the straight line control law, the phase domain is separated two domains, and the vector fields of the discontinuous dynamical system will be switched, and the two vector fields in the two domains are different. Owing to such discontinuity, any discontinuous dynamical system possesses more complicated dynamical behavior than the corresponding continuous dynamical system.

3.2 Transversality conditions

In phase plane, the vectors are introduced by

$$\mathbf{x} \stackrel{\Delta}{=} (x,\dot{x})^{\mathrm{T}} \equiv (x,y)^{\mathrm{T}} \text{ and } \mathbf{F} \stackrel{\Delta}{=} (y,F)^{\mathrm{T}}. \tag{3.5}$$

From Eq.(3.2), the control logic generates a discontinuous boundary in the system. To analyze dynamics of the system, the domains and boundary for three motion states are defined as follows. The two domains are

$$\Omega_1 = \{(x,y) | ax+by > c\} \text{ and } \Omega_2 = \{(x,y) | ax+by < c\}. \tag{3.6}$$

The separation boundary $\partial \Omega_{\alpha\beta} = \bar{\Omega}_\alpha \cap \bar{\Omega}_\beta$ ($\alpha, \beta = 1, 2$) is defined as

$$\partial \Omega_{12} = \partial \Omega_{21} = \bar{\Omega}_1 \cap \bar{\Omega}_2 = \{(x,y) | \varphi_{12}(x,y) \equiv ax+by-c = 0\}. \tag{3.7}$$

The domains and boundary are sketched in Fig.3.2. The boundary is depicted by a dotted straight line, governed by Eq.(2.2). The two domains are shaded. The arrows crossing the boundary indicate the passable directions of flows to the boundary. If a flow of the motion in phase space is in domain Ω_α ($\alpha = 1, 2$), the vector fields in such a domain are continuous. However, if a flow of motion from a domain Ω_α ($\alpha \in \{1,2\}$) switches into another domain Ω_β ($\beta \in \{1,2\}, \beta \neq \alpha$) through the boundary $\partial \Omega_{\alpha\beta}$, the vector field in domain Ω_α ($\alpha \in \{1,2\}$) will be changed into the one in domain Ω_β ($\beta = \{1,2\}$) accordingly. Because of the discontinuity, the flow may not pass over the boundary under a certain condition. In other words, the flow may slide along the boundary, which is called the sliding flow. As discussed in Luo (2006), the sliding motion on the boundary has an equilibrium point $(E,0)$ where $E = c/a$. Based on the equilibrium point, the parabolicity and hyperbolicity in vicinity of such an equilibrium can be discussed.

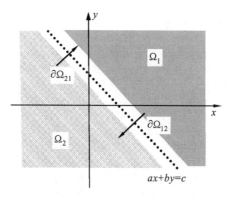

Fig. 3.2 Sub-domains and boundary of the discontinuous dynamical system ($a > 0$ and $b > 0$)

From Eq.(3.2), with initial condition $(x_i^{(0)}, \dot{x}_i^{(0)})$ on the boundary, the displacement and velocity for sliding are given by

$$\left.\begin{aligned}x^{(0)} &= \frac{c}{a} + \frac{1}{a}(ax_i^{(0)} - c)\exp[-\frac{a}{b}(t-t_i)],\\ y^{(0)} &= -\frac{1}{b}(ax_i^{(0)} - c)\exp[-\frac{a}{b}(t-t_i)].\end{aligned}\right\} \qquad (3.8)$$

The equations of motion can be further described as

$$\dot{\mathbf{x}} = \mathbf{F}^{(\lambda)}(\mathbf{x},t), \text{ for } \lambda \in \{0,\alpha\}, \qquad (3.9)$$

where

$$\left.\begin{aligned}\mathbf{F}^{(\alpha)}(\mathbf{x},t) &= (y, F_\alpha(\mathbf{x},t))^{\mathrm{T}}, \text{ in } \Omega_\alpha (\alpha \in \{1,2\}),\\ \mathbf{F}^{(0)}(\mathbf{x},t) &= (y, -\frac{a}{b}y)^{\mathrm{T}}, \text{ for sliding on } \partial\Omega_{\alpha\beta} \ (\alpha,\beta \in \{1,2\}),\\ \mathbf{F}^{(0)}(\mathbf{x},t) &= [\mathbf{F}^{(\alpha)}(\mathbf{x},t), \mathbf{F}^{(\beta)}(\mathbf{x},t)], \text{ for non-sliding on } \partial\Omega_{\alpha\beta},\end{aligned}\right\} \qquad (3.10)$$

$$F_\alpha(\mathbf{x},t) = -2d_\alpha y - c_\alpha x + A_0\cos(\Omega t + \phi) + b_\alpha. \qquad (3.11)$$

From the theory of discontinuous dynamical systems in Luo (2005a, 2006, 2008), for a sliding motion on $\partial\Omega_{\alpha\beta}$ with the corresponding normal vector $\mathbf{n}_{\partial\Omega_{\alpha\beta}}$ pointing to domain Ω_α (i.e., $\mathbf{n}_{\partial\Omega_{\alpha\beta}} \to \Omega_\alpha$), the necessary and sufficient conditions for sliding motions on the switching boundary are

$$\left.\begin{aligned}G^{(0,\alpha)}(\mathbf{x}_m, t_{m-}) &= \mathbf{n}_{\partial\Omega_{\alpha\beta}}^{\mathrm{T}} \cdot \mathbf{F}^{(\alpha)}(\mathbf{x}_m, t_{m-}) < 0,\\ G^{(0,\beta)}(\mathbf{x}_m, t_{m-}) &= \mathbf{n}_{\partial\Omega_{\alpha\beta}}^{\mathrm{T}} \cdot \mathbf{F}^{(\beta)}(\mathbf{x}_m, t_{m-}) > 0,\end{aligned}\right\} \qquad (3.12)$$

where $\alpha, \beta \in \{1,2\}$ and $\alpha \neq \beta$ with

3.2 Transversality conditions

$$\mathbf{n}_{\partial\Omega_{\alpha\beta}} = \nabla\varphi_{\alpha\beta} = \left(\frac{\partial\varphi_{\alpha\beta}}{\partial x}, \frac{\partial\varphi_{\alpha\beta}}{\partial y}\right)^{\mathrm{T}}_{(x_m, y_m)}. \tag{3.13}$$

$\nabla = (\partial/\partial x, \partial/\partial y)^{\mathrm{T}}$ is the Hamilton operator. Note that t_m is switching time for the motion to the switching boundary and $t_{m\pm} = t_m \pm 0$ reflects the responses in domains rather than on the boundary.

From Luo (2005a, 2006, 2008), the necessary and sufficient conditions of a motion switchable to the boundary $\partial\Omega_{\alpha\beta}$ with $\mathbf{n}_{\partial\Omega_{\alpha\beta}} \to \Omega_\alpha$ are as in Eq.(2.20), ie.,

$$\left.\begin{array}{l} G^{(0,\alpha)}(\mathbf{x}_m, t_{m-}) = \mathbf{n}^{\mathrm{T}}_{\partial\Omega_{\alpha\beta}} \cdot \mathbf{F}^{(\alpha)}(\mathbf{x}_m, t_{m-}) < 0 \\ G^{(0,\beta)}(\mathbf{x}_m, t_{m+}) = \mathbf{n}^{\mathrm{T}}_{\partial\Omega_{\alpha\beta}} \cdot \mathbf{F}^{(\beta)}(\mathbf{x}_m, t_{m+}) < 0 \end{array}\right\} \text{from } \Omega_\alpha \to \Omega_\beta, \\ \left.\begin{array}{l} G^{(0,\alpha)}(\mathbf{x}_m, t_{m+}) = \mathbf{n}^{\mathrm{T}}_{\partial\Omega_{\beta\alpha}} \cdot \mathbf{F}^{(\alpha)}(\mathbf{x}_m, t_{m+}) > 0 \\ G^{(0,\beta)}(\mathbf{x}_m, t_{m-}) = \mathbf{n}^{\mathrm{T}}_{\partial\Omega_{\beta\alpha}} \cdot \mathbf{F}^{(\beta)}(\mathbf{x}_m, t_{m-}) > 0 \end{array}\right\} \text{from } \Omega_\beta \to \Omega_\alpha. \tag{3.14}$$

As in Luo (2005a, b, 2008), the grazing motion to the separation boundary $\partial\Omega_{\alpha\beta}$ with $\mathbf{n}_{\partial\Omega_{\alpha\beta}} \to \Omega_\alpha$ is from Luo Eqs. (2.54)~(2.56), i.e.,

$$\left.\begin{array}{l} G^{(0,\alpha)}(\mathbf{x}_m, t_{m\pm}) = \mathbf{n}^{\mathrm{T}}_{\partial\Omega_{\alpha\beta}} \cdot \mathbf{F}^{(\alpha)}(\mathbf{x}_m, t_{m\pm}) = 0, \\ G^{(1,\alpha)}(\mathbf{x}_m, t_{m\pm}) = \mathbf{n}^{\mathrm{T}}_{\partial\Omega_{12}} \cdot D\mathbf{F}^{(\alpha)}(\mathbf{x}_m, t_{m\pm}) > 0, \\ G^{(1,\beta)}(\mathbf{x}_m, t_{m\pm}) = \mathbf{n}^{\mathrm{T}}_{\partial\Omega_{21}} \cdot D\mathbf{F}^{(\beta)}(\mathbf{x}_m, t_{m\pm}) < 0, \end{array}\right\} \tag{3.15}$$

where

$$D\mathbf{F}^{(\alpha)}(\mathbf{x}, t) = \left(F_\alpha(\mathbf{x}, t), \nabla F_\alpha(\mathbf{x}, t) \cdot \mathbf{F}^{(\alpha)}(\mathbf{x}, t) + \frac{\partial F_\alpha(\mathbf{x}, t)}{\partial t}\right)^{\mathrm{T}}. \tag{3.16}$$

The conditions presented herein are valid only for straight line boundary.
Substitution of Eq.(3.7) into Eq.(3.13) gives

$$\mathbf{n}_{\partial\Omega_{12}} = \mathbf{n}_{\partial\Omega_{21}} = (a, b)^{\mathrm{T}}. \tag{3.17}$$

From the forgoing equation, the normal vector always points to the domain Ω_1 (i.e., $\mathbf{n}_{\partial\Omega_{12}} \to \Omega_1$). Therefore,

$$\left.\begin{array}{l} G^{(0,\alpha)}(\mathbf{x}_m, t_m) = \mathbf{n}^{\mathrm{T}}_{\partial\Omega_{\alpha\beta}} \cdot \mathbf{F}^{(\alpha)}(\mathbf{x}_m, t_m) = ay_m + bF_\alpha(\mathbf{x}_m, t_m), \\ G^{(1,\alpha)}(\mathbf{x}_m, t_m) = \mathbf{n}^{\mathrm{T}}_{\partial\Omega_{\alpha\beta}} \cdot D\mathbf{F}^{(\alpha)}(\mathbf{x}_m, t_m) = aF_\alpha(\mathbf{x}_m, t_m) \\ \qquad + b\left[\nabla F_\alpha(\mathbf{x}, t) \cdot \mathbf{F}^{(\alpha)}(\mathbf{x}, t) + \dfrac{\partial F_\alpha(\mathbf{x}, t)}{\partial t}\right]_{(\mathbf{x}_m, t_m)}. \end{array}\right\} \tag{3.18}$$

From Eqs.(3.12) and (3.18), the conditions for sliding motion on the switching boundary are

$$G^{(0,1)}(\mathbf{x}_m, t_{m-}) < 0 \text{ and } G^{(0,2)}(\mathbf{x}_m, t_{m-}) > 0. \tag{3.19}$$

From Eqs.(3.14) and (3.18), the switching conditions for motions on the switching boundary are

$$\left.\begin{array}{l} G^{(0,1)}(\mathbf{x}_m,t_{m-}) < 0 \text{ and } G^{(0,2)}(\mathbf{x}_m,t_{m+}) < 0, \text{ from } \Omega_1 \to \Omega_2, \\ G^{(0,1)}(\mathbf{x}_m,t_{m+}) > 0 \text{ and } G^{(0,2)}(\mathbf{x}_m,t_{m-}) > 0, \text{ from } \Omega_2 \to \Omega_1. \end{array}\right\} \quad (3.20)$$

From the theory for discontinuous dynamical systems in Luo (2005a,b, 2008), the vanishing conditions for sliding motions on the separation boundary are

$$\left.\begin{array}{l} (-1)^\alpha G^{(0,\alpha)}(\mathbf{x}_m,t_{m-}) > 0 \text{ and } G^{(0,\beta)}(\mathbf{x}_m,t_{m\mp}) = 0 \text{ with} \\ (-1)^\beta G^{(1,\beta)}(\mathbf{x}_m,t_{m\mp}) < 0 \text{ from } \partial\Omega_{12} \to \Omega_\beta, (\alpha,\beta \in \{1,2\}). \end{array}\right\} \quad (3.21)$$

From Eq.(3.16), the onset condition of the sliding motion on the switching boundary is given by

$$\left.\begin{array}{l} (-1)^\alpha G^{(0,\alpha)}(\mathbf{x}_m,t_{m-}) > 0 \text{ and } G^{(0,\beta)}(\mathbf{x}_m,t_{m\pm}) = 0 \text{ with} \\ (-1)^\beta G^{(1,\beta)}(\mathbf{x}_m,t_{m\pm}) < 0 \text{ from } \Omega_\alpha \to \partial\Omega_{12}, (\alpha,\beta \in \{1,2\}). \end{array}\right\} \quad (3.22)$$

3.3 Mappings and predictions

In this section, switching planes and generic mappings will be introduced for mapping structures. For a sliding motion on the separation boundary, consider a sliding motion disappearing at time t_{i+1}. For $t_m \in [t_i, t_{i+1}]$, the solution of the sliding motion is given by

$$\left.\begin{array}{l} x_m = \dfrac{c}{a} + \dfrac{1}{a}(ax_i^{(0)} - c)\exp\left[-\dfrac{a}{b}(t_m - t_i)\right], \\ y_m = -\dfrac{1}{b}(ax_i^{(0)} - c)\exp\left[-\dfrac{a}{b}(t_m - t_i)\right]. \end{array}\right\} \quad (3.23)$$

The corresponding $G^{(0,\alpha)}$-function is

$$\begin{aligned} & G^{(0,\alpha)}(\mathbf{x}_m,t_{m-}) \\ & = ay_m + b[-2d_\alpha y_m - c_\alpha x_m + A_0\cos(\Omega t_{m-} + \phi) + b_\alpha]. \end{aligned} \quad (3.24)$$

For the non-sliding motion, once the initial condition is chosen on separation boundary, the solutions of Eq.(3.3) in all the domains Ω_α can be gotten, which are listed in Appendix. The basic solutions in Appendix will be applied for mapping construction. To construct basic mappings, switching planes in phase space will be introduced first. In phase plane, a trajectory in Ω_α, starting and ending at the switching boundary (i.e., from $\partial\Omega_{\beta\alpha}$ to $\partial\Omega_{\alpha\beta}$), are illustrated in Fig.3.3. The starting and ending points for mappings P_α in Ω_α are $(\mathbf{x}_i,t_i)$ and $(\mathbf{x}_{i+1},t_{i+1})$, respectively. The sliding mapping is P_0. Define switching planes as

3.3 Mappings and predictions

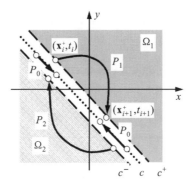

Fig. 3.3 Basic mappings of the discontinuous system ($a > 0$ and $b > 0$)

$$\left.\begin{aligned}
\Xi^0 &= \{(x_i, y_i, \Omega t_i) | \varphi_{\alpha\beta}(x_i, y_i) = c\}, \\
\Xi^1 &= \{(x_i^+, y_i^+, \Omega t_i) | \varphi_{\alpha\beta}(x_i^+, y_i^+) = c^+\}, \\
\Xi^2 &= \{(x_i^-, y_i^-, \Omega t_i) | \varphi_{\alpha\beta}(x_i^-, y_i^-) = c^-\},
\end{aligned}\right\} \quad (3.25)$$

where $c^- = \lim_{\delta \to 0}(c - \delta)$ and $c^+ = \lim_{\delta \to 0}(c + \delta)$ for an arbitrary small $\delta > 0$. Therefore, three mappings are defined as

$$P_0 : \Xi^0 \to \Xi^0, P_1 : \Xi^1 \to \Xi^1, P_2 : \Xi^2 \to \Xi^2. \quad (3.26)$$

From the switching planes and mappings, we have

$$\left.\begin{aligned}
P_0 &: (x_i, y_i, t_i) \to (x_{i+1}, y_{i+1}, t_{i+1}), \\
P_1 &: (x_i^+, y_i^+, t_i) \to (x_{i+1}^+, y_{i+1}^+, t_{i+1}), \\
P_2 &: (x_i^-, y_i^-, t_i) \to (x_{i+1}^-, y_{i+1}^-, t_{i+1}).
\end{aligned}\right\} \quad (3.27)$$

With Eq.(3.2), the governing equations for P_0 and $\alpha \in \{1, 2\}$ are

$$\left.\begin{aligned}
&-x_{i+1} + \frac{c}{a} + \frac{1}{a}(ax_i - c) \exp\left[-\frac{a}{b}(t_{i+1} - t_i)\right] = 0, \\
&(-1)^\beta G^{(0,\beta)}(\mathbf{x}_{i+1}, t_{i+1}) > 0, \\
&G^{(0,\alpha)}(\mathbf{x}_{i+1}, t_{i+1}) = 0, (-1)^\alpha G^{(1,\alpha)}(\mathbf{x}_{i+1}, t_{i+1}) < 0, \\
&G^{(0,1)}(\mathbf{x}_i, t_i) \times G^{(0,2)}(\mathbf{x}_i, t_i) \leqslant 0.
\end{aligned}\right\} \quad (3.28)$$

Because the two domains $\Omega_\alpha (\alpha \in \{1, 2\})$ are unbounded, from hypotheses in Luo (2005a, 2006), only three bounded motions can exist in domain $\Omega_\alpha (\alpha \in \{1, 2\})$. The corresponding governing equations of mapping P_α ($\alpha \in \{1, 2\}$) are obtained. With Eq.(3.2), the governing equations of each mapping P_λ ($\lambda \in \{0, 1, 2\}$) are expressed by

$$\left.\begin{array}{l} f_1^{(\lambda)}(x_i, \Omega t_i, x_{i+1}, \Omega t_{i+1}) = 0, \\ f_2^{(\lambda)}(x_i, \Omega t_i, x_{i+1}, \Omega t_{i+1}) = 0. \end{array}\right\} \quad (3.29)$$

From generic mappings, consider a generalized mapping structure for a periodic motion with sliding motion as

$$P = \underbrace{\left(P_2^{(k_{m2})} \circ P_1^{(k_{m1})} \circ P_0^{(k_{m0})}\right) \circ \cdots \circ \left(P_2^{(k_{12})} \circ P_1^{(k_{11})} \circ P_0^{(k_{10})}\right)}_{m\text{-terms}}, \quad (3.30)$$

where $k_{l\lambda} \in \{0,1\}$ for $l \in \{1, 2, \cdots, m\}$ and $\lambda \in \{0, 1, 2\}$. $P_\lambda^{(0)} = 1$ and $P_\lambda^{(k)} = P_\lambda \circ P_\lambda^{(k-1)}$. Note that the clockwise and counter-clockwise rotations of the order of the mapping P_λ in the complete mapping P in Eq.(3.30) will not change the periodic motion. However, only the initial conditions for such a periodic motion are different. For simplicity, the mapping notation in Eq.(3.30) is expressed by

$$P = P\underbrace{_{(2^{k_{m2}} 1^{k_{m1}} 0^{k_{m0}}) \cdots (2^{k_{12}} 1^{k_{11}} 0^{k_{10}})}}_{m\text{-terms}}.$$

For ($m = k_{12} = k_{11} = 1$ and $k_{10} = 0$), equation.(3.29) gives a mapping structure for the simplest, non-sliding periodic motion passing through the separation boundary. The procedure for prediction of periodic motions is presented through this periodic motion. The mapping structure for one of the simplest periodic motions is

$$P = P_2 \circ P_1 : \Xi^1 \to \Xi^2. \quad (3.31)$$

From the foregoing relation, we have

$$\left.\begin{array}{l} P_1 : (x_i^+, y_i^+, t_i) \to (x_{i+1}^+, y_{i+1}^+, t_{i+1}), \\ P_2 : (x_{i+1}^-, y_{i+1}^-, t_{i+1}) \to (x_{i+2}^-, y_{i+2}^-, t_{i+2}). \end{array}\right\} \quad (3.32)$$

Without sliding, $\mathbf{x}_{i+k}^+ = \mathbf{x}_{i+k}^- = \mathbf{x}_{i+k}$ ($k = 0, 1, 2$) exists. For a periodic motion of $\mathbf{y}_{i+2} = P\mathbf{y}_i$ where $\mathbf{y}_i = (x_i, y_i, \Omega t_i)^T$ during N-periods of excitation, the periodicity of the periodic motion is

$$x_{i+2} = x_i, y_{i+2} = y_i, \Omega t_{i+2} = \Omega t_i + 2N\pi. \quad (3.33)$$

Let $\mathbf{z}_i = (x_i, \Omega t_i)^T$ and $\mathbf{f}^{(\lambda)} = (f_1^{(\lambda)}, f_2^{(\lambda)})^T$. With Eq.(3.2), the governing equations for the simplest periodic motion are

$$\mathbf{f}^{(\alpha)}(\mathbf{z}_{i+\alpha-1}, \mathbf{z}_{i+\alpha}) = 0, \text{ for } \alpha = 1, 2. \quad (3.34)$$

With Eq.(3.33), equation (3.34) gives the switching sets for such a periodic motion. The existence of the periodic motion will be determined by the local stability analysis.

3.3 Mappings and predictions

Consider a mapping structure for a periodic motion with sliding motion ($m = k_{1\alpha} = 1$, $\alpha \in \{0,1,2\}$), the mapping structure for one of the simplest motions with sliding motion is

$$P = P_{210} \triangleq P_2 \circ P_1 \circ P_0. \tag{3.35}$$

For a sliding periodic motion of $\mathbf{y}_{i+3} = P\mathbf{y}_i$ where $\mathbf{y}_i = (x_i, y_i, \Omega t_i)^T$ during N-periods of excitation, the periodicity is given by

$$x_{i+3} = x_i, y_{i+3} = y_i, \Omega t_{i+3} = \Omega t_i + 2N\pi. \tag{3.36}$$

Again, with Eq.(3.2), the governing equations for such a periodic motion with sliding motion are

$$\mathbf{f}^{(\lambda)}(\mathbf{z}_{i+\lambda}, \mathbf{z}_{i+\lambda+1}) = 0, \text{ for } \lambda = 0, 1, 2. \tag{3.37}$$

For the two simple periodic motions, the mapping structures are sketched in Fig.3.4. Similarly, the periodic motion for the generalized mapping structure can be predicted through the corresponding governing equations.

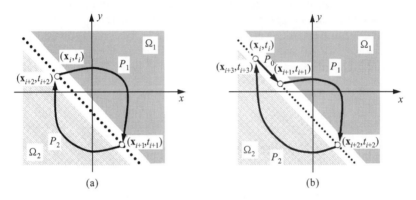

Fig. 3.4 Mapping structures: (a)$P_{21} = P_2 \circ P_1$ and (b) $P_{210} = P_2 \circ P_1 \circ P_0$

To determine the existence of periodic motions, consider the local stability and bifurcation through the eigenvalue analysis based on the corresponding Jacobian matrix. For a periodic motion relative to $\mathbf{z}_{i+\sum_{l=1}^{m}(k_{l2}+k_{l1}+k_{l0})} = P\mathbf{z}_i$, using Eq.(3.2), the Jacobian matrix is computed by

$$DP = \left[\frac{\partial(t_{i+\sum_{l=1}^{m}(k_{l2}+k_{l1}+k_{l0})}, x_{i+\sum_{l=1}^{m}(k_{l2}+k_{l1}+k_{l0})})}{\partial(t_i, x_i)} \right]_{(t_i, x_i)}$$
$$= \underbrace{(DP_2^{(k_{m2})} \cdot DP_1^{(k_{m1})} \cdot DP_0^{(k_{m0})}) \cdot \ \cdots \ \cdot (DP_2^{(k_{12})} \cdot DP_1^{(k_{11})} \cdot DP_0^{(k_{10})})}_{m\text{-terms}}, \tag{3.38}$$

where

$$DP_\lambda = \begin{bmatrix} \dfrac{\partial t_{v+1}}{\partial t_v} & \dfrac{\partial t_{v+1}}{\partial x_v} \\ \dfrac{\partial x_{v+1}}{\partial t_v} & \dfrac{\partial x_{v+1}}{\partial x_v} \end{bmatrix}, \tag{3.39}$$

for $\lambda \in \{0,1,2\}$ and $v \in \{i, i+1, \ldots, i + \sum_{l=1}^{m}(k_{l2}+k_{l1}+k_{l0}) - 1\}$, and the Jacobian matrix components $\partial t_{v+1}/\partial t_v$, $\partial t_{v+1}/\partial x_v$, $\partial x_{v+1}/\partial t_v$ and $\partial x_{v+1}/\partial x_v$ can be computed through Eqs.(3.28) and (3.29). Suppose the eigenvalues for the mapping structure of periodic motion are $\lambda_{1,2}$. The stable period-1 motion requires the eigenvalues be $|\lambda_\kappa| < 1$, ($\kappa \in \{1,2\}$). Once the aforementioned condition is not satisfied, the period-1 motion is unstable. If $|\lambda_{1\,(\text{or}\,2)}| = 1$ with complex numbers, the Neimark bifurcation occurs. If one of the two eigenvalues is -1 (i.e., $\lambda_{1(\text{or}2)} = -1$) and the other one is inside the unit circle, the period-doubling bifurcation occurs. If one of the two eigenvalues is $+1$ (i.e., $\lambda_{1(\text{or}2)} = +1$) and the second one is inside the unit circle, the first saddle-node bifurcation occurs. Without the local singularity involving with the discontinuity, the eigenvalue analysis can provide an adequate prediction. However, the eigenvalue analysis cannot work for the local singularity pertaining to the separation discontinuity. From the aforementioned causes, the onset, existence and disappearance of the sliding motion should be determined through the normal vector field criteria in Eqs.(3.21) and (3.22). The grazing bifurcation will be determined by Eq.(3.15).

Consider a set of parameters as

$$r_1 = 0.5, \ k_1 = 50, \ r_2 = 1.0, \ k_2 = 150, \ U_1 = 1,$$
$$U_2 = 1, \ Q_0 = 20, \ \Omega = 5, \ \phi = 0, \ b = 1, \ c = -3. \tag{3.40}$$

From the closed-form solutions in Appendix, bifurcation scenario will be presented. From the coppresponding mapping structures, the analytical prediction will be carried out by solving nonlinear algebraic equations, and the corresponding local stability analysis can be performed. Both analytical and numerical simulations results are presented in Fig.3.5 through the switching displacment and phase versus boundary displacement coefficient a, respectively. The black dots represent the numerical prediction of the switching responses on the separation boundary. The dark-gray hollow circular symbols give the analytical prediction of switching values on the separation boundary for stable periodic motions. The unstable periodic motions are given by the light-gray hollow circular symbols. The regions for chaotic and complex periodic motions are shaded. The grazing and sliding bifurcation conditions are used to determine the motion switching. The dashed lines are for the period-doubling bifurcation and grazing bifurcation. The dot-dashed lines are the sliding bifurcation, and the acronyms SB and GB represent the sliding bifurcation and grazing bifurcation. However, the acronym "PD" denotes the period-doubling bifurcation. Because the switching velocity and displacement on the switching boundary satisfy Eq.(3.2), the switching velocity versus excitation frequency will not be presented herein. The real parts and magnitudes of eigenvalues are respectively given in Fig.3.6(a) and (b), from which the stability and bifurcation can be determined. However, with the sliding and grazing bifurcations of the periodic motion, the eigenvalue analysis can-

3.3 Mappings and predictions 65

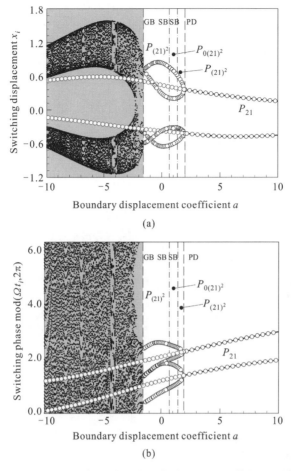

Fig. 3.5 Bifurcation scenario relative to boundary displacement coefficient a: (a) switching displacement and (b) switching phase. The black dots represent numerical simulations. The hollow circles give the analytical prediction of periodic motions ($r_1 = 0.5, k_1 = 50, r_2 = 1.0, k_2 = 150, U_1 = U_2 = 1, Q_0 = 150, \Omega = 5.0, \phi = 0, b = 1, c = -3$)

not provide adequate information. Therefore, the analytical conditions in Section 3.2 should be adopted. The thin and dark-gray hollow-circle curves are stable and unstable periodic motion of P_{21}, respectively. The thick solid curves give periodic motions of $P_{(21)^20}$ and $P_{(21)^2}$. For $\Omega \in (-10, -1.54)$, the complex and chaotic motions exist. The periodic motion of $P_{(21)^2}$ lies in $\Omega \in (-1.54, 0.66)$ and $(1.40, 1.92)$. The periodic motion of $P_{(21)^20}$ exists in the range of $\Omega \in (0.66, 1.40)$. The periodic motion of P_{21} is in the range of $\Omega \in (1.92, 10)$. Once the parameters change, the bifurcation scenario will be changed. Thus, the parameter maps for specific motions are also developed.

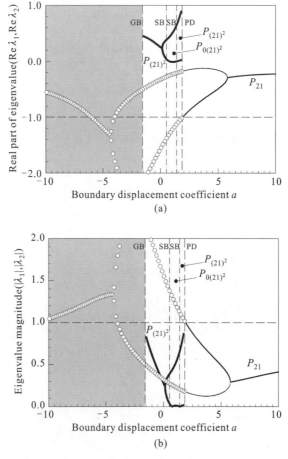

Fig. 3.6 Eigenvalue analysis of periodic motion for the boundary displacement coefficient a: (a) real part and (b) magnitude. The thin and hollow-circle curves are stable and unstable periodic motions of P_{21}, respectively. The thick solid curves give periodic motions of $P_{(21)^20}$ and $P_{(21)^2}$ ($r_1 = 0.5$, $k_1 = 50$, $r_2 = 1.0$, $k_2 = 150$, $U_1 = U_2 = 1$, $Q_0 = 150$, $\Omega = 5.0$, $\phi = 0$, $b = 1$, $c = -3$)

With the analytical conditions of the sliding and grazing bifurcations, the parameter map is developed from the stability and bifurcation conditions of periodic motions. The parameter map for the excitation amplitude versus the boundary displacement coefficient is given in Fig.3.7. The acronym "NM" denotes no motion that is touching the boundary. The "complex" or "chaos" simply means that in that area either a highly complex motion or a chaotic motion exists. Also, the complex sliding motion region means that there exists always sliding motion occurring with high number of mappings. The acronym "CSM" gives the "complex sliding motion". The other motion in the parameter map can be described from Eq.(3.30). Similarly, the parameter map for the excitation frequency versus amplitude is given in Fig.3.8.

3.4 Periodic and chaotic motions

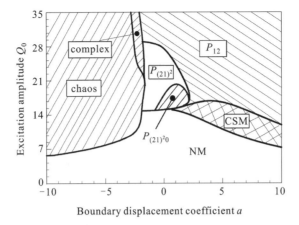

Fig. 3.7 Excitation amplitude versus boundary displacement coefficient ($r_1 = 0.5$, $k_1 = 50$, $r_2 = 1.0$, $k_2 = 150$, $U_1 = U_2 = 1$, $Q_0 = 150$, $\Omega = 5.0$, $\phi = 0$, $b = 1$, $c = -3$).

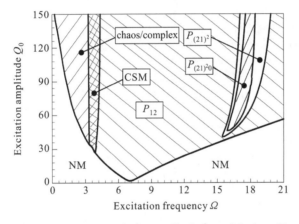

Fig. 3.8 Excitation frequency versus excitation amplitude ($r_1 = 0.5$, $k_1 = 50$, $r_2 = 1.0$, $k_2 = 150$, $U_1 = U_2 = 1$, $Q_0 = 150$, $\Omega = 5.0$, $\phi = 0$, $b = 1$, $c = -3$).

3.4 Periodic and chaotic motions

From analytical predictions, illustrations of periodic and chaotic motions are very important for a better understanding of periodic motions in such a discontinuous dynamical system under external excitations. The phase plane and displacement response for periodic motions will be presented. The normal vector field is very important to determine the grazing and sliding bifurcations, thus the normal vector fields versus displacement will be presented, and the normal vector field time-history will be given for illustrations of the transversality of the switching dynamical systems. For chaotic motions, the phase plane, Poincare mapping section, switching normal vector fields and the switching normal vector field product on the boundary will be presented along with the displacement. The closed-form solutions in Appendix will be used for numerical computations of motions for such a switching dynamical system. The initial conditions are selected from the analytical prediction.

For a periodic motion of P_{12} with parameters ($r_1 = 0.5, k_1 = 50, r_2 = 1, k_2 = 150, U_1 = U_2 = 1, \phi = 0, a = 1, b = 4, c = -3, Q_0 = 150, \Omega = 5$), an initial condition (i.e., $\Omega t_0 \approx 0.1875, x_0 \approx 1.1800$ and $y_0 \approx -1.2114$) is considered. The phase plane, normal vector field versus displacement, the time–histories of both displacement and normal vector fields are presented in Fig.3.9. The dark solid curves denote for real flows. The dashed curves represent the imaginary normal vector fields of the real flow. Such a periodic motion is one of the simplest periodic motions intersected with the separation boundary. For a starting point labeled with a black circle, it is observed that the normal vector fields $G^{(0,1)}$ and $G^{(0,2)}$ are negative (i.e., $G^{(0,1)} < 0$ and $G^{(0,2)} < 0$). For this switching system, the boundary normal vector always points to domain Ω_1. Therefore, from Eq.(3.20), the motion on the switching boundary will enter the domain Ω_2, as shown in Fig.3.9(a). In Fig.3.9(b), the dashed curves give the imaginary normal vector fields to the boundary for such a real flow. In other words, the imaginary normal vector field does not control the real flow, but records the other vector field changes in domains of the real flows. A detailed discussion can be referred to Luo (2005a, 2006). Once the motion in domain Ω_2 arrives at the separation boundary, it is observed that two normal vector fields are positive (i.e., $G^{(0,1)} > 0$ and $G^{(0,2)} > 0$). The motion on the boundary will switch into the domain Ω_1, as shown in Fig.3.9(a). The motion in domain Ω_1 returns back to the initial point. So the periodic motion is observed. The corresponding time-histories of displacement and normal vector fields are plotted in Fig.3.9(c) and (d), respectively. The periodicity of such a periodic motion is observed. This periodic motion on the boundary is always passable without any sliding motion.

For a periodic motion of P_{102} with parameters ($r_1 = 0.5$, $k_1 = 50$, $r_2 = 1$, $k_2 = 150$, $U_1 = U_2 = 1$, $\phi = 0$, $a = 4$, $b = 1$, $c = -3$, $Q_0 = 150$, $\Omega = 5$), from analytical prediction, an initial condition (i.e., $\Omega t_0 \approx 0.3337$, $x_0 \approx 0.0762$ and $y_0 \approx -3.3286$) is adopted. The phase plane, normal vector field versus displacement, the time-histories of both the displacement and normal vector fields are presented in Fig.3.10. The starting point is from a solid circle, and the normal vector fields are $G^{(0,1)} < 0$ and $G^{(0,2)} < 0$. From Eq.(3.20), the motion on the switching boundary will enter the domain Ω_2, as shown in Fig.3.10(a). In Fig.3.10(b), the imaginary and real normal vector fields are presented. Once the motion in domain Ω_2 arrives to the separation boundary, the two normal vector fields are $G^{(0,1)} < 0$ and $G^{(0,2)} > 0$. From Eq.(3.19), the motion should slide on the separation boundary. When the motion slides along the boundary, the normal vector fields will be changed with the location. For a moment, $G^{(0,1)} = 0$, $G^{(1,1)} > 0$ and $G^{(0,2)} > 0$ hold. From Eq.(3.22), the sliding motion on the boundary will vanish and the motion will switch into the domain Ω_1, which can be observed in Fig.3.10(a). The motion in domain Ω_1 returns back to the initial point. So the periodic motion is observed. The corresponding time responses of the displacement and normal vector fields are plotted in Fig.3.10(c) and (d), respectively. This periodic motion on the boundary shows a sliding motion exists.

3.4 Periodic and chaotic motions

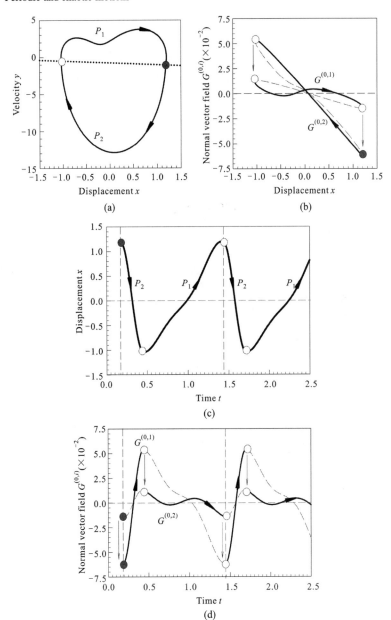

Fig. 3.9 Periodic motion of P_{12}: (a) phase plane, (b) normal vector field versus displacement, (c) displacement, and (d) the normal vector fields. Initial condition is $\Omega t_0 \approx 0.1875$, $x_0 \approx 1.1800$ and $y_0 \approx -1.2114$. The dashed curves represent the normal vector fields for the imaginary flows ($r_1 = 0.5$, $k_1 = 50$, $r_2 = 1.0$, $k_2 = 150$, $U_1 = 1$, $U_2 = 1$, $Q_0 = 150$, $\Omega = 5.0$, $\phi = 0$, $a = 1$, $b = 4$, $c = -3$)

70　　　　　　　　　　　　　　　　　　　　　　3 Transversality and Sliding Phenomena

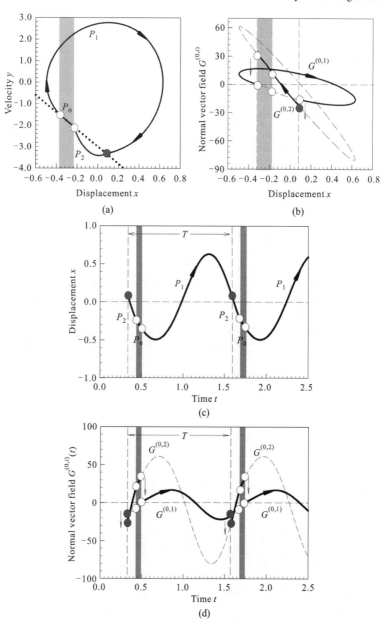

Fig. 3.10 Periodic motion of P_{102}: (a) phase plane, (b) normal vector field versus displacement, (c) displacement, and (d) the normal vector fields. Initial condition is $\Omega t_0 \approx 0.3337$, $x_0 \approx 0.0762$ and $y_0 \approx -3.3286$. The dashed curves represent the normal vector fields for the imaginary flow. The shaded area is for sliding motion ($r_1 = 0.5$, $k_1 = 50$, $r_2 = 1.0$, $k_2 = 150$, $U_1 = 1$, $U_2 = 1$, $Q_0 = 150$, $\Omega = 5.0$, $\phi = 0$, $a = 4$, $b = 1.0$, $c = -3$)

3.4 Periodic and chaotic motions

Consider a complex, periodic motion of $P_{(21)^201}$ with parameters ($r_1 = 0.5$, $k_1 = 50$, $r_2 = 1$, $k_2 = 150$, $U_1 = U_2 = 1$, $\phi = 0$, $a = 1$, $b = 4$, $c = -3$, $Q_0 = 150$, $\Omega = 5$). From analytical prediction, an initial condition (i.e., $\Omega t_0 \approx 0.8920$, $x_0 \approx -4.5103$, $y_0 \approx 0.9773$) is adopted. The phase plane, normal vector field versus displacement, the time-histories of the displacement and normal vector fields are presented in Fig.3.11. The starting point is from a solid circle, and the normal vector fields are $G^{(0,1)} > 0$ and $G^{(0,2)} > 0$. From Eq.(3.20), the motion on the switching boundary will enter the domain Ω_1, as shown in Fig.3.11(a). The normal vector fields are presented in Fig.3.11(b) as well. Once the motion in domain Ω_1 arrives to the separation boundary, the two normal vector fields ($G^{(0,1)} < 0$ and $G^{(0,2)} > 0$) are observed. From Eq.(3.19), the motion should slide on the separation boundary. When the motion slides along the boundary, the normal vector fields will be changed. For some moment, the conditions ($G^{(0,1)} = 0$, $G^{(1,1)} > 0$ and $G^{(0,2)} > 0$) are observed. From Eq.(3.22), the sliding motion on the boundary will vanish and such a motion will switch into the domain Ω_1, which can be observed in Fig.3.11(a). The motion in domain Ω_1 returns to the separation boundary with $G^{(0,1)} < 0$ and $G^{(0,2)} < 0$. From Eq.(3.20), the motion switches from domain Ω_1 into Ω_2. The motion in domain Ω_2 arrives to the switching boundary with $G^{(0,1)} > 0$ and $G^{(0,2)} > 0$. So the motion will switch from Ω_2 into Ω_1. Continuously, the motion in domain Ω_1 returns back the separation boundary with $G^{(0,1)} < 0$ and $G^{(0,2)} < 0$. So the motion will enter domain Ω_2. Finally the motion in domain Ω_2 returns back to the starting point, and such a periodic motion is achieved. The corresponding time-histories of displacement and normal vector fields are plotted in Fig.3.11(c) and (d), respectively.

Consider a chaotic motion of $P_{(21)^k}$. With parameters ($r_1 = 0.5$, $r_2 = 1$, $k_2 = 150$, $U_1 = U_2 = 1$, $Q_0 = 150$, $\Omega = 2.5$, $\phi = 0$, $a = 1$, $b = 4$, $c = -3$), an initial condition (i.e., $\Omega t_0 \approx 2.8192$, $x_0 \approx -0.9657$ and $y_0 \approx -7.8283$) is used. The phase plane, Poincare mapping section, switching normal vector fields, and switching normal vector field product are presented in Fig.3.12. In Fig.3.12(a), the trajectory in the phase plane is a chaotic pattern, and the corresponding switching points on the separation are depicted by the hollow circles, which will form the Poincare mapping section. In Fig.3.12(b), the Poincare mapping section of the chaotic motion is presented, which consists of two continuous parts relative to the separation boundary. The normal vector fields of the switching points on the separation boundary are presented in Fig.3.12(c). The solid dashed curves give the normal vector fields of $G^{(0,1)}(\mathbf{x}_k, t_k)$ and $G^{(0,2)}(\mathbf{x}_k, t_k)$, respectively. For the vicinity of $x_k = 1$, we have $G^{(0,1)} < 0$ and $G^{(0,2)} < 0$ for the chaotic motion switching from Ω_1 to Ω_2. However, in the vicinity of $x_k = -1$, $G^{(0,1)} > 0$ and $G^{(0,2)} > 0$, which implies that chaotic motion switches from Ω_2 to Ω_1. Such transversality of motion is expressed by the normal vector field product (i.e., $G^{(0,1)} \times G^{(0,2)}$), as shown in Fig.3.12(d). Therefore, this chaotic motion is always switchable because of $G^{(0,1)} \times G^{(0,2)} > 0$.

Consider a chaotic motion with parameters ($r_1 = 0.5$, $k_1 = 50$, $r_2 = 1$, $k_2 = 150$, $U_1 = U_2 = 1$, $Q_0 = 150$, $\Omega = 2.83$, $\phi = 0$, $a = 1$, $b = 1$, $c = -3$). The initial condition (i.e., $\Omega t_0 \approx 1.0205$, $x_0 \approx -1.5141$ and $y_0 \approx -1.4859$) is used. The chaotic motion is obtained, and the corresponding phase plane, Poincare mapping section, switching normal vector fields, and switching normal vector field product

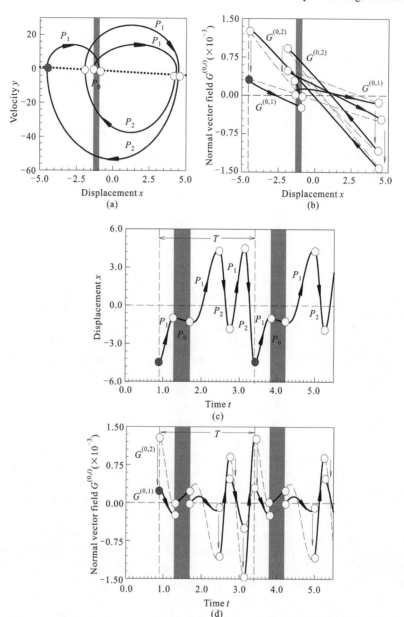

Fig. 3.11 Periodic motion of $P_{(21)^201}$: (a) phase plane, (b) normal vector field versus displacement, (c) displacement, and (d) the normal vector fields. Initial conditions are $\Omega t_0 \approx 0.8920$, $x_0 \approx -4.5103$ and $y_0 \approx 0.9773$. The dashed curves represent the normal vector fields for the imaginary flow. The shaded area is for sliding motion ($r_1 = 0.5$, $k_1 = 50$, $r_2 = 1.0$, $k_2 = 150$, $U_1 = 1$, $U_2 = 1$, $Q_0 = 150$, $\Omega = 2.5$, $\phi = 0$, $a = 1$, $b = 2.2$, $c = -3$)

3.4 Periodic and chaotic motions

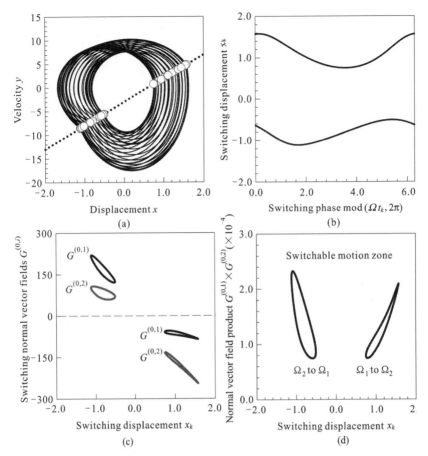

Fig. 3.12 Chaotic motion of $P_{(21)^k}$ ($k \to \infty$): (a) phase plane, (b) Poincare mapping section, (c) switching normal vector fields, and (d) switching normal vector field product. Initial conditions are $\Omega t_0 \approx 2.8192$, $x_0 \approx -0.9657$ and $y_0 \approx -7.8283$ ($r_1 = 0.5, k_1 = 50, r_2 = 1.0, k_2 = 150, U_1 = 1, U_2 = 1, Q_0 = 150, \Omega = 2.5, \phi = 0, a = -5, b = 1, c = -3$)

are presented in Fig.3.13. In Fig.3.13(a), the trajectory of this chaotic motion with the sliding is very chaotic. The Poincare mapping section of switching points on the boundary consists of several parts plus scattering points, as shown in Fig.3.13(b). It seems that such a strange attractor of chaotic motion is fragmentized. The normal vector fields for such a chaotic motion are presented in Fig.3.13(c). Again, the black and gray points represent the switching normal vector fields of $G^{(0,1)}(\mathbf{x}_k, t_k)$ and $G^{(0,2)}(\mathbf{x}_k, t_k)$. It is observed that the grazing bifurcation occurs in Ω_2, which is called the lower grazing (L-grazing). The grazing bifurcation in Ω_1 is called the upper grazing (U-grazing). The sliding motion vanishing is represented by the acronym "SMV". In the vicinity of $x_k = 2.5$, the chaotic motion on the boundary is switchable. However, for the vicinity of $x_k = -2.5$, the sliding motion parts exists in the chaotic motion. The grazing motion occurs around $x_k = -2$. To fur-

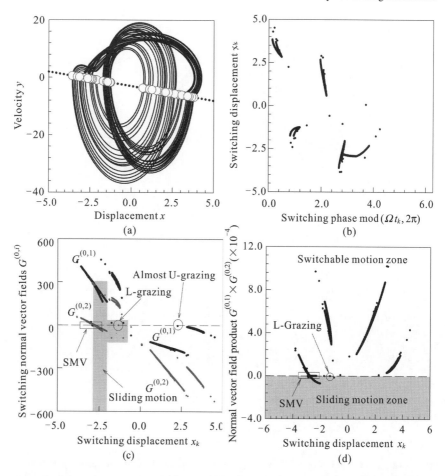

Fig. 3.13 Chaotic motion: (a) phase plane, (b) Poincare mapping section, (c) switching normal vector fields, and (d) switching normal vector field product. Initial conditions are $\Omega t_0 \approx 1.0205, x_0 \approx -1.5141$ and $y_0 \approx -1.4859$ ($r_1 = 0.5$, $k_1 = 50$, $r_2 = 1.0$, $k_2 = 150$, $U_1 = 1$, $U_2 = 1$, $Q_0 = 150$, $\Omega = 2.83$, $\phi = 0$, $a = 1$, $b = 1$, $c = -3$)

ther understand the transversality of chaotic motion, the normal vector field product is presented in Fig.3.13(d), and the switching motion zone and sliding motion are presented. If $G^{(0,1)} \times G^{(0,2)} > 0$, the switching points of the chaotic motion on the boundary are switchable. However, if $G^{(0,1)} \times G^{(0,2)} < 0$, the switching points of the chaotic motion on the boundary are non-switchable and the sliding motion will form on the boundary. If $G^{(0,1)} \times G^{(0,2)} = 0$, the sliding motion on the boundary will disappear. The grazing motion can be determined by isolated switching points of $G^{(0,1)} \times G^{(0,2)} = 0$. However, such a grazing motion to the boundary can be determined by the conditions in Eq.(3.15). Herein, such a condition will not be presented.

References

Aizerman, M. A., Pyatnitskii, E. S. (1974a), Foundation of a theory of discontinuous systems. 1, *Automatic and Remote Control*, **35**, pp. 1066-1079.

Aizerman, M. A., Pyatnitskii, E. S. (1974b), Foundation of a theory of discontinuous systems. 2, *Automatic and Remote Control*, **35**, pp. 1241-1262.

DeCarlo, R. A., Zak, S. H. and Matthews, G. P. (1988), Variable structure control of nonlinear multivariable systems: a tutorial, *Proceedings of the IEEE*, **76**, pp.212-232.

Filippov, A. F. (1964), Differential equations with discontinuous right-hand side, *American Mathematical Society Translations, Series 2*, **42**, pp. 199-231.

Filippov, A. F. (1988), *Differential Equations with Discontinuous Righthand Sides*, Dordrecht: Kluwer Academic Publishers.

Luo, A. C. J. (2005a), A theory for non-smooth dynamical systems on connectable domains, *Communication in Nonlinear Science and Numerical Simulation*, **10**, pp.1-55.

Luo, A. C. J. (2005b), Imaginary, sink and source flows in the vicinity of the separatrix of non-smooth dynamic system, *Journal of Sound and Vibration*, **285**, pp.443-456.

Luo, A. C. J. (2005c), The mapping dynamics of periodic motions for a three-piecewise linear system under a periodic excitation, *Journal of Sound and Vibration*, **283**, pp.723-748.

Luo, A. C. J. (2006), *Singularity and Dynamics on Discontinuous Vector Fields*, Amsterdam: Elsevier.

Luo, A. C. J. (2008), *Global Transversality, Resonance and Chaotic Dynamics*, Singapore, World Scientific.

Luo, A. C. J. and Chen, L. D. (2005), Periodic motion and grazing in a harmonically forced, piecewise, linear oscillator with impacts, *Chaos, Solitons and Fractals*, **24**, pp. 567-578.

Luo, A. C. J. and Gegg, B. C. (2005), On the mechanism of stick and non-stick periodic motion in a forced oscillator with dry-friction, ASME *Journal of Vibration and Acoustics*, **128**, pp.97-105.

Luo, A. C. J. and Gegg, B. C.(2006a), Stick and non-stick periodic motions in a periodically forced oscillator with dry-friction, *Journal of Sound and Vibration*, **291**, pp.132-168.

Luo, A. C. J. and Gegg, B. C. (2006b), Periodic motions in a periodically forced oscillator moving on an oscillating belt with dry friction, ASME *Journal of Computational and Nonlinear Dynamics*, **1**, pp.212-220.

Luo, A. C. J. and Gegg, B. C. (2006c), Dynamics of a periodically forced oscillator with dry friction on a sinusoidally time-varying traveling surface, *International Journal of Bifurcation and Chaos,***16**, pp.3539-3566.

Luo, A. C. J. and Rapp, B. M. (2007), Switching dynamics of a periodically forced discontinuous system with an inclined boundary, *Proceedings of IDETC'07*, 2007 ASME International Design Engineering Conferences and Exposition, September 4-7, 2007, Las Vegas, Nevada. IDETC2007-34863.

Luo, A. C. J. and Rapp, B. M. (2008), Sliding and transversal motions on an inclined boundary in a periodically forced discontinuous dynamical system, *Communications in Nonlinear Science and Numerical Simulation,* doi:10.1016/j.cnsns.2008.04.003.

Menon, S. and Luo, A. C. J. (2005), A global period-1 motion of a periodically forced, piecewise linear system, *International Journal of Bifurcation and Chaos*, **15**, pp.1945-1957.

Renzi, E. and Angelis, M. D. (2005), Optimal semi-active control and non-linear dynamics response of variable stiffness structures, *Journal of Vibration and Control*, **11**, pp.1253-1289.

Utkin, V. I. (1976), Variable structure systems with sliding modes, *IEEE Transactions on Automatic Control*, **AC-22**. pp. 212-222.

Chapter 4
A Frictional Oscillator on Time-varying Belt

In this chapter, an oscillator moving on a periodically traveling belt with dry friction is presented as a discontinuous dynamical system with a time-varying boundary. Such a time-varying boundary will cause the domain to change with time, on which the vector fields can be defined. From the theories in Chapter 2, analytical conditions for stick and non-stick motions on a periodically traveling belt in such an oscillator with dry friction will be developed, and the relative force product criteria will be obtained to predict such stick and non-stick motions. The grazing and stick (or sliding) bifurcations will be discussed for appearance and vanishing of regular motions in such an oscillator. Parameter maps for specific periodic motions will be obtained through mapping structures. The displacement, velocity and force responses of periodic motions will be presented to illustrate the analytical conditions of stick and non-stick motions.

4.1 Mechanical model

As in Luo and Gegg(2006a,b), consider a periodically forced oscillator attached to a fixed wall, which consists of a mass m, a spring of stiffness k and a damper of viscous damping coefficient r, as shown in Fig.4.1. The coordinate system $(\bar{x}, \bar{t})$ is absolute with a displacement $\bar{x}$ and time $\bar{t}$. The external force is $\bar{Q}_0 \cos \bar{\Omega}\bar{t}$ where $\bar{Q}_0$ and $\bar{\Omega}$ are the excitation amplitude and frequency, respectively. Since the mass

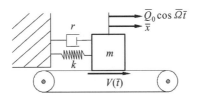

Fig. 4.1 Mechanical model for a friction oscillator on the time-varying belt

contacts the moving belt with friction, the mass can move along or rest on the belt. The belt travels with a time varying speed $\bar{V}(\bar{t})$, i.e.,

$$\bar{V}(\bar{t}) = \bar{V}_0 \cos(\omega\bar{t} + \beta) + \bar{V}_1, \tag{4.1}$$

where ω is the oscillation frequency of the traveling belt, and $\bar{V}_0$ is the oscillation amplitude of the traveling belt, and $\bar{V}_1$ is constant. Further, the kinetic friction force shown in Fig.4.2 is described as

$$\bar{F}_f(\dot{\bar{x}}) \begin{cases} = \mu_k F_N, & \dot{\bar{x}} \in (\bar{V}(t), \infty) \\ \in [-\mu_k F_N, \mu_k F_N], & \dot{\bar{x}} = \bar{V}(t) \\ = -\mu_k F_N, & \dot{\bar{x}} \in (-\infty, \bar{V}(t)) \end{cases} \tag{4.2}$$

where $\dot{\bar{x}} \triangleq d\bar{x}/dt$, μ_k and F_N are a dynamical friction coefficient and a normal force to the contact surface, respectively.

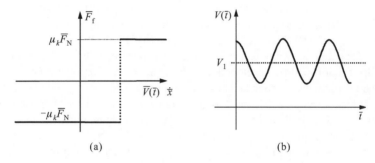

Fig. 4.2 (a) Friction forces and (b) the oscillation transport speed of the belt

For the mass moving with the same speed of the belt surface, the non-friction force acting on the mass in the x-direction is defined as

$$F_s = \bar{Q}_0 \cos \bar{\Omega}\bar{t} - r\bar{V}(\bar{t}) - k\bar{x}, \text{ for } \dot{\bar{x}} = \bar{V}(\bar{t}). \tag{4.3}$$

For the stick motion, this force cannot overcome the static friction force (i.e.,$|F_s| \leqslant \mu_k F_N$). Thus, the mass does not have any relative motion to the belt. In other words, the relative acceleration should be zero, i.e.,

$$\ddot{\bar{x}} = \dot{\bar{V}}(t) = -\bar{V}_0\omega\sin(\omega\bar{t} + \beta), \text{ for } \dot{\bar{x}} = \bar{V}(\bar{t}). \tag{4.4}$$

If $|F_s| > \mu_k F_N$, the non-friction force will overcome the static friction force on the mass and the non-stick motion will appear. For the non-stick motion, the total force acting on the mass is

$$F = \bar{Q}_0 \cos \bar{\Omega}\bar{t} - \mu_k F_N \text{sgn}(\dot{\bar{x}} - \bar{V}) - r\dot{\bar{x}} - k\bar{x}, \text{ for } \dot{\bar{x}} \neq \bar{V}, \tag{4.5}$$

4.1 Mechanical model

where sgn(·) is the sign function. Therefore, the equation of the non-stick motion for this oscillator with the dry-friction is

$$\ddot{\bar{x}} + 2\bar{d}\dot{\bar{x}} + \bar{c}\bar{x} = \bar{A}_0 \cos \bar{\Omega}\bar{t} - \tilde{F}_{\mathrm{f}} \operatorname{sgn}(\dot{\bar{x}} - \bar{V}(\bar{t})), \text{ for } \dot{\bar{x}} \neq \bar{V}(\bar{t}), \quad (4.6)$$

where $\bar{A}_0 = \bar{Q}_0/m, \bar{d} = r/2m, \bar{c} = k/m$ and $\tilde{F}_{\mathrm{f}} = \mu_k F_{\mathrm{N}}/m$. Introduce non-dimensional frequency and time

$$\left. \begin{array}{l} \Omega = \bar{\Omega}/\omega,\ t = \omega\bar{t},\ d = \bar{d}/\omega,\ c = \bar{c}/\omega^2,\ A_0 = \bar{A}_0/\omega^2, \\ F_{\mathrm{f}} = \tilde{F}_{\mathrm{f}}/\omega^2,\ V_0 = \bar{V}_0/\omega,\ V_1 = \bar{V}_1/\omega, V = \bar{V}/\omega, x = \bar{x}, \end{array} \right\} \quad (4.7)$$

$$V(t) = V_0 \cos(t + \beta) + V_1. \quad (4.8)$$

The phase constant β is used to synchronize the periodic force input and the velocity discontinuity after the modulus of time has been computed. The integration of Eq.(4.8) gives

$$X(t) = V_0 [\sin(t+\beta) - \sin(t_i + \beta)] + V_1 \cdot (t - t_i) + x_i, \text{ for } t > t_i, \quad (4.9)$$

which is the displacement response of the periodically time-varying traveling belt for $t = t_i, X(t_i) = x(t_i) \equiv x_i$.

Introduce the relative displacement, velocity, and acceleration as

$$\left. \begin{array}{l} z(t) = x(t) - X(t), \\ \dot{z}(t) = \dot{x}(t) - V(t), \\ \ddot{z}(t) = \ddot{x}(t) - \dot{V}(t). \end{array} \right\} \quad (4.10)$$

The non-friction force in Eq.(4.3) for $\dot{z} = 0$ becomes

$$F_s = A_0 \cos \Omega t - 2d[V(t) + \dot{z}(t)] - c[X(t) + z(t)] - \dot{V}(t). \quad (4.11)$$

Since the mass does not have any relative motion to the vibrating belt, the relative acceleration is zero, i.e.,

$$\ddot{x} = \dot{V}(t) = -V_0 \sin(t + \beta), \text{ for } \dot{x} = V(t), \quad (4.12)$$

or

$$\ddot{z} = 0, \text{ for } \dot{z} = 0. \quad (4.13)$$

For non-stick motion, the equation of this oscillator with friction becomes

$$\ddot{x} + 2d\dot{x} + cx = A_0 \cos \Omega t - F_{\mathrm{f}} \operatorname{sgn}(\dot{x} - V(t)), \text{ for } \dot{x} \neq V(t), \quad (4.14)$$

in the absolute frame, or

$$\ddot{z}(t) + 2d\dot{z}(t) + cz(t) = A_0 \cos \Omega t - F_{\mathrm{f}} \operatorname{sgn}(\dot{z}(t)) \\ - 2dV(t) - cX(t) - \dot{V}(t), \text{ for } \dot{z} \neq 0, \quad (4.15)$$

in the relative frame.

4.2 Analytical conditions

In this section, the sufficient and necessary conditions for passable, sliding and grazing motions to the time-varying boundary will be discussed. The corresponding physical interpretation will be given for a better understanding of mechanism for such motions.

4.2.1 Equations of motion

Because of the discontinuity, the phase plane partition for this friction oscillator is sketched in the absolute and relative frames in Fig.4.3. In the absolute frame, the separation boundary in phase plane is a curve determined by eliminating the parameter t in two parameter equations $X(t)$ and $V(t)$ in Eqs.(4.8) and (4.9). For $0 < V_0 < V_1$, the separation boundary is the prolate trochoid, as shown in Fig.4.3(a). Similarly, the separation boundary for $V_0 > V_1 > 0$ is the curtate trochoid. However, the boundary velocity in the relative frame is zero in Fig.4.3(b). In the relative frame, the criteria for motion to pass through the discontinuous boundary or not can be easily obtained from the theory of discontinuous dynamical systems as in Chapter 2. Therefore, the criteria for such a system will be developed based on the relative frame.

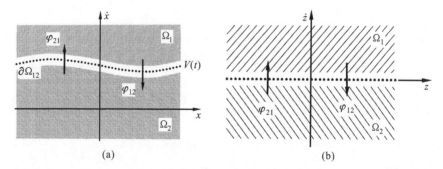

Fig. 4.3 Phase plane partition in (a) absolute and (b) relative phase planes for time varying boundary

In phase plane, one introduces two vectors

$$\mathbf{z} \triangleq (z,\dot{z})^{\mathrm{T}} \equiv (z,v)^{\mathrm{T}} \text{ and } \mathbf{F} \triangleq (v,F)^{\mathrm{T}}. \quad (4.16)$$

The corresponding regions and boundaries are expressed by

4.2 Analytical conditions

$$\left.\begin{array}{l}\Omega_1 = \{(z,v)| v \in (0,\infty)\}, \\ \Omega_2 = \{(z,v)| v \in (-\infty,0)\}, \\ \partial\Omega_{\alpha\beta} = \{(z,v)| \varphi_{\alpha\beta}(z,v) = v = 0\}. \end{array}\right\} \quad (4.17)$$

or

$$\left.\begin{array}{l}\Omega_1 = \{(x,\dot{x})| \dot{x} - V > 0\}, \\ \Omega_2 = \{(x,\dot{x})| \dot{x} - V < 0\}, \\ \partial\Omega_{\alpha\beta} = \{(x,\dot{x})| \varphi_{\alpha\beta}(x,\dot{x}) = \dot{x} - V(t) = 0\}. \end{array}\right\} \quad (4.18)$$

The subscripts $(\cdot)_{\alpha\beta}$ defines the boundary from Ω_α to Ω_β. The equations of motion in Eqs.(4.13) and (4.15) can be described as

$$\dot{\mathbf{z}} = \mathbf{F}_\lambda^{(\kappa)}(\mathbf{z},t), (\kappa, \lambda \in \{0,1,2\}), \quad (4.19)$$

where

$$\left.\begin{array}{l}\mathbf{F}_\alpha^{(\alpha)}(\mathbf{z},t) = (v, F_\alpha(\mathbf{z},t))^T \text{ in } \Omega_\alpha (\alpha \in \{1,2\}), \\ \mathbf{F}_\alpha^{(\beta)}(\mathbf{z},t) = (v, F_\beta(\mathbf{z},t))^T \text{ in } \Omega_\alpha (\alpha \neq \beta \in \{1,2\}), \\ \mathbf{F}_0^{(0)}(\mathbf{z},t) = (0,0)^T \text{ on } \partial\Omega_{\alpha\beta} \text{ for stick}, \\ \mathbf{F}_0^{(0)}(\mathbf{z},t) \in \left[\mathbf{F}_\alpha^{(\alpha)}(\mathbf{z},t), \mathbf{F}_\beta^{(\beta)}(\mathbf{z},t)\right] \text{ on } \partial\Omega_{\alpha\beta} \text{ for non-stick}. \end{array}\right\} \quad (4.20)$$

For the subscript and superscript (κ and λ) with non-zero values, they represent the two adjacent domains for $\alpha, \beta \in \{1,2\}$. $\mathbf{F}_\alpha^{(\alpha)}(\mathbf{z},t)$ is the true (or real) vector field in the α-domain. $\mathbf{F}_\alpha^{(\beta)}(\mathbf{z},t)$ is the fictitious (or imaginary) vector field in the α-domain, which is determined by the vector field in the β-domain. The detailed discussion can be referred to Luo (2005a, 2006). $\mathbf{F}_0^{(0)}(\mathbf{z},t)$ is the vector field on the separation boundary, and the discontinuity of the vector field for the entire system is presented through such an expression. $F_\alpha(\mathbf{z},t)$ is the scalar force in the α-domain. For the system in Eq.(4.15), we have the forces in the two domains as

$$F_\alpha(\mathbf{z},t) = A_0 \cos \Omega t - b_\alpha - 2d_\alpha [V(t) + \dot{z}(t)] \\ - c_\alpha [X(t) + z(t)] - \dot{V}(t). \quad (4.21)$$

Note that $b_1 = -b_2 = \mu g/\omega^2, d_\alpha = d$ and $c_\alpha = c$ for the model in Fig.4.1.

From Eq.(4.9), with increasing time t, the belt displacement will increase. However, the oscillator vibrates in the vicinity of equilibrium. The friction is dependent on the relative velocity between the oscillator and the belt. When the non-stick motion of the oscillator arrives to the discontinuous boundary, the particle of the belt that contacts the mass with the same velocity and displacement is different from the initial particle. To understand the stick and non-stick motions, the particle switching on the surface of the oscillating belt in the absolute phase plane is sketched in Fig.4.4. The particles (p_1, p_2, p_3) on the belts are represented by white, light-gray and gray circular symbols, respectively. The dark-gray circular symbol is the oscillator location. Because the oscillating velocity boundary is trochoid, when time $t \to \infty$, the selected belt particle will move to the infinity. However, the oscillator

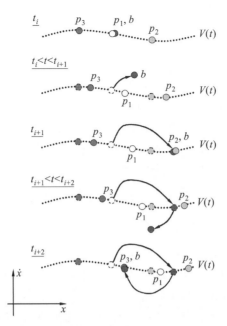

Fig. 4.4 Particles switching on the oscillating belt surface in absolute phase plane once the oscillator has the same speed as the oscillating belt at a moment t_i. The particles (p_1, p_2, p_3) on the belts are represented by white, light-gray and gray circular symbols, respectively. The dark-gray circular symbol is the oscillator location

will move around a certain location. To help one understand the particle switching on the translating belt, consider the oscillator with a belt particle p_1 at the same location at time t_i. Once the oscillator velocity is greater than the belt, the oscillator will move faster than the belt particle p_1 for $t \in (t_i, t_{i+1})$. When the oscillator has the same speed as the belt, the oscillator with the belt particle p_2 has the same location at t_{i+1}. Similarly, once the oscillator velocity is lower than the belt speed, the oscillator will move more slowly than the belt particle p_2 or move backwards with negative velocity for $t \in (t_{i+1}, t_{i+2})$. When the oscillator velocity is the same as the belt speed at time t_{i+2}, the oscillator with the belt particle p_3 has the same location. Such a belt particle switching can help understand the motion mechanisms in such an oscillator. This motion switching phenomena can exist extensively in mechanical systems. If the oscillator has the same speed with the translating belt in a time interval instead of a time point, then both the oscillator and the belt will move together during this time interval. In mathematics, such a motion is called *sliding flow*, but from a physics point of view, this motion is called *stick motion*. In Luo (2005a, 2006), the motion approaching the velocity boundary is called *passable motion* only if the motion can pass through the boundary and into another adjacent domain in phase space. Otherwise, such a motion is termed *non-passable motion*. For the non-passable motion, the sink and source motions to the boundary can be classified. The stick motion is a kind of the sink motion, which belongs to one of

4.2 Analytical conditions

the non-passable motions. If one is interested in this topic except for Chapter 2, the mathematical definitions for such motions to the discontinuous boundary can be referred to Luo (2005a, 2006).

4.2.2 Passable flows to boundary

Before presenting analytical conditions, the following functions are introduced in Luo (2007, 2008) as in Chapter 2

$$\begin{aligned}
G^{(0,\alpha)}(\mathbf{z},t_{m\pm}) &= \mathbf{n}_{\partial\Omega_{\alpha\beta}}^{\mathrm{T}} \cdot \left[\mathbf{F}_{\alpha}^{(\alpha)}(\mathbf{z},t_{m\pm}) - \mathbf{F}_{\alpha\beta}^{(0)}(\mathbf{z},t_{m\pm}) \right] \\
&= \mathbf{n}_{\partial\Omega_{\alpha\beta}}^{\mathrm{T}} \cdot \mathbf{F}_{\alpha}^{(\alpha)}(\mathbf{z},t_{m\pm}), \\
G^{(1,\alpha)}(\mathbf{z},t_{m\pm}) &= 2D\mathbf{n}_{\partial\Omega_{\alpha\beta}}^{\mathrm{T}} \cdot \left[\mathbf{F}_{\alpha}^{(\alpha)}(\mathbf{z},t_{m\pm}) - \mathbf{F}_{\alpha\beta}^{(0)}(\mathbf{z},t_{m\pm}) \right] \\
&\quad + \mathbf{n}_{\partial\Omega_{\alpha\beta}}^{\mathrm{T}} \cdot \left[D\mathbf{F}_{\alpha}^{(\alpha)}(\mathbf{z},t_{m\pm}) - D\mathbf{F}_{\alpha\beta}^{(0)}(\mathbf{z},t_{m\pm}) \right],
\end{aligned} \quad (4.22)$$

where $D = \frac{\partial}{\partial z}\dot{z} + \frac{\partial}{\partial w}\dot{w} + \frac{\partial}{\partial t}$. Notice that t_m represents the time for the motion on the velocity boundary and $t_{m\pm} = t_m \pm 0$ reflects the responses on the regions rather than the boundary. If the boundary $\partial\Omega_{\alpha\beta}$ in the relative frame is linear independent of time t (i.e., $D\mathbf{n}_{\partial\Omega_{\alpha\beta}}^{\mathrm{T}} = 0$), one obtains $D\mathbf{n}_{\partial\Omega_{\alpha\beta}}^{\mathrm{T}} \cdot \mathbf{F}_{\alpha\beta}^{(0)}(\mathbf{z},t_{m\pm}) + \mathbf{n}_{\partial\Omega_{\alpha\beta}}^{\mathrm{T}} \cdot D\mathbf{F}_{\alpha\beta}^{(0)}(\mathbf{z},t_{m\pm}) = 0$ due to $\mathbf{n}_{\partial\Omega_{\alpha\beta}}^{\mathrm{T}} \cdot \mathbf{F}_{\alpha\beta}^{(0)}(\mathbf{z},t_{m\pm}) = 0$. Thus, $\mathbf{n}_{\partial\Omega_{\alpha\beta}}^{\mathrm{T}} \cdot D\mathbf{F}_{\alpha\beta}^{(0)}(\mathbf{z},t_{m\pm}) = 0$. Equation (4.22) reduces to

$$G^{(1,\alpha)}(\mathbf{z}_m,t_{m\pm}) = \mathbf{n}_{\partial\Omega_{\alpha\beta}}^{\mathrm{T}} \cdot D\mathbf{F}_{\alpha}^{(\alpha)}(\mathbf{z}_m,t_{m\pm}). \quad (4.23)$$

For a general case, equation (4.22) instead of Eq.(4.23) will be used. Notice that $\mathbf{F}_{\alpha\beta}^{(0)}(\mathbf{z},t) = (0,0)^{\mathrm{T}}$.

For the boundary $\overrightarrow{\partial\Omega_{\alpha\beta}}$ with $\mathbf{n}_{\partial\Omega_{\alpha\beta}}^{\mathrm{T}} \to \Omega_\alpha$, the necessary and sufficient conditions for the non-stick motion (or called passable motion to boundary) are given by Eq.(2.20) (also see, Luo, 2005a)

$$\left.\begin{aligned}
G^{(0,\alpha)}(\mathbf{z}_m,t_{m-}) &= \mathbf{n}_{\partial\Omega_{\alpha\beta}}^{\mathrm{T}} \cdot \mathbf{F}_{\alpha}^{(\alpha)}(\mathbf{z}_m,t_{m-}) < 0 \\
G^{(0,\beta)}(\mathbf{z}_m,t_{m+}) &= \mathbf{n}_{\partial\Omega_{\alpha\beta}}^{\mathrm{T}} \cdot \mathbf{F}_{\beta}^{(\beta)}(\mathbf{z}_m,t_{m+}) < 0
\end{aligned}\right\} \text{from } \Omega_\alpha \to \Omega_\beta, \\
\left.\begin{aligned}
G^{(0,\beta)}(\mathbf{z}_m,t_{m-}) &= \mathbf{n}_{\partial\Omega_{\beta\alpha}}^{\mathrm{T}} \cdot \mathbf{F}_{\beta}^{(\beta)}(\mathbf{z}_m,t_{m-}) > 0 \\
G^{(0,\alpha)}(\mathbf{z}_m,t_{m+}) &= \mathbf{n}_{\partial\Omega_{\beta\alpha}}^{\mathrm{T}} \cdot \mathbf{F}_{\alpha}^{(\alpha)}(\mathbf{z}_m,t_{m+}) > 0
\end{aligned}\right\} \text{from } \Omega_\beta \to \Omega_\alpha, \quad (4.24)$$

or

$$L_{\alpha\beta}(t_{m-}) = G^{(0,\alpha)}(\mathbf{z}_m, t_{m-}) \times G^{(0,\beta)}(\mathbf{z}_m, t_{m+})$$
$$= \left[\mathbf{n}^T_{\partial\Omega_{\alpha\beta}} \cdot \mathbf{F}^{(\alpha)}_\alpha(\mathbf{z}_m, t_{m-})\right] \times \left[\mathbf{n}^T_{\partial\Omega_{\alpha\beta}} \cdot \mathbf{F}^{(\beta)}_\beta(\mathbf{z}_m, t_{m+})\right] > 0. \quad (4.25)$$

Note that $\mathbf{n}_{\partial\Omega_{\alpha\beta}} \to \Omega_\alpha$ means that the normal direction of $\partial\Omega_{\alpha\beta}$ points to domain Ω_α. Using the third equation of Eq.(4.17), the normal vector of the boundary $\partial\Omega_{12}$ or $\partial\Omega_{21}$ is

$$\mathbf{n}_{\partial\Omega_{12}} = \mathbf{n}_{\partial\Omega_{21}} = (0,1)^T. \quad (4.26)$$

Therefore, we have

$$\left.\begin{array}{l} G^{(0,\alpha)}(\mathbf{z},t) = \mathbf{n}^T_{\partial\Omega_{\alpha\beta}} \cdot \mathbf{F}^{(\alpha)}_\alpha(\mathbf{z},t) = F_\alpha(\mathbf{z},t), \alpha \in \{1,2\}, \\ G^{(1,\alpha)}(\mathbf{z},t) = \mathbf{n}^T_{\partial\Omega_{\alpha\beta}} \cdot D\mathbf{F}^{(\alpha)}_\alpha(\mathbf{z},t) = \nabla F_\alpha(\mathbf{z},t) \cdot \mathbf{F}^{(\alpha)}_\alpha(\mathbf{z},t) + \dfrac{\partial F_\alpha(\mathbf{z},t)}{\partial t}. \end{array}\right\} \quad (4.27)$$

The normal projection of the vector field of the oscillator to the discontinuous boundary is the total force per unit mass. The conditions in Eq.(4.24) or (4.25) become

$$\left.\begin{array}{l} \left.\begin{array}{l} G^{(0,\alpha)}(\mathbf{z}_m, t_{m-}) = F_\alpha(\mathbf{z}_m, t_{m-}) < 0 \\ G^{(0,\beta)}(\mathbf{z}_m, t_{m+}) = F_\beta(\mathbf{z}_m, t_{m+}) < 0 \end{array}\right\} \text{from } \Omega_\alpha \to \Omega_\beta, \\ \left.\begin{array}{l} G^{(0,\beta)}(\mathbf{z}_m, t_{m-}) = F_\beta(\mathbf{z}_m, t_{m-}) > 0 \\ G^{(0,\alpha)}(\mathbf{z}_m, t_{m+}) = F_\alpha(\mathbf{z}_m, t_{m+}) > 0 \end{array}\right\} \text{from } \Omega_\beta \to \Omega_\alpha. \end{array}\right\} \quad (4.28)$$

$$L_{\alpha\beta}(\mathbf{z}_m, t_{m\pm}) = G^{(0,\alpha)}(\mathbf{z}_m, t_{m-}) \times G^{(0,\beta)}(\mathbf{z}_m, t_{m+})$$
$$= F_\alpha(\mathbf{z}_m, t_{m-}) \times F_\beta(\mathbf{z}_m, t_{m+}) > 0. \quad (4.29)$$

The aforementioned conditions can be illustrated through the vector fields in the two domains, as shown in Fig.4.5. In Fig.4.5(a), the vector fields in the absolute frame are presented. The normal vector $\mathbf{n}_{\partial\Omega_{12}}$ is changed with time, and both the corresponding direction and magnitude vary with time and location. Thus, the corresponding normal component of vector fields will change with normal vector. However, in the relative frame, the normal vector of the boundary is constant. With time varying, the normal vector is invariant, and the vector fields and normal components of the two vector fields are sketched in Fig.4.5(b). The passable flow on the boundary requires that two normal components of two vector fields are of the same direction.

4.2 Analytical conditions

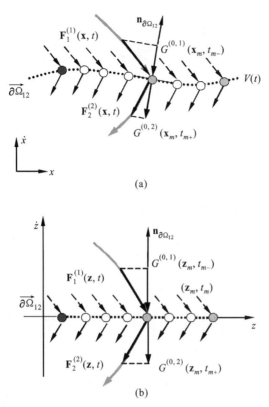

Fig. 4.5 Vector fields in the vicinity of the time-varying boundary for passable flows (a) in the absolute frame and (b) in the relative frame

4.2.3 Sliding flows on boundary

Consider the non-passable boundary $\widetilde{\partial\Omega}_{\alpha\beta}$ and passable boundary $\overrightarrow{\partial\Omega}_{\alpha\beta}$ on the velocity boundary in Fig.4.6. For a non-passable boundary $\widetilde{\partial\Omega}_{\alpha\beta}$ possessing non-zero measure in phase space, the critical initial and final states of the sliding flow are $(\Omega t_c, x_c, V_c)$ and $(\Omega t_f, x_f, V_f)$. Consider a sliding flow starting at $(\Omega t_i, x_i, V_i)$ and ending at $(\Omega t_{i+1}, x_{i+1}, V_{i+1})$, where $(\Omega t_{i+1}, x_{i+1}, V_{i+1}) \overset{\Delta}{=} (\Omega t_f, x_f, V_f)$. From Luo (2005a, 2006), the sliding flow (or called the stick motion in physics) through the real flow is guaranteed by

$$\begin{aligned}
L_{\alpha\beta}(t_{m-}) &= G^{(0,\alpha)}(\mathbf{z}_m, t_{m-}) \times G^{(0,\beta)}(\mathbf{z}_m, t_{m-}) \\
&= \left[\mathbf{n}_{\partial\Omega_{\alpha\beta}}^{\mathrm{T}} \cdot \mathbf{F}_\alpha^{(\alpha)}(\mathbf{z}_m, t_{m-})\right] \times \left[\mathbf{n}_{\partial\Omega_{\alpha\beta}}^{\mathrm{T}} \cdot \mathbf{F}_\beta^{(\beta)}(\mathbf{z}_m, t_{m+})\right] < 0, \\
&\text{for } t_m \in [t_i, t_{i+1}] \subseteq [t_c, t_f].
\end{aligned} \qquad (4.30)$$

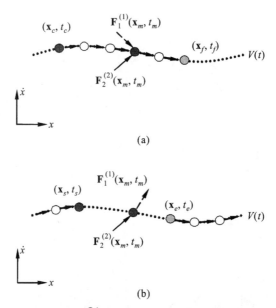

Fig. 4.6 (a) Non-passable boundary $\widetilde{\partial\Omega}_{12}$ with sliding flows and (b) passable boundary $\overrightarrow{\partial\Omega}_{12}$ with passable flows in the absolute frame

For the boundary $\partial\Omega_{\alpha\beta}$ with $\mathbf{n}_{\partial\Omega_{\alpha\beta}}^{\mathrm{T}} \to \Omega_\alpha$, the necessary and sufficient conditions for the stick motion (or called non-passable motion to the boundary) are also given in Luo (2005a, 2006), i.e.,

$$\left.\begin{aligned} G^{(0,\alpha)}(\mathbf{z}_m, t_{m-}) &= \mathbf{n}_{\partial\Omega_{\alpha\beta}}^{\mathrm{T}} \cdot \mathbf{F}_\alpha^{(\alpha)}(\mathbf{z}_m, t_{m-}) < 0, \\ G^{(0,\beta)}(\mathbf{z}_m, t_{m-}) &= \mathbf{n}_{\partial\Omega_{\alpha\beta}}^{\mathrm{T}} \cdot \mathbf{F}_\beta^{(\beta)}(\mathbf{z}_m, t_{m-}) > 0. \end{aligned}\right\} \qquad (4.31)$$

For a boundary $\overrightarrow{\partial\Omega}_{\alpha\beta}$ with the non-zero measure, the starting and ending states of the passable motion on the boundary are accordingly $(\Omega t_s, \mathbf{z}_s, V_s)$ and $(\Omega t_e, \mathbf{z}_e, V_e)$, as shown in Fig.4.5(b). For passable motion switching to non-stick motion on the boundary $\partial\Omega_{\alpha\beta}$ (i.e., from $\overrightarrow{\partial\Omega}_{\alpha\beta}$ to $\widetilde{\partial\Omega}_{\alpha\beta}$), we have $t_e = t_c$. Otherwise, $t_f = t_s$. Suppose the starting state is a switching state of the sliding flow, the switching condition of the sliding flows from $\overrightarrow{\partial\Omega}_{\alpha\beta}$ to $\widetilde{\partial\Omega}_{\alpha\beta}$ at $t_m = t_c$ is

$$\left.\begin{aligned} G^{(0,\alpha)}(\mathbf{z}_m, t_{m-}) &= \mathbf{n}_{\partial\Omega_{\alpha\beta}}^{\mathrm{T}} \cdot \mathbf{F}_\alpha^{(\alpha)}(\mathbf{z}_m, t_{m-}) < 0, \\ G^{(0,\beta)}(\mathbf{z}_m, t_{m\pm}) &= \mathbf{n}_{\partial\Omega_{\alpha\beta}}^{\mathrm{T}} \cdot \mathbf{F}_\beta^{(\beta)}(\mathbf{z}_m, t_{m\pm}) = 0, \\ G^{(1,\beta)}(\mathbf{z}_m, t_{m\pm}) &= \mathbf{n}_{\partial\Omega_{\alpha\beta}}^{\mathrm{T}} \cdot D\mathbf{F}_\beta^{(\beta)}(\mathbf{z}_m, t_{m\pm}) > 0, \end{aligned}\right\} \text{ for } \mathbf{n}_{\partial\Omega_{\alpha\beta}} \to \Omega_\alpha \qquad (4.32)$$

4.2 Analytical conditions

$$\left.\begin{aligned}G^{(0,\alpha)}(\mathbf{z}_m,t_{m\pm}) &= \mathbf{n}^T_{\partial\Omega_{\alpha\beta}} \cdot \mathbf{F}^{(\alpha)}_\alpha(\mathbf{z}_m,t_{m\pm}) = 0, \\ G^{(0,\beta)}(\mathbf{z}_m,t_{m-}) &= \mathbf{n}^T_{\partial\Omega_{\alpha\beta}} \cdot \mathbf{F}^{(\beta)}_\beta(\mathbf{z}_m,t_{m-}) > 0, \\ G^{(1,\alpha)}(\mathbf{z}_m,t_{m\pm}) &= \mathbf{n}^T_{\partial\Omega_{\alpha\beta}} \cdot D\mathbf{F}^{(\alpha)}_\alpha(\mathbf{z}_m,t_{m\pm}) < 0.\end{aligned}\right\} \text{ for } \mathbf{n}_{\partial\Omega_{\alpha\beta}} \to \Omega_\alpha \quad (4.33)$$

On $\widetilde{\partial\Omega}_{\alpha\beta}$ with $\mathbf{n}_{\partial\Omega_{\alpha\beta}} \to \Omega_\alpha$, a sliding flow vanishing from $\widetilde{\partial\Omega}_{\alpha\beta}$ and going into the domain Ω_γ ($\gamma \in \{\alpha,\beta\}$) at $t_m = t_f$ requires

$$\left.\begin{aligned}G^{(0,\alpha)}(\mathbf{z}_m,t_{m-}) &= \mathbf{n}^T_{\partial\Omega_{\alpha\beta}} \cdot \mathbf{F}^{(\alpha)}_\alpha(\mathbf{z}_m,t_{m-}) < 0, \\ G^{(0,\beta)}(\mathbf{z}_m,t_{m\pm}) &= \mathbf{n}^T_{\partial\Omega_{\alpha\beta}} \cdot \mathbf{F}^{(\beta)}_\beta(\mathbf{z}_m,t_{m\pm}) = 0, \\ G^{(1,\beta)}(\mathbf{z}_m,t_{m\pm}) &= \mathbf{n}^T_{\partial\Omega_{\alpha\beta}} \cdot D\mathbf{F}^{(\beta)}_\beta(\mathbf{z}_m,t_{m\pm}) < 0,\end{aligned}\right\} \quad (4.34)$$

$$\left.\begin{aligned}G^{(0,\beta)}(\mathbf{z}_m,t_{m-}) &= \mathbf{n}^T_{\partial\Omega_{\alpha\beta}} \cdot \mathbf{F}^{(\beta)}_\beta(\mathbf{z}_m,t_{m-}) > 0, \\ G^{(0,\alpha)}(\mathbf{z}_m,t_{m\mp}) &= \mathbf{n}^T_{\partial\Omega_{\alpha\beta}} \cdot \mathbf{F}^{(\alpha)}_\alpha(\mathbf{z}_m,t_{m\mp}) = 0, \\ G^{(1,\alpha)}(\mathbf{z}_m,t_{m\mp}) &= \mathbf{n}^T_{\partial\Omega_{\alpha\beta}} \cdot D\mathbf{F}^{(\alpha)}_\alpha(\mathbf{z}_m,t_{m\mp}) > 0.\end{aligned}\right\} \quad (4.35)$$

From Eqs.(4.32)~(4.35), the switching conditions for the sliding flows are summarized as

$$\left.\begin{aligned}&F_1(\mathbf{z}_m,\Omega t_{m-}) < 0 \text{ and } F_2(\mathbf{z}_m,\Omega t_{m\pm}) = 0, \\ &G^{(1,2)}(\mathbf{z}_m,t_{m\pm}) > 0,\end{aligned}\right\} \text{ for } \Omega_1 \to \widetilde{\partial\Omega}_{12}, \\ \left.\begin{aligned}&F_2(\mathbf{z}_m,\Omega t_{m-}) > 0 \text{ and } F_1(\mathbf{z}_m,\Omega t_{m\pm}) = 0, \\ &G^{(1,1)}(\mathbf{z}_m,t_{m\pm}) < 0,\end{aligned}\right\} \text{ for } \Omega_2 \to \widetilde{\partial\Omega}_{21}. \quad (4.36)$$

For a better understanding of the force characteristic of the stick motion (or sliding flow), the condition for the stick motion in Eq.(4.31) can be re-written by

$$F_1(\mathbf{z}_m,\Omega t_{m-}) < 0 \text{ and } F_2(\mathbf{z}_m,\Omega t_{m-}) > 0 \quad \text{on } \widetilde{\partial\Omega}_{12}. \quad (4.37)$$

With Eqs.(4.32) and (4.33), the onset (switching) conditions for the stick motion (or sliding flow) are

$$\left.\begin{aligned}&F_1(\mathbf{z}_m,\Omega t_{m-}) < 0 \text{ and } F_2(\mathbf{z}_m,\Omega t_{m\pm}) = 0, \\ &G^{(1,2)}(\mathbf{z}_m,t_{m\pm}) > 0,\end{aligned}\right\} \text{ for } \Omega_1 \to \widetilde{\partial\Omega}_{12}, \\ \left.\begin{aligned}&F_2(\mathbf{z}_m,\Omega t_{m-}) > 0 \text{ and } F_1(\mathbf{z}_m,\Omega t_{m\pm}) = 0, \\ &G^{(1,1)}(\mathbf{z}_m,t_{m\pm}) < 0,\end{aligned}\right\} \text{ for } \Omega_2 \to \widetilde{\partial\Omega}_{21}. \quad (4.38)$$

at $t_m = t_c$, and with Eqs.(4.34) and (4.35), the vanishing condition for the stick motion (or sliding flow) are re-written as at $t_m = t_f$

$$\left.\begin{array}{l} F_1(\mathbf{z}_m, \Omega t_{m-}) < 0 \text{ and } F_2(\mathbf{z}_m, \Omega t_{m\mp}) = 0, \\ G^{(1,2)}(\mathbf{z}_m, t_{m\mp}) < 0, \\ F_2(\mathbf{z}_m, \Omega t_{m-}) > 0 \text{ and } F_1(\mathbf{z}_m, \Omega t_{m\mp}) = 0, \\ G^{(1,1)}(\mathbf{z}_m, t_{m\mp}) > 0, \end{array}\right\} \begin{array}{l} \text{from } \widetilde{\partial \Omega}_{12} \to \Omega_2, \\ \\ \text{from } \widetilde{\partial \Omega}_{12} \to \Omega_1. \end{array} \right\} \quad (4.39)$$

A sketch of the stick motion (or sliding flow) on the discontinuous boundary is presented in Figs.4.7 and 4.8. In Fig. 4.7, the force conditions for the onset and vanishing of stick motion are sketched in the absolute frame through the vector fields of $\mathbf{F}_1^{(1)}(\mathbf{x},t)$ and $\mathbf{F}_2^{(2)}(\mathbf{x},t)$. The corresponding force conditions in the relative frame are presented in Fig.4.8 via the vector fields of $\mathbf{F}_1^{(1)}(\mathbf{z},t)$ and $\mathbf{F}_2^{(2)}(\mathbf{z},t)$. The vanishing condition for sliding flow along the velocity boundary is illustrated by $G^{(0,1)}(\mathbf{z}_m, t_{m-}) < 0$ (or $F_1(\mathbf{z}_m, t_{m-}) < 0$) and $G^{(0,2)}(\mathbf{z}_m, t_{m\pm}) = 0$ (or $F_2(\mathbf{z}_m, t_{m\pm}) = 0$) with $G^{(1,2)}(\mathbf{z}_m, t_{m\pm}) < 0$. The vanishing point is labeled by the grey circular symbol. However, the starting points of the sliding flow may not be the switching points from the passable flow to the sliding flow. The switching point satisfies the onset condition of the sliding flow in Eq.(4.36). When the flow arrives to the boundary, once Eq.(4.38) holds, the sliding flow along the discontinuous boundary will be formed. The onset of the sliding flow from the domain Ω_2 onto the sliding boundary $\widetilde{\partial \Omega}_{12}$ is

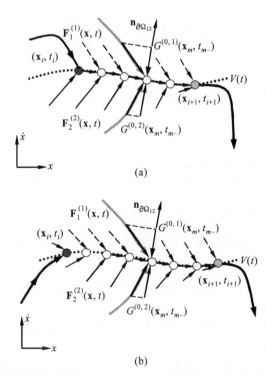

Fig. 4.7 Vector fields in the absolute frame: (a) vanishing and (b) appearance and vanishing of sliding flow on the time-varying boundary

4.2 Analytical conditions

labeled by the dark circular symbol in Fig.4.8(b). From Eq.(4.38), $F_1(\mathbf{z}_m, t_{m\pm}) = 0$ and $G^{(1,1)}(\mathbf{z}_m, t_{m\pm}) > 0$ with $F_2(\mathbf{z}_m, t_{m-}) < 0$ should hold for the onset of the sliding flow. Notice that the sliding flow in the relative motion is always at the origin.

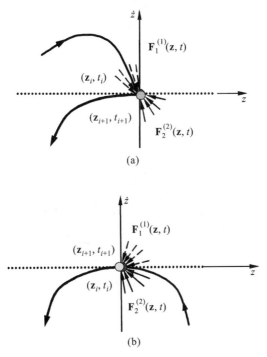

Fig. 4.8 Vector fields in the relative frame: (a) vanishing and (b) onset and vanishing of sliding flow

4.2.4 Grazing flows to boundary

Because of the discontinuity, the phase partition for this oscillator with friction is presented in the absolute and relative frames. Except for the flow passable at the boundary or sliding on the boundary, the flow in such a discontinuous system may be tangential to the boundary. Such a tangential flow is often called the grazing flow (or grazing motion). A sketch of grazing motions in the domain $\Omega_\alpha (\alpha = \{1,2\})$ is illustrated in Fig.4.9(a) and (b). In the absolute frame, the separation boundary is a curve varying with time. The corresponding vector fields for grazing flows are sketched by the arrows in two domains. However, the discontinuous boundary in the relative frame is constant. The corresponding grazing flow and vector fields are sketched in Fig.4.10(a) and (b). Because the boundary in the relative frame is a straight line, from Luo (2005a, 2006), the grazing flow is guaranteed by

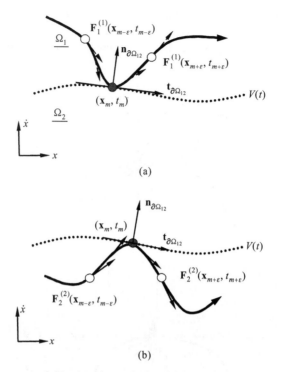

Fig. 4.9 Absolute vector fields of grazing motions in domains (a) Ω_1 and (b) Ω_2

$$\left.\begin{array}{l} \mathbf{n}_{\partial\Omega_{\alpha\beta}}^{\mathrm{T}} \cdot \mathbf{F}_{\alpha}^{(\alpha)}(\mathbf{z}_m, t_{m\pm}) = 0, \text{ for } \alpha = 1, 2, \\ G^{(1,1)}(\mathbf{z}_m, t_{m\pm}) = \mathbf{n}_{\partial\Omega_{\alpha\beta}}^{\mathrm{T}} \cdot D\mathbf{F}_1^{(1)}(\mathbf{z}_m, t_{m\pm}) > 0, \text{ and} \\ G^{(2,1)}(\mathbf{z}_m, t_{m\pm}) = \mathbf{n}_{\partial\Omega_{\alpha\beta}}^{\mathrm{T}} \cdot D\mathbf{F}_2^{(2)}(\mathbf{z}_m, t_{m\pm}) < 0. \end{array}\right\} \quad (4.40)$$

From Eqs.(4.26) and (4.27), the sufficient and necessary conditions for grazing flows are

$$\left.\begin{array}{l} G^{(0,\alpha)}(\mathbf{z}_m, t_{m\pm}) = F_\alpha(\mathbf{z}_m, t_{m\pm}) = 0, \alpha \in \{1, 2\}, \\ G^{(1,1)}(\mathbf{z}_m, t_{m\pm}) = \nabla F_1(\mathbf{z}_m, t_{m\pm}) \cdot \mathbf{F}_1^{(1)}(\mathbf{z}_m, t_{m\pm}) + \dfrac{\partial F_1(\mathbf{z}_m, t_{m\pm})}{\partial t_m} > 0, \\ G^{(1,2)}(\mathbf{z}_m, t_{m\pm}) = \nabla F_2(\mathbf{z}_m, t_{m\pm}) \cdot \mathbf{F}_2^{(2)}(\mathbf{z}_m, t_{m\pm}) + \dfrac{\partial F_2(\mathbf{z}_m, t_{m\pm})}{\partial t_m} < 0. \end{array}\right\} \quad (4.41)$$

The grazing conditions are also illustrated in Fig.4.10 (a) and (b), and the vector fields in Ω_1 and Ω_2 are expressed by the dashed and solid arrows, respectively. The condition in Eq.(4.41) for the grazing motion in Ω_α is presented via the vector field of $\mathbf{F}_\alpha^{(\alpha)}(t)$. In addition to $F_\alpha(\mathbf{z}_m, t_{m\pm}) = 0$, the sufficient condition requires $F_1(\mathbf{z}, t) < 0$ for time $t \in [t_{m-\varepsilon}, t_m)$ and $F_1(\mathbf{z}, t) > 0$ for time $t \in (t_m, t_{m+\varepsilon}]$ in Ω_1; and

4.3 Generic mappings and force product criteria

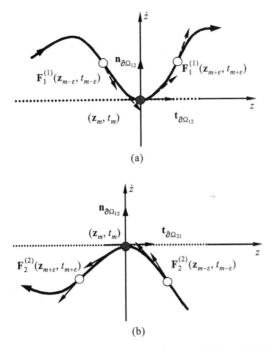

Fig. 4.10 Relative vector fields of grazing motions in domains (a) Ω_1 and (b) Ω_2

$F_2(\mathbf{z},t) > 0$ for $t \in [t_{m-\varepsilon}, t_m)$ and $F_2(\mathbf{z},t) < 0$ for $t \in (t_m, t_{m+\varepsilon}]$ in Ω_2. In other words, $G^{(1,1)}(\mathbf{z}_m, t_{m\pm}) > 0$ in Ω_1 and $G^{(1,2)}(\mathbf{z}_m, t_{m\pm}) < 0$ in Ω_2 are required.

4.3 Generic mappings and force product criteria

In this section, three basic mappings will be introduced. From such mappings, the force product criteria for stick and grazing flows to the boundary will be discussed.

4.3.1 Generic mappings

The direct integration of Eq.(4.8) with an initial condition $(t_i, z_i, 0)$ gives the sliding displacement (i.e., Eq.(4.9)). For a small δ-neighborhood of the sliding flow ($\delta \to 0$), substitution of Eqs.(4.8) and (4.9) into Eq.(4.21) gives the forces in the two domains $\Omega_\alpha (\alpha \in \{1,2\})$. For passable motions, select the initial condition on the velocity boundary (i.e., $\dot{x}_i = V_i$ and $x_i = X_i$) in the absolute frame. Based on Eq.(4.14), the basic solutions of the generalized discontinuous linear oscillator for a certain domain in Appendix will be used for construction of mappings. In absolute phase space, the trajectories in Ω_α starting and ending at the velocity discontinuity (i.e., from $\partial\Omega_{\beta\alpha}$ to $\partial\Omega_{\alpha\beta}$) are sketched in Fig.4.11. The starting and ending points for mappings P_α in Ω_α are (x_i, V_i, t_i) and $(x_{i+1}, V_{i+1}, t_{i+1})$, respectively. The sliding (or stick) mapping on the boundary is P_0. On the boundary $\partial\Omega_{\alpha\beta}$, letting $z_i = 0$, the switching planes are defined as

$$\begin{aligned}
\Xi^0 &= \{(x_i, \Omega t_i) | \dot{x}_i = V_i\}, \\
\Xi^1 &= \{(x_i, \Omega t_i) | \dot{x}_i = V_i^+\}, \\
\Xi^2 &= \{(x_i, \Omega t_i) | \dot{x}_i = V_i^-\},
\end{aligned} \quad (4.42)$$

where $V_i^- = \lim_{\delta \to 0}(V_i - \delta)$ and $V_i^+ = \lim_{\delta \to 0}(V_i + \delta)$ for an arbitrarily small $\delta > 0$. Therefore,

$$P_1 : \Xi^1 \to \Xi^1, P_2 : \Xi^2 \to \Xi^2, P_0 : \Xi^0 \to \Xi^0. \quad (4.43)$$

From the foregoing two equations, we have

$$\begin{aligned}
P_0 &: (x_i, V_i, t_i) \to (x_{i+1}, V_{i+1}, t_{i+1}), \\
P_1 &: (x_i, V_i^+, t_i) \to (x_{i+1}, V_{i+1}^+, t_{i+1}), \\
P_2 &: (x_i, V_i^-, t_i) \to (x_{i+1}, V_{i+1}^-, t_{i+1}).
\end{aligned} \quad (4.44)$$

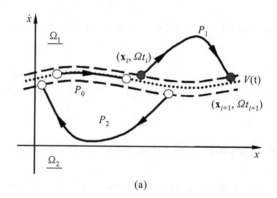

(a)

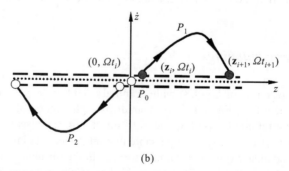

(b)

Fig. 4.11 Basic mappings in (a) absolute and (b) relative frames for the oscillator with a time-varying boundary

4.3 Generic mappings and force product criteria

From Eq.(4.9) and $F_\alpha(\mathbf{z}_{i+1},t_{i+1}) = 0$, with $z_{i+1} = z_i = 0$ and $\dot{z}_{i+1} = \dot{z}_i = 0$, the governing equations for P_0 and $\alpha \in \{1,2\}$ are

$$\left.\begin{array}{l} x_{i+1} - V_0\left[\sin(t_{i+1}+\beta_i) - \sin(t_i+\beta_i)\right] - V_1(t_{i+1}-t_i) - x_i = 0, \\ A_0 \cos\Omega t_{i+1} - b_\alpha - 2d_\alpha V_{i+1} - \dot{V}_{i+1} - c_\alpha x_{i+1} = 0. \end{array}\right\} \quad (4.45)$$

For this problem, the two domains Ω_α ($\alpha \in \{1,2\}$) are unbounded. From Luo (2005a, 2006), the flows of dynamical systems on the corresponding domains should be bounded. For any non-sliding flow, there are three possible stable motions in the two domains $\Omega_\alpha(\alpha \in \{1,2\})$, the governing equations of mapping $P_\alpha(\alpha \in \{1,2\})$ are obtained from the displacement and velocity in Appendix. Thus, the governing equations of mapping P_α ($\alpha \in \{0,1,2\}$) can be expressed by

$$\left.\begin{array}{l} f_1^{(\alpha)}(x_i,\Omega t_i, x_{i+1}, \Omega t_{i+1}) = 0, \\ f_2^{(\alpha)}(x_i,\Omega t_i, x_{i+1}, \Omega t_{i+1}) = 0. \end{array}\right\} \quad (4.46)$$

4.3.2 Sliding flows and fragmentation

Based on the switching conditions, the corresponding criteria can be developed by the product of normal vector fields as in Luo (2006). Two equations with four unknowns in Eq.(4.47) cannot give the unique solutions for the sliding flow. Once the initial state is given, the final state of the sliding flow is uniquely determined. Consider switching states $(\Omega t_c, X_c, V(t_c))$ and $(\Omega t_f, X_f, V(t_f))$ as the initial and final conditions of a sliding flow, respectively. For a time interval $[t_i, t_{i+1}] \subseteq [t_c, t_f]$ of any sliding flows, assume $(\Omega t_{i+1}, X_{i+1}, V(t_{i+1})) \equiv (\Omega t_f, X_f, V(t_f))$. Because of $z_{i+1} = z_i = 0$ and $\dot{z}_{i+1} = \dot{z}_i = 0$, we have

$$\begin{aligned} F_\alpha(\mathbf{z}_i, \Omega t_i) &\equiv F_\alpha(x_i, V(t_i), \Omega t_i) \\ &= A_0 \cos\Omega t_i - b_\alpha - 2d_\alpha V(t_i) - \dot{V}(t_i) - c_\alpha x_i. \end{aligned} \quad (4.47)$$

From Eqs.(4.30) and (4.47), the sliding flow on the boundary $\widetilde{\partial\Omega}_{\alpha\beta}$ for all $t_i \in [t_c, t_f)$ satisfies the following force product relation

$$\begin{aligned} L_{12}(t_i) &\equiv L_{12}(x_i, V(t_i), \Omega t_i) \\ &= F_1(x_i, V(t_i), \Omega t_i) \times F_2(x_i, V(t_i), \Omega t_i) < 0. \end{aligned} \quad (4.48)$$

The critical condition in Eq.(4.28) or (4.29) gives the *initial force product condition* at the critical time t_c, i.e.,

$$\begin{aligned} L_{12}(t_c) &\equiv L_{12}(x_c, V(t_c), \Omega t_c) \\ &= F_1(x_c, V(t_c), \Omega t_c) \times F_2(x_c, V(t_c), \Omega t_c) = 0, \end{aligned} \quad (4.49)$$

$$(-1)^\alpha G^{(1,\alpha)}(\mathbf{z}_c, t_{c\pm}) < 0, \text{ for } \alpha \in \{1,2\}. \tag{4.50}$$

If the initial condition satisfies Eqs.(4.49) and (4.50), the sliding flow is called the *critical sliding flow*. The condition in Eq.(4.39) for the sliding flow gives the *final force product condition* at the time $t_{i+1} = t_f$, i.e.,

$$\begin{aligned}L_{12}(t_f) &\equiv L_{12}(x_f, V_f, \Omega t_f) \\ &= F_1(x_f, V_f, \Omega t_f) \times F_2(x_f, V_f, \Omega t_f) = 0,\end{aligned} \tag{4.51}$$

$$(-1)^\alpha G^{(1,\alpha)}(\mathbf{z}_f, t_{f\pm}) < 0, \text{ for } \alpha \in \{1,2\}.$$

Once the force product of the sliding flow changes its sign at $t_m \in (t_i, t_{i+1})$, such a sliding flow will vanish. So the characteristic of the force product for the sliding flow should be further discussed. From Chapter 2, a set of the local maximum of the force product relative to the domains Ω_1 and Ω_2 is defined as

$$_{L\max}L_{12}(t_k) = \left\{ L_{12}(t_k) \left| \begin{array}{l} \forall t_k \in (t_i, t_{i+1}), \exists L_{12}(t_k) \leq 0, \\ \left.\dfrac{d}{dt}L_{12}(t)\right|_{(t_k, \mathbf{x}_k)} = 0 \text{ and } \left.\dfrac{d^2}{dt^2}L_{12}(t)\right|_{(t_k, \mathbf{x}_k)} < 0, \end{array} \right. \right\} \tag{4.52}$$

and the maximum force product is defined as

$$_{G\max}L_{12}(t_k) = \max_{t_k \in (t_i, t_{i+1})} \{L_{12}(t_k) | L_{12}(t_k) \in {_{L\max}L_{12}(t_k)}\}. \tag{4.53}$$

Once $L_{12}(t_k) \in {_{L\max}L_{12}(t_k)}$ is greater than zero, the sliding flow does not exist. Further, the sliding flow will be fragmentized. The corresponding critical condition is

$$_{G\max}L_{12}(t_k) = 0, \text{ for } t_k \in (t_i, t_{i+1}). \tag{4.54}$$

The foregoing condition is termed the *sliding fragmentation condition*. After fragmentation, the sliding flow for $t_i < t_{i+1} \leq t_{i+2} < t_{i+3}$ becomes two pieces from $(x_i, V(t_i), \Omega t_i)$ to $(x_{i+1}, V(t_{i+1}), \Omega t_{i+1})$ and $(x_{i+2}, V(t_{i+2}), \Omega t_{i+2})$ to $(x_{i+3}, V(t_{i+3}), \Omega t_{i+3})$, respectively. From Eq.(4.29), the passable motion between the two fragmentized sliding flows requires

$$L_{12}(t_k) > 0, \text{ for } t_k \in (t_{i+1}, t_{i+2}). \tag{4.55}$$

The fragmentation of the sliding flow is sketched in Fig.4.12 through the force product characteristics of the sliding flow in phase plane. The sliding, critical sliding and sliding fragmentation motions are depicted for given parameters and excitation amplitude $A_0^{(1)} < A_0^{(2)} < A_0^{(3)}$. The sliding flow vanishes from the boundary and goes into the domain Ω_1. Two points $\mathbf{x}_i$ and $\mathbf{x}_{i+1}$ represent the starting and vanishing switching displacements of the sliding flow, respectively. The point $\mathbf{x}_k \in (\mathbf{x}_i, \mathbf{x}_{i+1})$ represents a critical point for the fragmentation of the sliding flow. After the fragmentation, the critical point $\mathbf{x}_k$ splits into two new points $\mathbf{x}_{i+1}$ and $\mathbf{x}_{i+2}$, and the index of $\mathbf{x}_{i+1}$ will be shifted as $\mathbf{x}_{i+3}$. Suppose the force product $L_{12}(t_k)$ increases with

4.3 Generic mappings and force product criteria

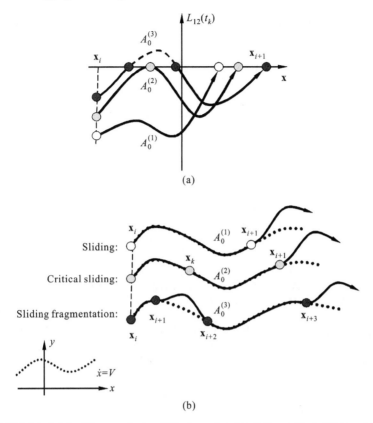

Fig. 4.12 (a) A sketch of force product and (b) phase plane for sliding, critical sliding and sliding fragmentation motions under given parameters and $A_0^{(1)} < A_0^{(2)} < A_0^{(3)}$. The ending point of the sliding flow vanishes on the boundary and goes to the domain Ω_1

increasing excitation amplitude. The flows from the sliding to fragmented sliding flows are given for increasing A_0 from $A_0^{(1)}$ to $A_0^{(3)}$. The passable portion of the sliding fragmentation needs $L_{12}(t) > 0$ as in Eq.(4.55) for $t \in U_n = (t_j^n, t_{j+1}^n) \subset (t_i, t_{i+1})$, where $L_{12}(t_k) = 0$ for $t_k = t_j^n, t_{j+1}^n$. However, for the sliding portion, the force products keep $L_{12}(t_k) < 0$ for $t_k \in (t_i, t_{i+1}) \setminus \bigcup_n U_n$. Otherwise, suppose the force products at the potential merging points of the two sliding flows decrease with decreasing excitation. The merging of the two sliding flows can be observed for decreasing A_0 from $A_0^{(3)}$ to $A_0^{(1)}$. Once the merging or fragmentation of a sliding flow occurs, the old sliding flow will be destroyed. For a sliding flow on $\widetilde{\partial \Omega}_{\alpha\beta}$ with starting and ending points $(x_i, V_i, \Omega t_i)$ and $(x_{i+1}, V_{i+1}, \Omega t_{i+1})$, consider $\Delta = \sqrt{\delta_{x_i}^2 + \delta_{t_i}^2 + \delta_{V_i}^2}$ with $\delta_{V_i} = V_{i+1} - V_i$, $\delta_{x_i} = x_{i+1} - x_i$ and $\delta_{t_i} = \Omega t_{i+1} - \Omega t_i \geq 0$. Suppose the starting and ending points satisfy the initial and final force product conditions. As $\Delta \to 0$, the following force product condition is termed the *onset* condition for the sliding flow.

$$\left. \begin{array}{l} L_{12}(t_k) = 0, \text{ for } k \in \{i, i+1\}, \\ (x_{i+1}, V_{i+1}, \Omega t_{i+1}) = \lim_{\delta \to 0} (x_i + \delta_{x_i}, V_i + \delta_{V_i}, \Omega t_i + \delta_{t_i}). \end{array} \right\} \quad (4.56)$$

The detailed discussion can be referred to Gegg, Luo and Suh (2009).

4.3.3 Grazing flows

If the grazing for two mappings of passable motions occurs at the final state $(x_{i+1}, V_{i+1}, t_{i+1})$, from Eq.(4.29), the grazing conditions based on mappings are obtained. With Eq.(4.21), the grazing condition in Eq.(4.42) becomes

$$\left. \begin{array}{l} A_0 \cos \Omega t_{i+1} - b_\alpha - 2d_\alpha V_{i+1} - c_\alpha x_{i+1} - \dot{V}_{i+1} = 0, \quad (\alpha \in \{1,2\}), \\ -2d_\alpha \dot{V}_{i+1} - c_\alpha V_{i+1} - \ddot{V}_{i+1} - A_0 \Omega \sin \Omega t_{i+1} > 0, \quad (\alpha = 1), \\ -2d_\alpha \dot{V}_{i+1} - c_\alpha V_{i+1} - \ddot{V}_{i+1} - A_0 \Omega \sin \Omega t_{i+1} < 0, \quad (\alpha = 2). \end{array} \right\} \quad (4.57)$$

The grazing conditions for the two non-sliding mappings can be illustrated. The grazing conditions in Eq.(4.57) are given through the forces and jerks. Hence, both the initial and final switching sets of the two non-sliding mappings will vary with system parameters. Because the grazing characteristics of the two non-sliding mappings are different, illustrations of grazing conditions for the two mappings will be presented separately.

To ensure the initial switching sets to be passable, from Luo (2006), the initial switching sets of mapping $P_\alpha (\alpha \in \{1,2\})$ should satisfy the following condition as in Luo and Gegg (2005b, 2006a,b).

$$L_{12}(t_m) = F_1(\mathbf{z}_m, t_{m\mp}) \times F_2(\mathbf{z}_m, t_{m\pm}) > 0. \quad (4.58)$$

The condition in Eq.(4.58) guarantees that the initial switching sets of mapping P_α ($\alpha \in \{1,2\}$) is passable to the discontinuous boundary (i.e., $\dot{x}_i = V_i$). The force product for the initial switching sets is also illustrated to ensure the non-sliding mapping exists. The force conditions for the final switching sets of mapping $P_\alpha (\alpha \in \{1,2\})$ are presented in Eq.(4.57).

The grazing conditions are given by Eqs.(4.46) and (4.57). Three equations plus an inequality with four unknowns require one of the unknown be given. For instance, the initial displacement or phase of mapping $P_\alpha (\alpha \in \{1,2\})$ can be selected to specific values from Eq.(4.57). Therefore, three equations with three unknowns will give the grazing conditions. Namely, the initial switching phase, the final switching phase and displacement of mapping $P_\alpha (\alpha \in \{1,2\})$ will be determined by Eqs. (4.46) and (4.57). From the inequality of Eq.(4.57), the critical value for $\text{mod}(\Omega t_{i+1}, 2\pi)$ is introduced by

4.4 Periodic motions

$$\Theta_\alpha^{cr} = \arcsin\left(-\frac{\gamma_\alpha}{A_0\Omega}\right), \qquad (4.59)$$

where $\gamma_\alpha = c_\alpha V_{i+1} + 2d_\alpha \dot{V}_{i+1} + \ddot{V}_{i+1}$ ($\alpha = 1, 2$) and the superscript "cr" represents a critical value relative to grazing and $\alpha \in \{1, 2\}$. From the second equation of Eq.(4.57), the final switching phase for mapping P_1 has the following six cases:

$$\left.\begin{aligned}
&\mathrm{mod}\,(\Omega t_{i+1}, 2\pi) \in (\pi + |\Theta_1^{cr}|, 2\pi - |\Theta_1^{cr}|) \subset (\pi, 2\pi), \\
&\quad\text{for } 0 < \gamma_1 < A_0\Omega; \\
&\mathrm{mod}\,(\Omega t_{i+1}, 2\pi) \in (\pi - \Theta_1^{cr}, 2\pi] \cup [0, \Theta_1^{cr}), \\
&\quad\text{for } \gamma_1 < 0 \text{ and } A_0\Omega > |\gamma_1|; \\
&\mathrm{mod}\,(\Omega t_{i+1}, 2\pi) \in (\pi, 2\pi), \text{ for } \gamma_1 = 0;
\end{aligned}\right\} \qquad (4.60)$$

$$\left.\begin{aligned}
&\mathrm{mod}\,(\Omega t_{i+1}, 2\pi) \in \{\varnothing\}, &&\text{for } \gamma_1 > 0 \text{ and } A_0\Omega \leqslant \gamma_1; \\
&\mathrm{mod}\,(\Omega t_{i+1}, 2\pi) \in [0, 2\pi], &&\text{for } \gamma_1 < 0 \text{ and } A_0\Omega < |\gamma_1|; \\
&\mathrm{mod}\,(\Omega t_{i+1}, 2\pi) \in [0, 2\pi]/\{\pi/2\}, &&\text{for } \gamma_1 < 0 \text{ and } A_0\Omega = |\gamma_1|.
\end{aligned}\right\} \qquad (4.61)$$

The parameter characteristics of grazing for mapping P_2 will be presented as follows. Similarly, from the third equation of Eq.(4.57), the six cases of the final switching phase for mapping P_2 are

$$\left.\begin{aligned}
&\mathrm{mod}\,(\Omega t_{i+1}, 2\pi) \in [0, \pi + |\Theta_2^{cr}|) \cup (2\pi - |\Theta_2^{cr}|, 2\pi], \\
&\quad\text{for } 0 < \gamma_2 < A_0\Omega; \\
&\mathrm{mod}\,(\Omega t_{i+1}, 2\pi) \in (\Theta_2^{cr}, \pi - \Theta_2^{cr}) \subset (0, \pi), \\
&\quad\text{for } \gamma_2 < 0 \text{ and } A_0\Omega > |\gamma_2|; \\
&\mathrm{mod}\,(\Omega t_{i+1}, 2\pi) \in (0, \pi), \text{ for } \gamma_2 = 0;
\end{aligned}\right\} \qquad (4.62)$$

$$\left.\begin{aligned}
&\mathrm{mod}\,(\Omega t_{i+1}, 2\pi) \in [0, 2\pi], &&\text{for } \gamma_2 > 0 \text{ and } A_0\Omega < \gamma_2; \\
&\mathrm{mod}\,(\Omega t_{i+1}, 2\pi) \in [0, 2\pi]/\left\{\frac{3\pi}{2}\right\}, &&\text{for } \gamma_2 > 0 \text{ and } A_0\Omega = \gamma_2; \\
&\mathrm{mod}\,(\Omega t_{i+1}, 2\pi) \in \{\varnothing\}, &&\text{for } \gamma_2 < 0 \text{ and } A_0\Omega < |\gamma_2|.
\end{aligned}\right\} \qquad (4.63)$$

The more details can be referred to Gegg, Suh and Luo (2008).

4.4 Periodic motions

In this section, periodic motions for such oscillator will be presented through mapping structures. The analytical and numerical predictions will be presented through mapping structures and force product. The local stability and bifurcation of periodic motions will be presented and the parameter maps will be given for such an oscillator with a time-varying boundary.

4.4.1 Mapping structures

A generalized mapping structure for periodic motions with stick as in Luo and Gegg (2005a,b) is given by

$$P = \underbrace{\left(P_2^{(k_{m2})} \circ P_1^{(k_{m1})} \circ P_0^{(k_{m0})}\right) \circ \cdots \circ \left(P_2^{(k_{12})} \circ P_1^{(k_{11})} \circ P_0^{(k_{10})}\right)}_{m\text{-terms}}, \qquad (4.64)$$

where $k_{l\alpha} \in \{0,1\}$ for $l \in \{1,2,\ldots,m\}$ and $\alpha \in \{0,1,2\}$, $P_\alpha^{(k)} = P_\alpha \circ P_\alpha^{(k-1)}$ and $P_\alpha^{(0)} = 1$ as discussed in Chapter 3. The clockwise and counter–clockwise rotations of the order of the mapping P_α in the entire mapping group P in Eq.(4.64) will not change the periodic motion, while only the initial conditions for such a periodic motion are different. The mapping notation in Eq.(4.64) will be expressed by

$$P = P_{\underbrace{(2^{k_{m2}} 1^{k_{m1}} 0^{k_{m0}}) \cdots (2^{k_{12}} 1^{k_{11}} 0^{k_{10}})}_{m\text{-terms}}}.$$

Consider $m = k_{12} = k_{11} = 1$ and $k_{10} = 0$. Equation (4.64) gives a mapping structure for one of the simplest non-stick periodic motions, as shown in Fig.4.13. The mapping structure of such a simplest periodic motion is

$$P = P_2 \circ P_1 : \Xi^+ \to \Xi^-. \qquad (4.65)$$

From Eq.(4.42), the detailed relationship for such a mapping structure is given by

$$\left.\begin{array}{l} P_1 : (x_i, V^+(t_i), \Omega t_i) \to (x_{i+1}, V^+(t_{i+1}), \Omega t_{i+1}), \\ P_2 : (x_{i+1}, V^-(t_{i+1}), \Omega t_{i+1}) \to (x_{i+2}, V^-(t_{i+2}), \Omega t_{i+2}). \end{array}\right\} \qquad (4.66)$$

Let $\mathbf{y}_{i+2} = P\mathbf{y}_i$ where $\mathbf{y}_i = (x_i, \Omega t_i)^T$ during N-periods of excitation, the periodicity for such a mapping structure is

$$x_{i+2} = x_i, \Omega t_{i+2} = \Omega t_i + 2N\pi. \qquad (4.67)$$

The governing equations for such a simplest periodic motion are given by four algebraic equations, i.e.,

$$\left.\begin{array}{l} f_1^{(1)}(x_i, \Omega t_i, x_{i+1}, \Omega t_{i+1}) = 0, \\ f_2^{(1)}(x_i, \Omega t_i, x_{i+1}, \Omega t_{i+1}) = 0, \\ f_1^{(2)}(x_{i+1}, \Omega t_{i+1}, x_{i+2}, \Omega t_{i+2}) = 0, \\ f_2^{(2)}(x_{i+1}, \Omega t_{i+1}, x_{i+2}, \Omega t_{i+2}) = 0. \end{array}\right\} \qquad (4.68)$$

With Eq.(4.67), equation (4.68) gives the initial switching sets for such a periodic motion. The existence of the periodic motion without stick will be determined by the local stability analysis.

4.4 Periodic motions

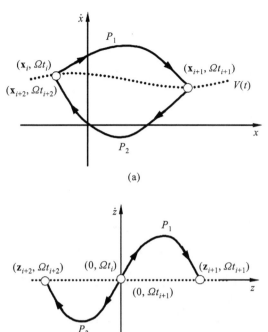

Fig. 4.13 Non-stick periodic motion of a linear oscillator with dry-friction in (a) absolute and (b) relative phase planes

Consider the mapping structure for one of the simplest periodic motions with sticking ($m = k_{1\alpha} = 1$, $\alpha \in \{0,1,2\}$), the mapping structure for a periodic motion with stick is

$$P = P_{210} \overset{\Delta}{=} P_2 \circ P_1 \circ P_0. \tag{4.69}$$

The periodic motion based on the foregoing mapping is sketched in Fig.4.14. For the stick periodic motion $\mathbf{y}_{i+3} = P\mathbf{y}_i$ where $\mathbf{y}_i = (x_i, \Omega t_i)^T$ during N-periods of excitation, the periodicity is

$$x_{i+3} = x_i, \Omega t_{i+3} = \Omega t_i + 2N\pi. \tag{4.70}$$

The six governing equations for such a periodic motion with stick are

$$\left. \begin{array}{l} f_1^{(\alpha)}(x_{i+\alpha}, \Omega t_{i+\alpha}, x_{i+\alpha+1}, \Omega t_{i+\alpha+1}) = 0, \\ f_2^{(\alpha)}(x_{i+\alpha}, \Omega t_{i+\alpha}, x_{i+\alpha+1}, \Omega t_{i+\alpha+1}) = 0 \end{array} \right\} \tag{4.71}$$

for $\alpha \in \{0,1,2\}$. Similarly, the periodic motion relative to the generalized mapping structure can be predicted through the corresponding governing equations. For periodic motions with stick, the traditional local stability analysis may not be useful. Based on such a reason, the onset, existence and disappearance of the stick

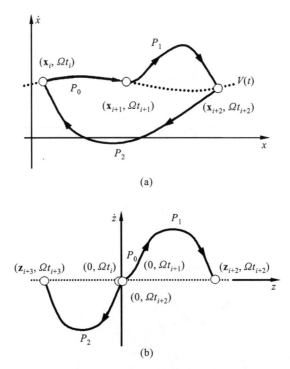

Fig. 4.14 A stick periodic motion of a linear oscillator with dry-friction in (a) absolute and (b) relative phase planes

motion should be determined by the force criteria in Eqs.(4.32)∼(4.39). The grazing bifurcation will be determined by Eq.(4.40) or (4.41). For the general case of the nonlinear friction-induced oscillators, the eigenvalue analysis cannot be done. However, the force criteria can be embedded in the numerical algorithm to detect the switching, non-stick and grazing motions. For comparison, the traditional eigenvalue analysis will be also presented in illustrations.

4.4.2 Illustrations

Consider system parameters ($\Omega = 1, V_0 = 0.25, V_1 = 1, d_1 = 1, d_2 = 0, \ b_1 = -b_2 = 0.5, c_1 = c_2 = 30$) for illustrations. Using the mapping structure in Eq.(4.64), all the periodic motions for the investigated range of excitation amplitudes can be determined analytically through the governing equations similar to Eqs. (4.70) and (4.71). For nonlinear discontinuous systems, the analytical prediction will become difficult to be completed. However, the force criteria developed in Section 2 will be very easily used to numerically predict the sliding flows and grazing bifurcations in nonlinear, non-smooth dynamical systems. The numerical simulation is based on

4.4 Periodic motions

the closed form solutions in Appendix. In addition, the numerical prediction can be also carried out through the traditional numerical integration algorithm of differential equations, but, the motion at the discontinuous boundary should be treated specially.

The switching phase and displacement varying with excitation amplitude are predicted analytically and numerically, as illustrated in Fig.4.15. The diamond sym-

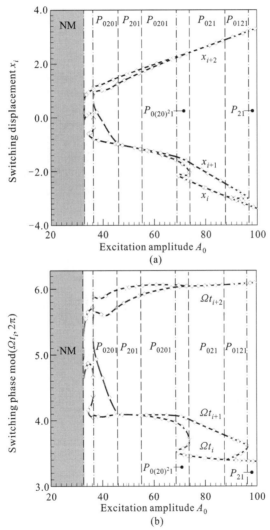

Fig. 4.15 Analytical and numerical predictions of switching sets varying with excitation amplitude: (a) switching displacement and (b) switching phase. ($\Omega = 1$, $V_0 = 0.25$, $V_1 = 1, d_1 = 1$, $d_2 = 0$, $b_1 = -b_2 = 0.5, c_1 = c_2 = 30$). The dashed and dash-dotted vertical lines are stick and grazing bifurcations, respectively. The diamond symbols and the solid curves represent the analytical and numerical predictions, respectively

bols and the solid curves represent the analytical and numerical predictions, respectively. The analytical and numerical predictions of periodic motions are in a very good agreement for this non-smooth dynamical system. The dashed and dash-dot vertical lines denote the boundaries for the onset and vanishing of the sliding flow (or stick motion), and for the grazing bifurcation, respectively. The *sliding* bifurcation denotes the appearance or disappearance of a stick motion, thus the mapping structure is switched. The regions between two adjacent boundary lines are labeled with corresponding mapping structures. The switching phases and displacements are also labeled for the corresponding mapping. The *grazing* bifurcation refers to the motion of the mass, tangential to the velocity boundary, or the relative velocity of the mass and belt for this grazing bifurcation equals zero. The regions between two lines are labeled by mapping structures for $A_0 \in (29.85, 100.0)$. If the excitation amplitude is less than the lowest boundary, no periodic motion intersects with the velocity boundary (i.e., $y = V(t)$). The periodic motion for P_{20} exists in the range of $A_0 \in (31.66, 33.64)$. The periodic motion for P_{201} lies in range of $A_0 \in (39.20, 57.75)$. The periodic motion for P_{21} is in the range of $A_0 \in (88.75, 100.0)$. The periodic motion for $P_{2(01)^2}$ is in the range of $A_0 \in (79.15, 84.09)$ and $(38.45, 39.22)$. The periodic motion for P_{0201} is in the range of $A_0 \in (70.20, 79.10)$, $(57.80, 63.50)$ and $(33.65, 38.40)$. The periodic motion for $P_{(02)^201}$ is in the range of $A_0 \in (63.55, 79.15)$. The periodic motion for $P_{(20)^2}$ is in the range of $A_0 \in (31.66, 33.64)$. Finally, the periodic motion for $P_{(20)^k}$ ($k = 1, 2, 3$) is in the range of $A_0 \in (29.85, 30.16)$. The grazing bifurcations occur at the following approximate excitation amplitudes $\{88.755, 39.220, 63.550, 39.200, 31.660, 29.850\}$. For $A_0 \approx 38.45$, the sliding bifurcation for mapping P_1 exists, and the other critical values are for the sliding bifurcations of mapping P_2. The onset and vanishing of stick periodic motions occur at the approximate excitation amplitudes $\{30.160, 30.170, 31.670, 32.450, 32.650, 33.650, 38.450, 57.800, 70.200, 79.150, 84.010\}$. Excitation amplitudes for specific periodic motions, grazing and sliding bifurcations are summarized in Table 4.1.

Table 4.1 The summary of excitation frequencies for specific motions ($\Omega = 1, V_0 = 0.25, V_1 = 1, d_1 = 1, d_2 = 0, b_1 = -b_2 = 0.5, c_1 = c_2 = 30$)

Mapping structures	Excitation amplitude	Grazing bifurcation	Sliding bifurcation
P_{21}	(96.6616, 100.0)	96.6616~96.6615	—
P_{0121}	(88.0794, 96.6615)	96.6616~96.6615	88.0794
P_{021}	(73.6610, 88.0793)	—	88.0793, 73.6610
$P_{(02)^201}$	(68.7450, 73.6609)	68.7450~68.7449	73.6609
P_{0201}	(54.5117, 68.7450) (36.5593, 45.8571)	68.7450~68.7449	54.5117, 45.8571, 36.5593
P_{201}	(45.8572, 54.5117)	—	54.5117, 45.8572
$P_{2(01)^2}$	(79.15, 84.09) (38.45, 39.22)	39.220~39.215	84.090, 79.150, 38.450
P_{20}	(30.17, 31.65)	—	31.670, 30.170
$P_{(20)^2}$	(34.5756, 36.4695)	34.5756~34.5755	36.5592
$P_{(20)^k}$ ($k = 1, 2, 3, ...$)	(29.85, 30.16)	29.850~29.845	30.150

4.4 Periodic motions

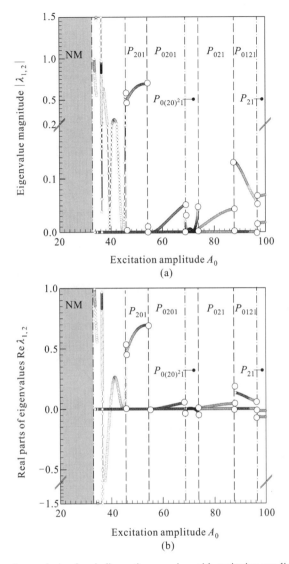

Fig. 4.16 Eigenvalue analysis of periodic motions varying with excitation amplitude: (a) magnitude and (b) real part. ($\Omega = 1$, $V_0 = 0.25$, $V_1 = 1$, $d_1 = 1$, $d_2 = 0$, $b_1 = -b_2 = 0.5$, $c_1 = c_2 = 30$). The dashed and dash-dotted vertical lines are stick and grazing bifurcations, respectively

The eigenvalue analysis of the analytical solutions based on the mapping structure can be completed very easily from Appendix. The magnitude and real parts of eigenvalues for all the analytical solutions of periodic motions are presented in Fig.4.16(a) and (b), respectively. The corresponding mapping structures are also labeled. From the local stability analysis of periodic motions, all the periodic motions are always stable before a new periodic motion appears. Therefore, the traditional

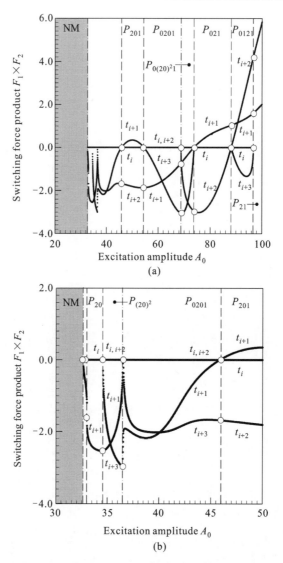

Fig. 4.17 Force products at switching points for periodic motion: (a) the entire range ($A_0 \in [20, 100]$), (b) zoomed area ($A_0 \in [39.9, 50]$) ($\Omega = 1$, $V_0 = 0.25$, $V_1 = 1$, $d_1 = 1$, $d_2 = 0$, $b_1 = -b_2 = 0.5$, $c_1 = c_2 = 30$). The dashed and dash-dotted vertical lines are stick and grazing bifurcations, respectively

eigenvalue analysis tells us that those periodic motions should exist in the wide range of parameters. However, before the local stability conditions are destroyed, the existing periodic motion already switches to a new periodic motion due to a global event, resulting from the stick and grazing bifurcations. Once the stick exists in periodic motions, the traditional eigenvalue analysis for the motion switching

4.4 Periodic motions 105

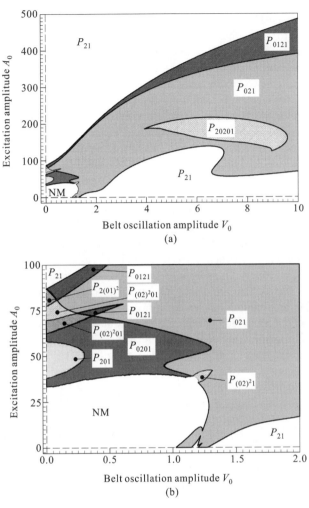

Fig. 4.18 The (V_0, A_0) parameter map for periodic motions: (a) the large range, (b) zoomed area ($\Omega = 1$, $V_1 = 1$, $d_1 = 1$, $d_2 = 0$, $b_1 = -b_2 = 0.5$, $c_1 = c_2 = 30$). The acronym "NM" represents no motion interacting with the discontinuous boundary (color plot in the book end)

cannot work well. In Fig.4.16, from the non-stick motion to the stick motion, one of the eigenvalues for the periodic motion is zero. From a stick motion to an adjacent stick motion, the eigenvalue analysis cannot provide the further information to find any signature of the motion switching. Since the necessary and sufficient conditions were given for the passable and non-passable motions in non-smooth dynamical systems in Luo (2005a,b), the corresponding conditions for the stick and non-stick motion in this specific problem can be presented through the relative force product, as similar to Luo and Gegg (2005c,d). Therefore, force products at switching points for periodic motions are shown in Fig. 4.17. If the stick motion exists, the switching points for the vanishing of the sliding in periodic motion always possesses a zero force product from the criteria in Eq.(4.48). For the onsets of the stick and grazing, the corresponding force products are zero. Before and after such onsets, the force product should not be zero. Therefore, the grazing and sliding bifurcation can be detected through the force product. The more discussion on grazing and sliding appearance can be referred to Luo and Gegg (2007a, b). From the force product, it is very clear to observe the grazing bifurcation and the onset and vanishing of the stick motions. This force product computation does not need any closed-form solution of non-smooth dynamical system, thus the computation of force products for nonlinear, non-smooth dynamical systems can be carried out to determine the corresponding switching of stick and non-stick periodic motions. Without the eigenvalue analysis, the closed-form solutions are not necessary. Furthermore, the methodology presented herein can be used in nonlinear, non-smooth dynamical systems.

Based on the force product criteria of the stick and non-stick motions, the parameter maps for excitation amplitude versus belt oscillation amplitude, constant belt speed, belt oscillation frequency are presented in Figs.4.18~4.20 for given other parameters, respectively. In addition, friction force effects on periodic motion for such a dynamical system are discussed. The parameter map of friction force versus excitation amplitude is presented in Fig.4.21. The acronym "NM" represents no periodic motion interacting with the discontinuous boundary. The white areas represent the non-stick, periodic motion relative to P_{21} and the non-motion intersected with discontinuous velocity boundary condition. The stick, periodic motion relative to P_{021} is filled by the gray color. The blue, yellow, red and green colors represent periodic motions relative to $P_{0201}, P_{201}, P_{(21)^2}$ and $P_{(20)^2}$, respectively. In addition, the white-left-diagonal, white-right-diagonal, gray-left-diagonal, gray-right-diagonal, blue-left-diagonal shaded areas represent the periodic motions pertaining to $P_{(02)^201}, P_{(02)^21}, P_{2(01)^2}, P_{(02)^201}$ and P_{0121}, respectively. The boundaries in the parameter maps are the grazing bifurcations and the onsets of stick motions. Both the belt constant speed and the belt oscillation amplitude versus excitation amplitude give the two completely different parameter maps. If $V_1 = 0$, the oscillation belt travels with a cosine wave speed. If $V_1 < 0$, the oscillation belt moves backward compared to the case of $V_1 > 0$. Consider $\Omega = \bar{\Omega}/\omega = 1$ as an example to demonstrate the parameter map for periodic motion. Let the frequency ratio $\Omega \equiv m/n$ where integers m and n are irreducible and the periodic motion can be obtained. The mapping structure will be strongly dependent on the two integers m and n. Herein, such a case will not be discussed. If the frequency ratio Ω is irrational number, quasi-periodic motions will be obtained. The mapping structures will be endless.

From the parameter map in Fig.4.18, for $V_0 > 1.5$ and $A_0 > 100$, the periodic motions of $P_{21}, P_{021}, P_{0121}$ and $P_{(20)^21}$ exist. For $A_0 \in (30, 100)$ and $V_0 \in (0, 1.5)$, the periodic motions become very complicated, and the corresponding mapping structures are labeled there. For $A_0 \in (0, 30)$ and $V_0 \in (0, 1)$, no any periodic motion intersected with the velocity boundary can be founded. The zoomed parameter map is illustrated in Fig.4.18(b). For a given oscillation of the belt, the parameter map for excitation amplitude versus the non-oscillation speed is given in Fig.4.19 with $V_0 \in (-10, 10)$ and $A_0 \in (0, 500)$. The complicated periodic motions exist on the

4.4 Periodic motions

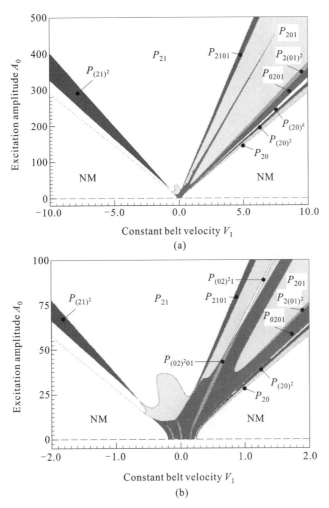

Fig. 4.19 The (V_1, A_0) parameter map for periodic motions: (a) the large range, (b) zoomed area ($\Omega = 1$, $V_0 = 0.25$, $d_1 = 1$, $d_2 = 0$, $b_1 = -b_2 = 0.5$, $c_1 = c_2 = 30$). The acronym "NM" represents no motion interacting with the discontinuous boundary (color plot in the book end)

side of $V_0 > 0$ because of $V_1 > 0$. For $V_1 < 0$, the complicated motions should be on the side of $V_0 < 0$. Both of the parameter maps will be symmetric. Such symmetry of periodic motions in non-smooth dynamical systems can be referred to Luo (2005c,d, 2006). The periodic motion associated with P_{21} takes a large region of the parameter map. On the right side of the parameter map, before the periodic motions intersected with the velocity boundary disappear, the periodic motions pertaining to $P_{(20)^k}(k = 1, 2, 3, \ldots)$ exist. However, on the left side of the parameter map, before periodic motions disappear, the periodic motion is P_{21}. The periodic motion relative to $P_{(21)^2}$ seems a period-2 motion exists. In fact, it does not have any char-

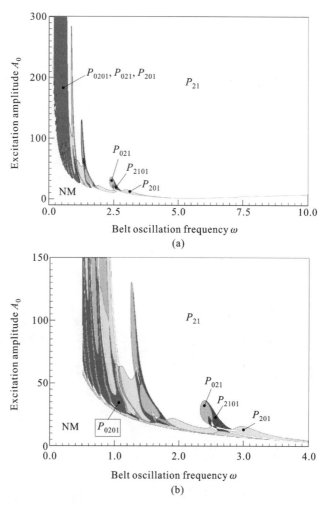

Fig. 4.20 The (A_0, ω) parameter map for periodic motions: (a) the large range, (b) zoomed area ($\Omega = 1$, $V_0 = 0.25$, $V_1 = 1$, $d_1 = 1$, $d_2 = 0$, $b_1 = -b_2 = 0.5$, $c_1 = c_2 = 30$). The acronym "NM" represents no motion interacting with the discontinuous boundary (color plot in the book end)

acteristics of period doubling motion. In the vicinity of $V_1 = 0$, the periodic motions are relative to mapping structures of $P_{201}, P_{(02)^2 21}, P_{0201}$ and P_{20}. The parameter map of the excitation amplitude with the belt oscillation frequency is given in Fig.4.20 for $\omega \in (0, 10)$ and $A_0 \in (0, 300)$. Because of the frequency ratio $\Omega = 1$, this parameter map is for excitation amplitude and frequency, similar to the parameter map given in Luo and Gegg (2005b). That parameter map is based on the belt speed being constant. For the high oscillation frequency $\omega > 5.0$, only the periodic motion of P_{21} exists. For an oscillation frequency less than 0.4 (i.e., $\omega < 0.4$), the periodic motion intersected with the velocity boundary cannot be obtained. For $\omega \in (0.4, 5)$,

4.4 Periodic motions

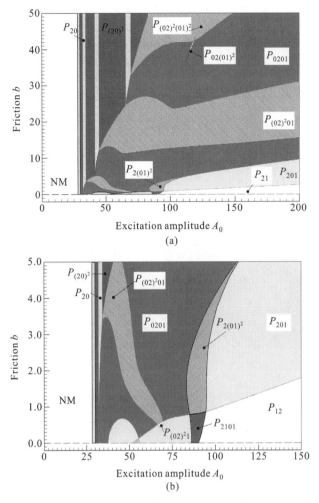

Fig. 4.21 The (A_0, b) parameter map for periodic motions: (a) the large range, (b) zoomed area ($\Omega = 1$, $V_0 = 0.25, V_1 = 1$, $d_1 = 1$, $d_2 = 0$, $b_1 = -b_2 = b, c_1 = c_2 = 30$). The acronym "NM" represents no motion interacting with the discontinuous boundary (color plot in the book end)

the complicated periodic motions exist. With decreasing ω, the parameter map of periodic motions becomes more complex. The detailed parameters for $\omega \in (0,4)$ and $A_0 \in (0,150)$ are shown in Fig.4.20(b). The strange domains in the parameter map are observed. The parameter map for friction force and excitation amplitude is presented in Fig.4.21 for $b \in (0,50)$ and $A_0 \in (0,200)$. With increasing friction force b, the periodic motion with stick will appear. With increasing excitation amplitude, the more non-stick periodic motions will exist, and finally the motion relative to P_1 will disappear. For $A_0 < 30$, the periodic motion intersected with the veloc-

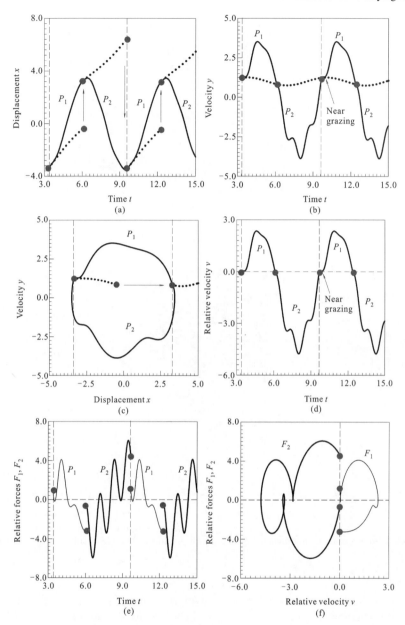

Fig. 4.22 Responses of non-stick periodic motion relative to mapping P_{21} ($A_0 = 99$): (a) absolute displacement, (b) absolute velocity, (c) phase trajectory, (d) relative velocity, (e) relative force, and (f) relative force versus velocity ($\Omega = 1$, $V_0 = 0.25$, $V_1 = 1$, $d_1 = 1$, $d_2 = 0$, $b_1 = -b_2 = 0.5$, $c_1 = c_2 = 30$). The solid and dotted curves represent the oscillator's responses and belt responses, respectively. The dark-gray symbols denote switching points for the passable motion

4.4 Periodic motions

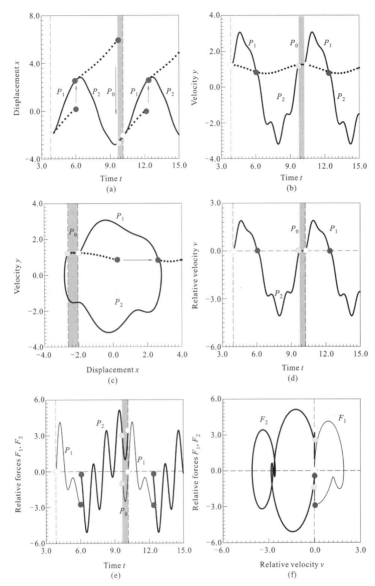

Fig. 4.23 Responses of periodic motion with stick, pertaining to mapping P_{021} ($A_0 = 80$): (a) absolute displacement, (b) absolute velocity, (c) phase trajectory, (d) relative velocity, (e) relative force, and (f) relative force versus velocity ($\Omega = 1$, $V_0 = 0.25$, $V_1 = 1$, $d_1 = 1, d_2 = 0$, $b_1 = -b_2 = 0.5$, $c_1 = c_2 = 30$). The solid and dotted curves represent the oscillator's responses and belt responses, respectively. The dark-gray, light-gray and white circular symbols denote the switching points relative to passable motion, stick motion onset and vanishing, respectively The shade area is for stick motion.

ity boundary cannot be obtained. For the lower excitation amplitude, the periodic motions pertaining to $P_{(20)^k}$ ($k = 1, 2, 3, \ldots$) exist.

4.5 Numerical simulations

From the analytical prediction, a periodic motion of the friction-induced oscillator can be directly obtained. The same parameters will be used for numerical simulations (i.e., $\Omega = 1$, $V_0 = 0.25$, $V_1 = 1$, $d_1 = 1$, $d_2 = 0$, $b_1 = -b_2 = 0.5$, $c_1 = c_2 = 30$). In all plots of numerical simulations, the solid and dotted curves are relative to the responses of the oscillator and the oscillating belts. The dark-gray, light-gray and white symbols represent the passable motion, starting and vanishing of the stick motion, respectively. The non-stick, periodic motion is presented in Fig.4.22 for $(x_i, \dot{x}_i, \Omega t_i) \approx (-3.3510, 1.2311, 3.3750)$ with $A_0 = 99$. The absolute displacement and velocity responses associated with mapping $P_{21} = P_2 \circ P_1$ are presented in Fig.4.22(a) and (b). The belt displacement has a jump because of the belt particle switching, as discussed in Fig.4.3. It is clearly observed that the motion switches in the two domains with a passable motion. In Fig.4.22(c), it is shown that the absolute phase trajectory of the non-stick periodic motion for the oscillator interacting with dry-friction is presented. To confirm the criteria in the relative frame, the relative velocity and force responses, and their relationship are illustrated in Fig.4.22(d)~(f), respectively. From Fig.4.22(d), the relative velocity response shows this non-stick periodic motion almost possessing the grazing bifurcation. The relative force responses show that the relative forces have the same sign at the switching point, which satisfy the conditions for passable motion conditions in Eq.(4.29).

Consider an excitation amplitude $A_0 = 80$ with the initial conditions $(x_i, \dot{x}_i, \Omega t_i) \approx (-2.0348, 1.2478, 3.8999)$ for a periodic motion with stick. The absolute displacement, velocity and phase trajectories for P_{021} are shown in Fig.4.23(a)~(c), respectively. The periodic motion with sliding flow is clearly observed. The shaded areas in all plots are relative to the stick motion (or sliding flow). Since the force criteria are developed from the relative motion, the relative velocity and force responses are presented in Fig.4.23(d) and (e). In Eq.(4.37), $F_1(\mathbf{z}_{m-}, t) \times F_2(\mathbf{z}_{m-}, t) < 0$ is for the stick motion. The stick motion disappears from the velocity boundary once one of the two forces becomes zero (see, Eq.(4.38)). In phase plane, the sliding flow moves along the velocity boundary until Eq.(4.38) is satisfied. Phase trajectories in absolute phase plane are shown in Fig.4.24. Such phase trajectories show complicated periodic motions in the friction oscillator with an periodically time-varying, traveling belt.

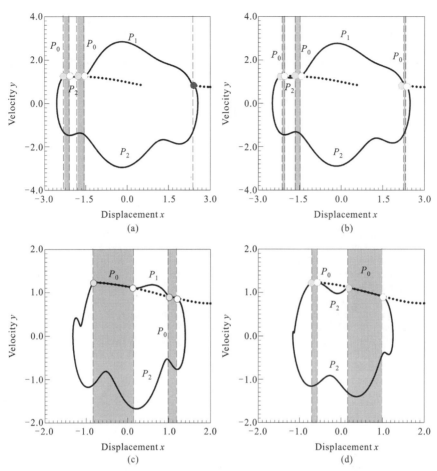

Fig. 4.24 Phase trajectories in absolute phase plane: (a) $P_{(02)^2 1}$ ($A_0 = 73$), (b) $P_{(02)^2 01}$ ($A_0 = 70$), (c) P_{0201} ($A_0 = 38$) and (d) P_{0202} ($A_0 = 35$) ($\Omega = 1$, $V_0 = 0.25$, $V_1 = 1$, $d_1 = 1$, $d_2 = 0$, $b_1 = -b_2 = 0.5$, $c_1 = c_2 = 30$). The solid and dotted curves represent trajectories for the oscillator and transport belt, respectively. The dark-gray, light-gray and white circular symbols denote the switching points relative to passable motion, stick motion onset and vanishing, respectively. The shade area is for stick motion.

References

Gegg, B. C., Luo, A. C. J. and Suh, S. (2009), Sliding motions on a periodically time-varying boundary for a friction-induced oscillator, *Journal of Vibration and Control,* in press.

Gegg, B. C., Luo, A. C. J. and Suh, S. (2008), Grazing bifurcations of a harmonically excited oscillator moving on a time-varying, translation belt, *Nonlinear Analysis: Real World Application,* **9**, pp.2156-2174.

Luo, A. C. J. (2005a), A theory for non-smooth dynamical systems on connectable domains, *Communication in Nonlinear Science and Numerical Simulation,* **10**, pp.1-55.

Luo, A. C. J. (2005b), Imaginary, sink and source flows in the vicinity of the separatrix of non-smooth dynamic system, *Journal of Sound and Vibration*, **285,** pp.443-456.

Luo, A. C. J. (2005c), The mapping dynamics of periodic motions for a three-piecewise linear system under a periodic excitation, *Journal of Sound and Vibration*, **283**, pp.723-748.

Luo, A. C. J. (2005d), On the symmetry of motions in non-smooth dynamical systems with two constraints, *Journal of Sound and Vibration*, **273**, 1118-1126.

Luo, A. C. J. (2006), *Singularity and Dynamics on Discontinuous Vector Fields*, Amsterdam: Elsevier.

Luo, A. C. J. (2007), Differential geometry of flows in nonlinear dynamical systems, Proceedings of IDECT'07, ASME International Design Engineering Technical Conferences, September 4-7, 2007, Las Vegas, Nevada, USA. DETC2007-84754.

Luo, A. C. J. (2008), *Global Transversality, Resonance and Chaotic Dynamics,* Singapore: World Scientific.

Luo, A. C. J. and Gegg, B. C. (2005a), On the mechanism of stick and non-stick periodic motion in a forced oscillator with dry-friction, ASME *Journal of Vibration and Acoustics*, **128**, pp.97-105.

Luo, A. C. J. and Gegg, B. C. (2005b), Stick and non-stick periodic motions in a periodically forced oscillator with dry-friction, *Journal of Sound and Vibration,* **291**, pp.132-168.

Luo, A. C. J. and Gegg, B. C. (2005c), Grazing phenomena in a periodically forced, friction-induced, linear oscillator, *Communications in Nonlinear Science and Numerical Simulation*, **11**, pp.777-802.

Luo, A. C. J. and Gegg, B. C. (2006a), Periodic motions in a periodically forced oscillator moving on an oscillating belt with dry friction, ASME *Journal of Computational and Nonlinear Dynamics*, **1**, pp.212-220.

Luo, A. C. J. and Gegg, B. C. (2006b), Dynamics of a periodically excited oscillator with dry friction on a sinusoidally time-varying, traveling surface, *International Journal of Bifurcation and Chaos*, **16**, pp.3539-3566.

Luo, A. C. J. and Gegg, B. C. (2007), An analytical prediction of sliding motions along discontinuous boundary in non-smooth dynamical systems, *Nonlinear Dynamics*, **49**, pp.401-424.

Chapter 5
Two Oscillators with Impacts and Stick

In this chapter, impact and stick motions of two oscillators at the moving boundary will be of great interest. The dynamics mechanism of the impacting chatter with stick at the moving boundary will be presented from the local singularity theory of discontinuous dynamical systems. The analytical conditions for the onset and vanishing of stick motions will be developed, and the condition for maintaining stick motion will be achieved as well. Analytical prediction of periodic motions relative to impacting chatter with and without stick in two dynamical oscillators will be presented through the mapping structure. The corresponding local stability and bifurcation analyses will be completed, and the grazing and stick conditions will be adopted for the existence of periodic motions. Numerical simulations will be given to illustrate periodic impacting chatters and stick in two oscillators.

5.1 Physical problem

In this section, the interaction on two oscillators at the moving boundary will be introduced through a gear transmission system. This problem exists extensively in engineering. For a better understanding of the physical mechanism, a theory for discontinuous dynamical systems on time-varying domains will be adopted to describe such a physical problem.

5.1.1 Introduction to problem

As in Luo and O'Connor (2007a,b), the dynamics of gear transmission systems is discussed. Such a problem originates from a physical problem of one-stage gear rattling, as shown in Fig.5.1. In Fig.5.1(a), the physical problem is sketched and a possible sequence of impacts is presented in Fig.5.1(b). For a gear transmission system, the driving gear with the pitch radius of R_1 is subject to a torque $T_0 +$

$T_1 \cos \Omega t$, and the angular displacement is φ_1. The driven gear with a pitch radius of R_2 has an angular displacement φ_2. The maximum clearance between the two teeth of the driven gear is d. The relative distance between two teeth is

$$z(\varphi_1, \varphi_2, t) = \varphi_1 R_1 - \varphi_2 R_2. \tag{5.1}$$

The corresponding linear displacements are defined as $x^{(1)} = R_1 \varphi_1$ and $x^{(2)} = R_2 \varphi_2$ for the driving and driven gears, respectively. In Fig.5.1(b), impact occurs on each side of the driven gear when $z = -d/2$ and $d/2$.

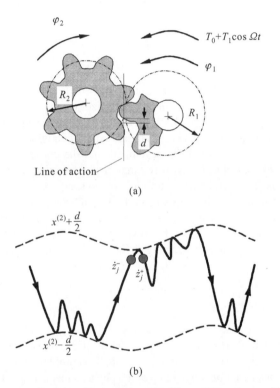

Fig. 5.1 One-stage gear chattering: (a) physical model; (b) possible sequence of impacts

To model the aforementioned gear transmission system, consider a periodically forced oscillator confined between the teeth of a second oscillator, as shown in Fig.5.2. Interaction between the two gears causes impacting and sticking at time-varying boundary. Since the gears are supported by shafts, each gear m_i ($i = 1, 2$) is connected to a spring and damper. The spring stiffness of k_i represents the torsional stiffness of shafts, and the damping of r_i is from lubricating fluids. The free-flying gap between the two teeth of the driven gear is d. The external force $B_0 + A_0 \cos \Omega t$ acts on the driving gear m_1, where A_0 and Ω are amplitude and frequency, respectively. B_0 is from constant torque. The displacements of each mass measured from

5.1 Physical problem

their equilibriums are expressed by $x^{(1)}$ and $x^{(2)}$. Impacts between the two gears possess a restitution coefficient of e. The equilibrium of the driving gear is at the center of the two teeth from the driven gear placed at equilibrium. This mechanical model gives a dynamical system consisting of two oscillators and the motions of the two oscillators are defined in two time-varying domains. The interaction of the two oscillators occurs at the time-varying boundaries. The dynamical behaviors of such dynamical systems are much richer than those in the time-independent domains, and it is much more difficult to figure out such dynamical behaviors. Such impact chattering and stick phenomena in the dynamical systems with impact and stick is of great interest. For such a system, there are three motion regions: (i) free–flight motion and (ii) two stick motions at the left and right ends of the driven gears. The subscripts of $x^{(1)}$ and $x^{(2)}$ should be added (i.e., $x^{(i)}_\alpha$, $i = 1, 2$, $\alpha = 1, 2, 3$) from now on.

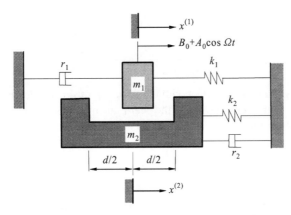

Fig. 5.2 Two oscillators with impacts and stick on time-varying boundaries

5.1.2 Equations of motion

Without any interaction between two gear oscillators (i.e., free-flight motion), from Newton's second law, the equations of motion are

$$\ddot{x}^{(i)}_2 + 2\zeta^{(i)}_2 \dot{x}^{(i)}_2 + \left(\omega^{(i)}_2\right)^2 x^{(i)}_2 = b^{(i)}_2 + Q^{(i)}_2 \cos \Omega t, \quad \text{for } i = 1, 2, \qquad (5.2)$$

where the constant coefficients are defined by

$$\left.\begin{array}{l} \zeta^{(i)}_2 = \dfrac{r_i}{2m_i},\ \omega^{(i)}_2 = \sqrt{\dfrac{k_i}{m_i}}, i = 1, 2, \\[6pt] b^{(1)}_2 = \dfrac{B_0}{m_1},\ Q^{(1)}_2 = \dfrac{A_0}{m_1}, b^{(2)}_2 = 0, Q^{(2)}_2 = 0 \end{array}\right\} \qquad (5.3)$$

for a mechanical model in Fig.5.2. The subscript $(\cdot)_2$ denotes the region of free-flight motion, and the superscript $(\cdot)^{(i)}$ represents the first mass ($i = 1$) and the second mass ($i = 2$). For $i = 1$ or 2, $\bar{i}$ is defined as

$$\bar{i} = \begin{cases} 2, & \text{if } i = 1, \\ 1, & \text{if } i = 2. \end{cases} \tag{5.4}$$

At $|x_2^{(i)} - x_2^{(\bar{i})}| = d/2$, impacts between the two masses occur if $\dot{x}_2^{(i)} \neq \dot{x}_2^{(\bar{i})}$. From the momentum conservation and the simple impact law, velocities after impact for the two gears are

$$\dot{x}_2^{(i)+} = I_1^{(i)} \dot{x}_2^{(i)-} + I_2^{(i)} \dot{x}_2^{(\bar{i})-}, \tag{5.5}$$

where the superscripts "−" and "+" represent *before* and *after* impact, and the corresponding coefficients are

$$\left. \begin{aligned} I_1^{(1)} &= \frac{m_1 - m_2 e}{m_1 + m_2}, \quad I_2^{(1)} = \frac{(1+e)m_2}{m_1 + m_2}, \\ I_1^{(2)} &= \frac{m_2 - m_1 e}{m_1 + m_2}, \quad I_2^{(2)} = \frac{(1+e)m_1}{m_1 + m_2}. \end{aligned} \right\} \tag{5.6}$$

If the first mass maintains contact with the left or right side of the second mass and they move together, such a motion in the gear transmission system is called the *stick motion*. To identify regions of stick motion at the left and right ends of the driven gear, indices ($\alpha = 1$ and 3) are used. Once the two oscillators stick together, the equations of motion are

$$\ddot{x}_\alpha^{(i)} + 2\zeta_\alpha^{(i)} \dot{x}_\alpha^{(i)} + \left(\omega_\alpha^{(i)}\right)^2 x_\alpha^{(i)} = b_\alpha^{(i)} + Q_\alpha^{(i)} \cos \Omega t \quad \text{for } i = 1, 2 \text{ and } \alpha = 1, 3, \tag{5.7}$$

where the constant coefficients are

$$\left. \begin{aligned} Q_\alpha^{(i)} &= \frac{A_0}{m_1 + m_2}, \quad \zeta_\alpha^{(i)} = \frac{r_1 + r_2}{2(m_1 + m_2)}, \quad \omega_\alpha^{(i)} = \sqrt{\frac{k_1 + k_2}{m_1 + m_2}}, \\ b_\alpha^{(1)} &= \frac{B_0}{m_1 + m_2} \pm \frac{k_2 d}{2(m_1 + m_2)}, \quad b_\alpha^{(2)} = \frac{B_0}{m_1 + m_2} \mp \frac{k_1 d}{2(m_1 + m_2)}. \end{aligned} \right\} \tag{5.8}$$

From a physical point of view, there is a pair of internal forces for such a stick motion. The sign convention for the positive direction of internal forces is opposite to all other forces and is expressed as

$$\left. \begin{aligned} f_\alpha^{(1)} &= -m_1 \ddot{x}_\alpha^{(1)} - r_1 \dot{x}_\alpha^{(1)} - k_1 x_\alpha^{(1)} + B_0 + A_0 \cos \Omega t, \\ f_\alpha^{(2)} &= -m_2 \ddot{x}_\alpha^{(2)} - r_2 \dot{x}_\alpha^{(2)} - k_2 x_\alpha^{(2)}. \end{aligned} \right\} \quad \text{for } \alpha = 1, 3 \tag{5.9}$$

From the Newton's third law, the two internal forces are with the same magnitude and the opposite direction, i.e.,

$$f_\alpha^{(1)} = -f_\alpha^{(2)}. \tag{5.10}$$

Consider the second mass to be a base reference as in Fig.5.2. In region of $\alpha = 1$, $f_\alpha^{(1)} > 0$ and $f_\alpha^{(2)} < 0$, but in region of $\alpha = 3$, $f_\alpha^{(1)} < 0$ and $f_\alpha^{(2)} > 0$. The stick motion vanishing requires

$$f_\alpha^{(i)} = 0, \text{ for } i = 1, 2. \tag{5.11}$$

The stick condition for two gear oscillators, which represents a compressive force, is given by

$$f_\alpha^{(i)} \operatorname{sgn}(x_\alpha^{(i)} - x_\alpha^{(\bar{i})}) > 0 \quad \text{for } i = 1, 2 \text{ and } \alpha = 1, 3. \tag{5.12}$$

Furthermore, the condition for stick vanishing is given by

$$f_\alpha^{(i)} \operatorname{sgn}(x_\alpha^{(i)} - x_\alpha^{(\bar{i})}) = 0. \tag{5.13}$$

In the region of $\alpha = 2$, the two masses do not contact each other and $f_2^{(i)} = 0$. This means that the oscillators are independent of each other and exist in the state of free-flight motion.

5.2 Domains and vector fields

The domains varying with time for such dynamical system cause more difficulty to analyze the dynamical behavior of such a system. To help one understand such impact chatter and stick on the time-varying boundary, in this section, absolute and relative descriptions will be given. The corresponding time-varying domain and boundaries will be developed for the absolute and relative descriptions of such a dynamical system.

5.2.1 Absolute motion description

Before the stick motion appears in such a dynamical system, impact between the two masses occurs upon contact. For this case, only one domain exists with two boundaries of flow barrier. From the definition of Luo (2006), such a boundary for $\dot{x}^{(i)} \neq \dot{x}^{(\bar{i})}$ ($i = 1, 2$) possesses a permanent flow barrier in the prescribed dynamical system. In other words, the motion never passes through such a non-passable boundary (i.e., the oscillators cannot enter the regions of stick motion). The free-flight motion region $\alpha = 2$ is now defined as a domain by

$$\Omega_2^{(i)} = \left\{ (x^{(i)}, \dot{x}^{(i)}) \; \middle| \; \begin{array}{l} x^{(i)} \in \left(x^{(\bar{i})}(t_m) - \dfrac{d}{2}, x^{(\bar{i})}(t_m) + \dfrac{d}{2} \right) \\ t_m \in (0, \infty). \end{array} \right\} \tag{5.14}$$

The two boundaries for the impacting chatter are

$$
\begin{aligned}
{}^R\partial\Omega^{(i)}_{2\infty} &= \left\{ \left(x^{(i)},\dot{x}^{(i)}\right) \middle| \begin{array}{l} {}^R\varphi^{(i)}_{2\infty} \equiv x^{(i)} - x^{(\bar{i})}(t_m) - \dfrac{d}{2} = 0 \\ \dot{x}^{(i)} \neq \dot{x}^{(\bar{i})}(t_m),\ t_m \in (0,\infty) \end{array} \right\}, \\
{}^L\partial\Omega^{(i)}_{2\infty} &= \left\{ \left(x^{(i)},\dot{x}^{(i)}\right) \middle| \begin{array}{l} {}^L\varphi^{(i)}_{2\infty} \equiv x^{(i)} - x^{(\bar{i})}(t_m) + \dfrac{d}{2} = 0 \\ \dot{x}^{(i)} \neq \dot{x}^{(\bar{i})}(t_m),\ t_m \in (0,\infty) \end{array} \right\},
\end{aligned}
\tag{5.15}
$$

where the subscript "∞" implies that the boundary is non-passable. The free-flight motions exists in domain Ω_2, and impacts or impacting chatters between the two gears occur at such a boundary with the permanent flow barrier, as defined in Eq.(5.15). If the stick motion in such a two–oscillator system exists, then there are three regions for motions of the two masses (i.e., a free-flight motion region and two stick motion regions). Thus, the phase space can be partitioned into three domains. To develop the mathematical model, the notations ${}^Lx^{(i)}_{2\pm}$ and ${}^Rx^{(i)}_{2\pm}$ ($i=1,2$) are used for the onset and vanishing of the stick motion at the left and right sides of the ith oscillator gear in domain $\Omega^{(i)}_2$. The "−" and "+" represent just before the onset and just after vanishing of stick, respectively.

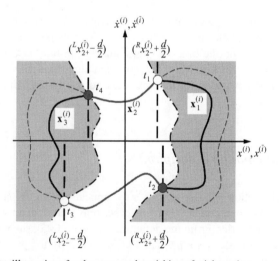

Fig. 5.3 Boundary illustrations for the onset and vanishing of stick motion

The trajectories in phase space are sketched for two oscillators in Fig.5.3. The state vector $\mathbf{x}^{(i)}_\alpha = (x^{(i)}_\alpha, \dot{x}^{(i)}_\alpha)^{\text{T}}$ with $i=1,2$ and $\alpha=1,2,3$ defines the position and velocity for a given oscillator. The two thick solid curves between the paralleled dash-dotted curves represent the free-flight motion trajectory of the ith oscillator, and the thick dash-dotted curves represent the trajectory for the left and right sides of the

5.2 Domains and vector fields

$\bar{i}^{\text{th}}$ oscillator. The times t_1 and t_2 herein represent switching times for the onset and vanishing of stick motion on the right side, and the vertical dashed lines represent the boundary locations at t_m by $^Rx_{2-}^{(\bar{i})} + d/2$ and $^Rx_{2+}^{(\bar{i})} + d/2$. The solid black curves show the trajectories of stick motion in phase space, and the thin dashed curves are the imaginary trajectories of the ith oscillator without stick or impact. In addition, without stick and impact between the two oscillators, the motion of the $\bar{i}^{\text{th}}$ oscillator will follow the thick, dash-dotted curves in the state space.

One is already familiar to the particle impact phenomena in university physics. However, once the particles become oscillators or dynamical systems, the stick phenomena between two oscillators will occur. In particle physics and molecular dynamics, one likes to describe a particle in molecule as a free-particle. If one consider such a particle as an oscillator rather than a freely moving particle, it can be very easy to explain the motion mechanism of a molecule. In this book, we will not discuss such an issue. However, the materials and methodology presented in this book is applicable to such problems. For instance, such an idea applying to the Fermi accelerator was reported (e.g., Luo and Guo, 2008).

For the onset and vanishing of stick motion, the boundaries of domain Ω_2 can be determined by

$$\left.\begin{aligned} x^{(i)} - {}^Rx_{2\pm}^{(\bar{i})} &= \frac{d}{2}, \\ x^{(i)} - {}^Lx_{2\pm}^{(\bar{i})} &= -\frac{d}{2}, \\ \dot{x}^{(i)} = {}^L\dot{x}_{2\pm}^{(\bar{i})}, \dot{x}^{(i)} &= {}^R\dot{x}_{2\pm}^{(\bar{i})}. \end{aligned}\right\} \tag{5.16}$$

Because the stick motion requires $\dot{x}^{(i)} = \dot{x}^{(\bar{i})}$, and the critical values are defined by $^Lx_{2\pm}^{(\bar{i})}$ and $^Rx_{2\pm}^{(\bar{i})}$. The absolute phase space of the ith gear is partitioned as

$$\Omega^{(i)} = \bigcup_{\alpha=1}^{3} \Omega_{\alpha}^{(i)} \bigcup \partial \Omega_{12}^{(i)} \bigcup \partial \Omega_{21}^{(i)} \bigcup \partial \Omega_{23}^{(i)} \bigcup \partial \Omega_{32}^{(i)}, \tag{5.17}$$

where the three sub-domains for such a gear system are defined as

$$\left.\begin{aligned} \Omega_1^{(i)} &= \left\{ \left(x^{(i)}, \dot{x}^{(i)}\right) \middle| \begin{array}{l} x^{(i)} \in \left({}^Rx_2^{(\bar{i})}(t_m) + \frac{d}{2}, \infty \right) \\ t_m \in (0, \infty) \end{array} \right\}, \\ \Omega_2^{(i)} &= \left\{ \left(x^{(i)}, \dot{x}^{(i)}\right) \middle| \begin{array}{l} x^{(i)} \in \left({}^Lx_2^{(\bar{i})}(t_m) - \frac{d}{2}, {}^Rx_2^{(\bar{i})}(t_m) + \frac{d}{2} \right) \\ t_m \in (0, \infty) \end{array} \right\}, \\ \Omega_3^{(i)} &= \left\{ \left(x^{(i)}, \dot{x}^{(i)}\right) \middle| \begin{array}{l} x^{(i)} \in \left(-\infty, {}^Lx_2^{(\bar{i})}(t_m) - \frac{d}{2} \right) \\ t_m \in (0, \infty) \end{array} \right\}. \end{aligned}\right\} \tag{5.18}$$

Within the domain $\Omega_\alpha^{(i)}$ ($\alpha = 1, 3$), $x^{(i)} = x^{(\bar{i})} \pm d/2$ and $\dot{x}^{(i)} = \dot{x}^{(\bar{i})}$. The corresponding separation boundaries for stick motion are defined as

$$\begin{aligned}
\partial\Omega_{12}^{(i)} &= \bar{\Omega}_1^{(i)} \cap \bar{\Omega}_2^{(i)} \\
&= \left\{ (x^{(i)}, \dot{x}^{(i)}) \;\middle|\; \begin{array}{l} \varphi_{12}^{(i)} \equiv \dot{x}^{(i)} - {}^R\dot{x}_{2+}^{(\bar{i})}(t_m) = 0 \\ x^{(i)} - {}^Rx_{2+}^{(\bar{i})}(t_m) = \dfrac{d}{2},\; t_m \in (0, \infty) \end{array} \right\}, \\
\partial\Omega_{21}^{(i)} &= \bar{\Omega}_1^{(i)} \cap \bar{\Omega}_2^{(i)} \\
&= \left\{ (x^{(i)}, \dot{x}^{(i)}) \;\middle|\; \begin{array}{l} \varphi_{21}^{(i)} \equiv \dot{x}^{(i)} - {}^R\dot{x}_{2-}^{(\bar{i})}(t_m) = 0 \\ x^{(i)} - {}^Rx_{2-}^{(\bar{i})}(t_m) = \dfrac{d}{2},\; t_m \in (0, \infty) \end{array} \right\},
\end{aligned} \quad (5.19a)$$

$$\begin{aligned}
\partial\Omega_{23}^{(i)} &= \bar{\Omega}_2^{(i)} \cap \bar{\Omega}_3^{(i)} \\
&= \left\{ (x^{(i)}, \dot{x}^{(i)}) \;\middle|\; \begin{array}{l} \varphi_{23}^{(i)} \equiv \dot{x}^{(i)} - {}^L\dot{x}_{2+}^{(\bar{i})}(t_m) = 0 \\ x^{(i)} - {}^Lx_{2+}^{(\bar{i})}(t_m) = -\dfrac{d}{2},\; t_m \in (0, \infty) \end{array} \right\}, \\
\partial\Omega_{32}^{(i)} &= \bar{\Omega}_2^{(i)} \cap \bar{\Omega}_3^{(i)} \\
&= \left\{ (x^{(i)}, \dot{x}^{(i)}) \;\middle|\; \begin{array}{l} \varphi_{32}^{(i)} \equiv \dot{x}^{(i)} - {}^L\dot{x}_{2-}^{(\bar{i})}(t_m) = 0 \\ x^{(i)} - {}^Lx_{2-}^{(\bar{i})}(t_m) = -\dfrac{d}{2},\; t_m \in (0, \infty) \end{array} \right\},
\end{aligned} \quad (5.19b)$$

where $\bar{\Omega}_\alpha^{(i)}$ is the closure of $\Omega_\alpha^{(i)}$ for $i = 1, 2$ and $\alpha = 1, 2, 3$. In a similar fashion, for the $\bar{i}^{\text{th}}$ gear, the corresponding domains (i.e., $\Omega_3^{(\bar{i})}$, $\Omega_2^{(\bar{i})}$ and $\Omega_1^{(\bar{i})}$) and boundaries (i.e., $\partial\Omega_{32}^{(\bar{i})}$, $\partial\Omega_{23}^{(\bar{i})}$, $\partial\Omega_{21}^{(\bar{i})}$ and $\partial\Omega_{12}^{(\bar{i})}$) can be defined. The partitions of phase plane for impacting chatter and stick motions in the absolute frame are sketched in Fig.5.4, and the domains and boundaries are presented. In Fig.5.4(a), the shaded domain $\Omega_2^{(i)}$ is sketched for the free-flight motion with chatter impacting of the two gears. The two non-passable boundaries for the chatter impacts are presented by the dash-dot curves, labeled by ${}^L\partial\Omega_{2\infty}^{(i)}$ and ${}^R\partial\Omega_{2\infty}^{(i)}$. The impacting time t_m labels the impact location on the boundary, and such boundaries are determined by the left and right sides of the $\bar{i}^{\text{th}}$ oscillator. For the motion continuity of such a gear transmission system, a transport law should be used. Herein, the transport law is the simple impact law. In Fig.5.4(b), the two domains for the stick motion are $\Omega_1^{(i)}$ and $\Omega_3^{(i)}$. The sub-domain for free-flight motion is still $\Omega_2^{(i)}$. The boundaries for the onset and vanishing of stick motion are sketched by the two dash-dot lines, and the switching time t_m marks the locations for the appearance and disappearance of stick motions by dashed lines. The hollow and solid circular symbols represent the onset and vanishing of stick motion, respectively. For these boundaries, under certain conditions, the motion can pass through the boundary from one domain to an adjacent domain (i.e., the oscillators can enter the regions of stick motion). No transport law is needed for motion continuation.

To investigate the motion mechanism of a two-oscillator system, the following vectors are introduced:

5.2 Domains and vector fields

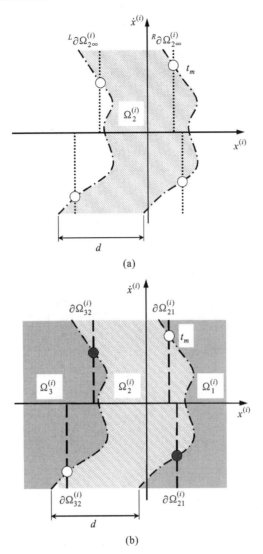

Fig. 5.4 Phase space partition: (a) impacting chatter; (b) stick motion

$$\left.\begin{array}{l}\mathbf{x}_\alpha^{(i)} = (x_\alpha^{(i)}, \dot{x}_\alpha^{(i)})^{\mathrm{T}} = (x_\alpha^{(i)}, y_\alpha^{(i)})^{\mathrm{T}}, \\ \mathbf{F}_\alpha^{(i)} = (\dot{x}_\alpha^{(i)}, F_\alpha^{(i)})^{\mathrm{T}} = (y_\alpha^{(i)}, F_\alpha^{(i)})^{\mathrm{T}}. \end{array}\right\} \quad (5.20)$$

Using Eq.(5.20), equations (5.2) and (5.7) give

$$\dot{\mathbf{x}}_\alpha^{(i)} = \mathbf{F}_\alpha^{(i)}(\mathbf{x}_\alpha^{(i)}, t), \text{ for } i = 1, 2 \text{ and } \alpha = 1, 2, 3 \quad (5.21)$$

where

$$F_\alpha^{(i)} = -2\zeta_\alpha^{(i)} \dot{x}_\alpha^{(i)} - (\omega_\alpha^{(i)})^2 x_\alpha^{(i)} + b_\alpha^{(i)} + Q_\alpha^{(i)} \cos \Omega t. \tag{5.22}$$

At boundaries $^L\partial\Omega_{2\infty}^{(i)}$ and $^R\partial\Omega_{2\infty}^{(i)}$ with $i = 1, 2$, the impact relation in Eq.(5.5) is used when $\dot{x}_\alpha^{(1)} \neq \dot{x}_\alpha^{(2)}$. The impacting chatters with and without stick are sketched in Fig.5.5(a) and (b), respectively. As in Luo (1995) (or Han et al, 1995), consider an impacting chatter with m-impacts on the left boundary and n-impacts on the right boundary as sketched in Fig.5.5(a). In addition to the impacting chatter motion, the impacting chatter with stick is also important and is sketched in Fig.5.5(b). The stick boundaries are represented by the two dashed lines. However, the boundaries labeled $^L\partial\Omega_{2\infty}^{(i)}$ and $^R\partial\Omega_{2\infty}^{(i)}$ for impacting chatter are non-passable, and the velocity jump is

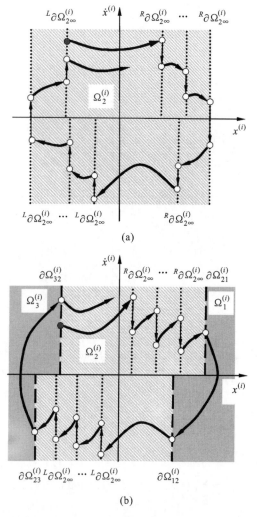

Fig. 5.5 Phase Plane: (a) impacting chatter; (b) impacting chatter with stick motion

5.2 Domains and vector fields

caused by impact. The trajectories with impacting chatter or impacting chatter with stick are depicted by the solid curves. At boundary $\partial\Omega^{(i)}_{\alpha\beta}$ ($\alpha, \beta = 1,2,3$, $\alpha \neq \beta$), the stick motion can begin when $\dot{x}^{(1)}_\alpha = \dot{x}^{(2)}_\alpha$.

5.2.2 Relative motion description

Although the description of motion based on the absolute coordinate systems is very intuitive, it is very difficult to develop analytical conditions because the boundaries vary with time. Hence, two relative variables are introduced herein, i.e.,

$$z^{(i)}_\alpha = x^{(i)}_\alpha - x^{(\bar{i})}_{\bar{\alpha}} \text{ and } v^{(i)}_\alpha = \dot{z}^{(i)}_\alpha = \dot{x}^{(i)}_\alpha - \dot{x}^{(\bar{i})}_{\bar{\alpha}}. \tag{5.23}$$

From domain definitions, $\alpha = 1,2,3$ corresponds to $\bar{\alpha} = 3,2,1$. From the foregoing equation, for $i = 1,2$ and $\alpha = 1,2,3$, the equations of motion are

$$\ddot{z}^{(i)}_\alpha + 2\zeta^{(i)}_\alpha \dot{z}^{(i)}_\alpha + (\omega^{(i)}_\alpha)^2 z^{(i)}_\alpha = b^{(i)}_\alpha + Q^{(i)}_\alpha \cos\Omega t \\ -\ddot{x}^{(\bar{i})}_{\bar{\alpha}} - 2\zeta^{(\bar{i})}_\alpha \dot{x}^{(\bar{i})}_{\bar{\alpha}} - (\omega^{(\bar{i})}_\alpha)^2 x^{(\bar{i})}_{\bar{\alpha}}, \tag{5.24}$$

$$\ddot{x}^{(\bar{i})}_{\bar{\alpha}} + 2\zeta^{(\bar{i})}_{\bar{\alpha}} \dot{x}^{(\bar{i})}_{\bar{\alpha}} + (\omega^{(\bar{i})}_{\bar{\alpha}})^2 x^{(\bar{i})}_{\bar{\alpha}} = b^{(\bar{i})}_{\bar{\alpha}} + Q^{(\bar{i})}_{\bar{\alpha}} \cos\Omega t. \tag{5.25}$$

In a similar fashion, two more vectors are introduced as follows:

$$\left. \begin{array}{l} \mathbf{z}^{(i)}_\alpha = (z^{(i)}_\alpha, \dot{z}^{(i)}_\alpha)^\mathrm{T} = (z^{(i)}_\alpha, v^{(i)}_\alpha)^\mathrm{T}, \\ \mathbf{g}^{(i)}_\alpha = (\dot{z}^{(i)}_\alpha, g^{(i)}_\alpha)^\mathrm{T} = (v^{(i)}_\alpha, g^{(i)}_\alpha)^\mathrm{T}. \end{array} \right\} \tag{5.26}$$

From Eqs.(5.24) and (5.25), for $i = 1,2$ and $\alpha = 1,2,3$ with $\bar{\alpha} = 3,2,1$, the equations of motion become

$$\left. \begin{array}{l} \dot{\mathbf{z}}^{(i)}_\alpha = \mathbf{g}^{(i)}_\alpha(\mathbf{z}^{(i)}_\alpha, \mathbf{x}^{(\bar{i})}_{\bar{\alpha}}, t), \\ \dot{\mathbf{x}}^{(\bar{i})}_{\bar{\alpha}} = \mathbf{F}^{(\bar{i})}_{\bar{\alpha}}(\mathbf{x}^{(\bar{i})}_{\bar{\alpha}}, t), \end{array} \right\} \tag{5.27}$$

where

$$g^{(i)}_\alpha(\mathbf{z}^{(i)}_\alpha, \mathbf{x}^{(\bar{i})}_{\bar{\alpha}}, t) = -2\zeta^{(i)}_\alpha \dot{z}^{(i)}_\alpha - (\omega^{(i)}_\alpha)^2 z^{(i)}_\alpha + b^{(i)}_\alpha + Q^{(i)}_\alpha \cos\Omega t \\ -\ddot{x}^{(\bar{i})}_{\bar{\alpha}} - 2\zeta^{(\bar{i})}_\alpha \dot{x}^{(\bar{i})}_{\bar{\alpha}} - (\omega^{(\bar{i})}_\alpha)^2 x^{(\bar{i})}_{\bar{\alpha}}. \tag{5.28}$$

Because the stick motion requires the relative motion to vanish between the two gears, the domains $\Omega^{(i)}_1$ and $\Omega^{(i)}_3$ become two points in relative phase space. In the relative frame, the sub-domains in Eqs.(5.14) and (5.18) can be expressed by

$$\left.\begin{aligned}\Omega_1^{(i)} &= \left\{(z^{(i)}, \dot{z}^{(i)}) \,\Big|\, z^{(i)} = \frac{d}{2}, \dot{z}^{(i)} = 0\right\}, \\ \Omega_2^{(i)} &= \left\{(z^{(i)}, \dot{z}^{(i)}) \,\Big|\, z^{(i)} \in (-\frac{d}{2}, \frac{d}{2})\right\}, \\ \Omega_3^{(i)} &= \left\{(z^{(i)}, \dot{z}^{(i)}) \,\Big|\, z^{(i)} = -\frac{d}{2}, \dot{z}^{(i)} = 0\right\}.\end{aligned}\right\} \quad (5.29)$$

In the relative frame, the impacting chatter boundaries in Eq.(5.15) become

$$\left.\begin{aligned}{}^R\partial\Omega_{2\infty}^{(i)} &= \left\{(z^{(i)}, \dot{z}^{(i)}) \,\Big|\, {}^R\varphi_{2\infty}^{(i)} \equiv z^{(i)} - \frac{d}{2} = 0\right\}, \\ {}^L\partial\Omega_{2\infty}^{(i)} &= \left\{(z^{(i)}, \dot{z}^{(i)}) \,\Big|\, {}^L\varphi_{2\infty}^{(i)} \equiv z^{(i)} + \frac{d}{2} = 0\right\}.\end{aligned}\right\} \quad (5.30)$$

Through their subsets, such boundary sets become

$$\left.\begin{aligned}{}^R\partial\Omega_{2\infty}^{(i)} &= {}^R_+\partial\Omega_{2\infty}^{(i)} \cup {}^R_-\partial\Omega_{2\infty}^{(i)}, \\ {}^L\partial\Omega_{2\infty}^{(i)} &= {}^L_+\partial\Omega_{2\infty}^{(i)} \cup {}^L_-\partial\Omega_{2\infty}^{(i)},\end{aligned}\right\} \quad (5.31)$$

where

$$\left.\begin{aligned}{}^R_+\partial\Omega_{2\infty}^{(i)} &= \left\{(\mathbf{z}^{(i)}, \dot{\mathbf{z}}^{(i)}) \,\Bigg|\, \begin{array}{l}{}^R\varphi_{2\infty}^{(i)} \equiv z^{(i)} - \frac{d}{2} = 0 \\ \dot{z}^{(i)} \in (0, \infty)\end{array}\right\}, \\ {}^R_-\partial\Omega_{2\infty}^{(i)} &= \left\{(z^{(i)}, \dot{z}^{(i)}) \,\Bigg|\, \begin{array}{l}{}^R\varphi_{2\infty}^{(i)} \equiv z^{(i)} - \frac{d}{2} = 0 \\ \dot{z}^{(i)} \in (-\infty, 0)\end{array}\right\}, \\ {}^L_+\partial\Omega_{2\infty}^{(i)} &= \left\{(z^{(i)}, \dot{z}^{(i)}) \,\Bigg|\, \begin{array}{l}{}^L\varphi_{2\infty}^{(i)} \equiv z^{(i)} + \frac{d}{2} = 0 \\ \dot{z}^{(i)} \in (0, \infty)\end{array}\right\}, \\ {}^L_-\partial\Omega_{2\infty}^{(i)} &= \left\{(z^{(i)}, \dot{z}^{(i)}) \,\Bigg|\, \begin{array}{l}{}^L\varphi_{2\infty}^{(i)} \equiv z^{(i)} + \frac{d}{2} = 0 \\ \dot{z}^{(i)} \in (-\infty, 0)\end{array}\right\}.\end{aligned}\right\} \quad (5.32)$$

The stick boundaries become two points, which are expressed by

5.3 Mechanism of stick and grazing

$$\begin{aligned}
\partial\Omega_{12}^{(i)} &= \left\{ (z^{(i)}, \dot{z}^{(i)}) \,\Big|\, z^{(i)} - \frac{d}{2} = 0, \varphi_{12}^{(i)} \equiv \dot{z}^{(i)} = 0_+ \right\}, \\
\partial\Omega_{21}^{(i)} &= \left\{ (z^{(i)}, \dot{z}^{(i)}) \,\Big|\, z^{(i)} - \frac{d}{2} = 0, \varphi_{21}^{(i)} \equiv \dot{z}^{(i)} = 0_- \right\}, \\
\partial\Omega_{32}^{(i)} &= \left\{ (z^{(i)}, \dot{z}^{(i)}) \,\Big|\, z^{(i)} + \frac{d}{2} = 0, \varphi_{23}^{(i)} \equiv \dot{z}^{(i)} = 0_+ \right\}, \\
\partial\Omega_{23}^{(i)} &= \left\{ (z^{(i)}, \dot{z}^{(i)}) \,\Big|\, z^{(i)} + \frac{d}{2} = 0, \varphi_{23}^{(i)} \equiv \dot{z}^{(i)} = 0_- \right\}.
\end{aligned} \qquad (5.33)$$

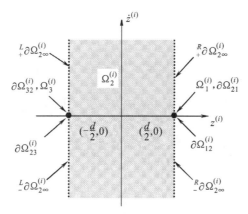

Fig. 5.6 Relative phase plane partition

The boundaries in the relative frame are independent of time. The phase partition in relative phase space is illustrated in Fig.5.6, where the stick boundaries and domains are presented by the two large dots. From the relative description of the two-oscillator systems, the analytical conditions for stick and impacting chatters are presented in the next section.

5.3 Mechanism of stick and grazing

In this section, the necessary and sufficient conditions for stick onsets, grazing, and stick motions for the two-oscillator system will be developed through equations of motion in the relative frame. To help one understand the analytical conditions, the corresponding physical explanations of such analytical conditions will be presented herein.

5.3.1 Analytical conditions

The motion and interaction for two gears at a discontinuous boundary can be described mathematically. Before presenting the analytical conditions for onset and vanishing of stick motion, the following functions from Chapter 2 (or Luo, 2007, 2008a,b,c) are introduced as

$$G_i^{(0,\alpha)}(\mathbf{z}_\alpha^{(i)}, \mathbf{x}_{\bar{\alpha}}^{(\bar{i})}, t_{m\pm})$$
$$= \mathbf{n}_{\partial\Omega_{\alpha\beta}^{(i)}}^{\mathrm{T}} \cdot [\mathbf{g}_\alpha^{(i)}(\mathbf{z}_\alpha^{(i)}, \mathbf{x}_{\bar{\alpha}}^{(\bar{i})}, t_{m\pm}) - \mathbf{g}_{\alpha\beta}^{(0)}(\mathbf{z}_{\alpha\beta}^{(i)}, t_m)], \tag{5.34}$$

$$G_i^{(1,\alpha)}(\mathbf{z}_\alpha^{(i)}, \mathbf{x}_{\bar{\alpha}}^{(\bar{i})}, t_{m\pm})$$
$$= 2D\mathbf{n}_{\partial\Omega_{\alpha\beta}^{(i)}}^{\mathrm{T}} \cdot [\mathbf{g}_\alpha^{(i)}(\mathbf{z}_\alpha^{(i)}, \mathbf{x}_{\bar{\alpha}}^{(\bar{i})}, t_{m\pm}) - \mathbf{g}_{\alpha\beta}^{(0)}(\mathbf{z}_{\alpha\beta}^{(i)}, t_m)]$$
$$+ \mathbf{n}_{\partial\Omega_{\alpha\beta}^{(i)}}^{\mathrm{T}} \cdot [D\mathbf{g}_\alpha^{(i)}(\mathbf{z}_\alpha^{(i)}, \mathbf{x}_{\bar{\alpha}}^{(\bar{i})}, t_{m\pm}) - D\mathbf{g}_{\alpha\beta}^{(0)}(\mathbf{z}_{\alpha\beta}^{(i)}, t_m)], \tag{5.35}$$

where $D = \dot{z}\partial()/\partial z + \dot{v}\partial()/\partial v + \partial()/\partial t$. $\mathbf{g}_{\alpha\beta}^{(0)}(\mathbf{z}_{\alpha\beta}^{(i)}, t_m) = (0,0)^{\mathrm{T}}$ for stick boundary. For impact chatter, $\mathbf{g}_{\alpha\beta}^{(0)}(\mathbf{z}_{\alpha\beta}^{(i)}, t_m) = (0, \pm\infty)^{\mathrm{T}}$. The function in Eq.(5.34) is the normal component of the difference between the vector fields on a domain and on a boundary. The time change-rate of this normal component of the vector field difference is given in Eq.(5.35). The vector field $\mathbf{g}_\alpha^{(i)}(\mathbf{z}_\alpha^{(i)}, \mathbf{x}_{\bar{\alpha}}^{(\bar{i})}, t_{m\pm})$ is for a flow of the ith oscillator in domain $\Omega_\alpha^{(i)}$, and the vector field $\mathbf{g}_{\alpha\beta}^{(0)}(\mathbf{z}_{\alpha\beta}^{(i)}, t_m)$ is for a flow on the boundary $\partial\Omega_{\alpha\beta}^{(i)}$. The normal vector $\mathbf{n}_{\partial\Omega_{\alpha\beta}^{(i)}}$ of the boundary $\partial\Omega_{\alpha\beta}^{(i)}$ is computed by

$$\mathbf{n}_{\partial\Omega_{\alpha\beta}^{(i)}} = \nabla\varphi_{\alpha\beta}^{(i)} = \left(\frac{\partial\varphi_{\alpha\beta}^{(i)}}{\partial z^{(i)}}, \frac{\partial\varphi_{\alpha\beta}^{(i)}}{\partial v^{(i)}}\right)^{\mathrm{T}}, \tag{5.36}$$

where $\nabla = (\partial/\partial z, \partial/\partial v)^{\mathrm{T}}$. Owing to $\mathbf{n}_{\partial\Omega_{\alpha\beta}^{(i)}}^{\mathrm{T}} \cdot \mathbf{g}_{\alpha\beta}^{(0)}(\mathbf{z}_{\alpha\beta}^{(i)}, t_m) = 0$, the total derivative gives

$$D\mathbf{n}_{\partial\Omega_{\alpha\beta}^{(i)}}^{\mathrm{T}} \cdot \mathbf{g}_{\alpha\beta}^{(0)}(\mathbf{z}_{\alpha\beta}^{(i)}, t_m) + \mathbf{n}_{\partial\Omega_{\alpha\beta}^{(i)}}^{\mathrm{T}} \cdot D\mathbf{g}_{\alpha\beta}^{(0)}(\mathbf{z}_{\alpha\beta}^{(i)}, t_m) = 0 \tag{5.37}$$

If the boundary $\partial\Omega_{\alpha\beta}^{(i)}$ is linear independent of time t, we obtain $D\mathbf{n}_{\partial\Omega_{\alpha\beta}^{(i)}}^{\mathrm{T}} = 0$. Therefore, Eq. (5.37) becomes

$$\mathbf{n}_{\partial\Omega_{\alpha\beta}^{(i)}}^{\mathrm{T}} \cdot D\mathbf{g}_{\alpha\beta}^{(0)}(\mathbf{z}_{\alpha\beta}^{(i)}, t_m) = 0. \tag{5.38}$$

5.3 Mechanism of stick and grazing

Equations (5.34) and (5.35) reduce to

$$\left.\begin{array}{l} G_i^{(0,\alpha)}(\mathbf{z}_\alpha^{(i)}, \mathbf{x}_{\bar{\alpha}}^{(\bar{i})}, t_{m\pm}) = \mathbf{n}_{\partial\Omega_{\alpha\beta}^{(i)}}^T \cdot \mathbf{g}_\alpha^{(i)}(\mathbf{z}_\alpha^{(i)}, \mathbf{x}_{\bar{\alpha}}^{(\bar{i})}, t_{m\pm}), \\ G_i^{(1,\alpha)}(\mathbf{z}_\alpha^{(i)}, \mathbf{x}_{\bar{\alpha}}^{(\bar{i})}, t_{m\pm}) = \mathbf{n}_{\partial\Omega_{\alpha\beta}^{(i)}}^T \cdot D\mathbf{g}_\alpha^{(i)}(\mathbf{z}_\alpha^{(i)}, \mathbf{x}_{\bar{\alpha}}^{(\bar{i})}, t_{m\pm}). \end{array}\right\} \quad (5.39)$$

For a general case, Eqs.(5.34) and (5.35) instead of Eq.(5.39) will be used. From Luo (2005, 2006), the grazing motion (i.e., the motion tangential to the boundary) is guaranteed by

$$\left.\begin{array}{l} G_i^{(0,\alpha)}(\mathbf{z}_\alpha^{(i)}, \mathbf{x}_{\bar{\alpha}}^{(\bar{i})}, t_{m\pm}) = \mathbf{n}_{\partial\Omega_{\alpha\beta}^{(i)}}^T \cdot \mathbf{g}_\alpha^{(i)}(\mathbf{z}_\alpha^{(i)}, \mathbf{x}_{\bar{\alpha}}^{(\bar{i})}, t_{m\pm}) = 0 \\ \qquad\qquad \text{for } \alpha, \beta = 1, 2 \text{ or } 2, 3 \\ (-1)^i G_i^{(1,\alpha)}(\mathbf{z}_\alpha^{(i)}, \mathbf{x}_{\bar{\alpha}}^{(\bar{i})}, t_{m\pm}) = (-1)^i \mathbf{n}_{\partial\Omega_{\alpha\beta}^{(i)}}^T \cdot D\mathbf{g}_\alpha^{(i)}(\mathbf{z}_\alpha^{(i)}, \mathbf{x}_{\bar{\alpha}}^{(\bar{i})}, t_{m\pm}) > 0 \\ \qquad\qquad \text{for } \mathbf{n}_{\partial\Omega_{\alpha\beta}^{(i)}} \to \Omega_\beta^{(i)} \text{ and} \\ (-1)^i G_i^{(1,\alpha)}(\mathbf{z}_\alpha^{(i)}, \mathbf{x}_{\bar{\alpha}}^{(\bar{i})}, t_{m\pm}) = (-1)^i \mathbf{n}_{\partial\Omega_{\alpha\beta}^{(i)}}^T \cdot D\mathbf{g}_\alpha^{(i)}(\mathbf{z}_\alpha^{(i)}, \mathbf{x}_{\bar{\alpha}}^{(\bar{i})}, t_{m\pm}) < 0 \\ \qquad\qquad \text{for } \mathbf{n}_{\partial\Omega_{\alpha\beta}^{(i)}} \to \Omega_\alpha^{(i)} \end{array}\right\} \quad (5.40)$$

where

$$D\mathbf{g}_\alpha^{(i)}(\mathbf{z}_\alpha^{(i)}, \mathbf{x}_{\bar{\alpha}}^{(\bar{i})}, t) = \left(g_\alpha^{(i)}(\mathbf{z}_\alpha^{(i)}, \mathbf{x}_{\bar{\alpha}}^{(\bar{i})}, t), \right.$$
$$\left. \nabla g_\alpha^{(i)}(\mathbf{z}_\alpha^{(i)}, \mathbf{x}_{\bar{\alpha}}^{(\bar{i})}, t) \cdot \mathbf{g}_\alpha^{(i)}(\mathbf{z}_\alpha^{(i)}, \mathbf{x}_{\bar{\alpha}}^{(\bar{i})}, t) + \frac{\partial g_\alpha^{(i)}(\mathbf{z}_\alpha^{(i)}, \mathbf{x}_{\bar{\alpha}}^{(\bar{i})}, t)}{\partial t} \right)^T. \quad (5.41)$$

For stick motions in domain $\Omega_\beta^{(i)}$ ($\beta = 1, 3$), the condition for a flow from domain $\Omega_2^{(i)}$ to such a stick domain $\Omega_\beta^{(i)}$ ($\beta = 1, 3$) is very important. From Luo(2005,2006), the passable motion to the boundary $\partial\Omega_{\alpha\beta}^{(i)}$ for an impacting chatter motion into the stick motion is guaranteed by

$$_iL_{\alpha\beta}^{(0,0)}(t_m) = G_i^{(0,\alpha)}(\mathbf{z}_\alpha^{(i)}, \mathbf{x}_{\bar{\alpha}}^{(\bar{i})}, t_{m-}) \times G_i^{(0,\beta)}(\mathbf{z}_\beta^{(i)}, \mathbf{x}_{\bar{\beta}}^{(\bar{i})}, t_{m+}) > 0. \quad (5.42)$$

In other words, the condition for stick motion can be expressed by

$$\left.\begin{array}{l}(-1)^{i}G_{i}^{(0,2)}(\mathbf{z}_{2}^{(i)},\mathbf{x}_{2}^{(\bar{i})},t_{m-})<0,\\ (-1)^{i}G_{i}^{(0,1)}(\mathbf{z}_{1}^{(i)},\mathbf{x}_{3}^{(\bar{i})},t_{m+})<0,\end{array}\right\} \text{on } \partial\Omega_{21}^{(i)},$$

$$\left.\begin{array}{l}(-1)^{i}G_{i}^{(0,2)}(\mathbf{z}_{2}^{(i)},\mathbf{x}_{2}^{(\bar{i})},t_{m-})>0,\\ (-1)^{i}G_{i}^{(0,3)}(\mathbf{z}_{3}^{(i)},\mathbf{x}_{1}^{(\bar{i})},t_{m+})>0,\end{array}\right\} \text{on } \partial\Omega_{23}^{(i)}.$$

(5.43)

Once the stick motion exists in domain $\Omega_{\beta}^{(i)}$ ($\beta = 1, 3$), the vanishing condition of stick motion requires

$$\left.\begin{array}{l}G_{i}^{(0,1)}(\mathbf{z}_{1}^{(i)},\mathbf{x}_{3}^{(\bar{i})},t_{m-})=0,\\ (-1)^{i}G_{i}^{(1,1)}(\mathbf{z}_{1}^{(i)},\mathbf{x}_{3}^{(\bar{i})},t_{m-})>0,\\ (-1)^{i}G_{i}^{(1,2)}(\mathbf{z}_{2}^{(i)},\mathbf{x}_{2}^{(\bar{i})},t_{m+})>0,\end{array}\right\} \text{on } \partial\Omega_{12}^{(i)},$$

$$\left.\begin{array}{l}G_{i}^{(0,3)}(\mathbf{z}_{3}^{(i)},\mathbf{x}_{1}^{(\bar{i})},t_{m-})=0,\\ (-1)^{i}G_{i}^{(1,3)}(\mathbf{z}_{3}^{(i)},\mathbf{x}_{1}^{(\bar{i})},t_{m-})<0,\\ (-1)^{i}G_{i}^{(1,2)}(\mathbf{z}_{2}^{(i)},\mathbf{x}_{2}^{(\bar{i})},t_{m+})<0,\end{array}\right\} \text{on } \partial\Omega_{32}^{(i)}.$$

(5.44)

From Eqs.(5.30) and (5.32), the normal vector of the impacting chatter boundaries $^{L}\partial\Omega_{2\infty}^{(i)}$ and $^{R}\partial\Omega_{2\infty}^{(i)}$ is expressed by

$$\mathbf{n}_{^{L}\partial\Omega_{2\infty}^{(i)}} = \mathbf{n}_{^{R}\partial\Omega_{2\infty}^{(i)}} = (1,0)^{\mathrm{T}}. \tag{5.45}$$

Therefore, equations (5.34) and (5.35) give

$$\left.\begin{array}{l}G_{i}^{(0,2)}(\mathbf{z}_{2}^{(i)},\mathbf{x}_{2}^{(\bar{i})},t)=\mathbf{n}_{\partial\Omega_{2\infty}^{(i)}}^{\mathrm{T}}\cdot\mathbf{g}_{2}^{(i)}(\mathbf{z}_{2}^{(i)},\mathbf{x}_{2}^{(\bar{i})},t)=v_{2}^{(i)},\\ G_{i}^{(1,2)}(\mathbf{z}_{2}^{(i)},\mathbf{x}_{2}^{(\bar{i})},t)=\mathbf{n}_{\partial\Omega_{2\infty}^{(i)}}^{\mathrm{T}}\cdot D\mathbf{g}_{2}^{(i)}(\mathbf{z}_{2}^{(i)},\mathbf{x}_{2}^{(\bar{i})},t)=g_{2}^{(i)}(\mathbf{z}_{2}^{(i)},\mathbf{x}_{2}^{(\bar{i})},t).\end{array}\right\}$$

(5.46)

From Eq.(5.40), the analytical conditions for grazing motions on the impacting chatter boundary are

$$\left.\begin{array}{l}v_{2}^{(i)}=0, (-1)^{i}g_{2}^{(i)}(\mathbf{z}_{2}^{(i)},\mathbf{x}_{2}^{(\bar{i})},t_{m\pm})>0 \text{ on } ^{R}\partial\Omega_{2\infty}^{(i)},\\ v_{2}^{(i)}=0, (-1)^{i}g_{2}^{(i)}(\mathbf{z}_{2}^{(i)},\mathbf{x}_{2}^{(\bar{i})},t_{m\pm})<0 \text{ on } ^{L}\partial\Omega_{2\infty}^{(i)}.\end{array}\right\}$$

(5.47)

For instance if $i = 1$, the conditions in Eq.(5.47) give the relative velocity $v_{2}^{(1)}(t_{m}) = 0$ and the relative acceleration $g_{2}^{(1)}(\mathbf{z}_{2}^{(1)},\mathbf{x}_{2}^{(2)},t_{m\pm}) < 0$ at the right side boundary $^{R}\partial\Omega_{2\infty}^{(1)}$ (i.e., $z_{2}^{(1)} = d/2$). Because $g_{2}^{(1)} < 0$ for $t > t_{m}$, the relative velocity $v_{2}^{(1)}(t) < 0$ and the

5.3 Mechanism of stick and grazing

relative displacement $z_2^{(1)}(t) < d/2$, which means that the motion remains in $\Omega_2^{(1)}$. Such a phenomenon is called the grazing of the motion to the boundary ${}^R\partial\Omega_{2\infty}^{(1)}$. However, for the stick boundaries $\partial\Omega_{\alpha\beta}^{(i)}$, the normal vector is

$$\mathbf{n}_{\partial\Omega_{23}^{(i)}} = \mathbf{n}_{\partial\Omega_{12}^{(i)}} = (0,1)^{\mathrm{T}}. \tag{5.48}$$

The corresponding G-functions are

$$G_i^{(0,\alpha)}(\mathbf{z}_\alpha^{(i)}, \mathbf{x}_{\bar\alpha}^{(\bar i)}, t) = \mathbf{n}^{\mathrm{T}}_{\partial\Omega_{\alpha\beta}^{(i)}} \cdot \mathbf{g}_\alpha^{(i)}(\mathbf{z}_\alpha^{(i)}, \mathbf{x}_{\bar\alpha}^{(\bar i)}, t) = g_\alpha^{(i)}(\mathbf{z}_\alpha^{(i)}, \mathbf{x}_{\bar\alpha}^{(\bar i)}, t),$$

$$G_i^{(1,\alpha)}(\mathbf{z}_\alpha^{(i)}, \mathbf{x}_{\bar\alpha}^{(\bar i)}, t) = \mathbf{n}^{\mathrm{T}}_{\partial\Omega_{\alpha\beta}^{(i)}} \cdot D\mathbf{g}_\alpha^{(i)}(\mathbf{z}_\alpha^{(i)}, \mathbf{x}_{\bar\alpha}^{(\bar i)}, t)$$

$$= \nabla g_\alpha^{(i)}(\mathbf{z}_\alpha^{(i)}, \mathbf{x}_{\bar\alpha}^{(\bar i)}, t) \cdot \mathbf{g}_\alpha^{(i)}(\mathbf{z}_\alpha^{(i)}, \mathbf{x}_{\bar\alpha}^{(\bar i)}, t) + \frac{\partial g_\alpha^{(i)}(\mathbf{z}_\alpha^{(i)}, \mathbf{x}_{\bar\alpha}^{(\bar i)}, t)}{\partial t}. \tag{5.49}$$

From Eq.(5.27), $g_\alpha^{(i)}(\mathbf{z}_\alpha^{(i)}, t)$ is equivalent to $g_\alpha^{(i)}(\mathbf{z}_\alpha^{(i)}, \mathbf{x}_{\bar\alpha}^{(\bar i)}, t)$ because $\mathbf{x}_\alpha^{(\bar i)}$ is a function of time and solved by the second equation in Eq.(5.27), so one obtains

$$G_i^{(1,\alpha)}(\mathbf{z}_\alpha^{(i)}, \mathbf{x}_{\bar\alpha}^{(\bar i)}, t) = -2\zeta_\alpha^{(i)}\ddot{z}_\alpha^{(i)} - (\omega_\alpha^{(i)})^2 \dot{z}_\alpha^{(i)} - Q_\alpha^{(i)}\Omega\sin\Omega t$$
$$- \dddot{x}_{\bar\alpha}^{(\bar i)} - 2\zeta_\alpha^{(i)}\ddot{x}_{\bar\alpha}^{(\bar i)} - (\omega_\alpha^{(i)})^2 \dot{x}_{\bar\alpha}^{(\bar i)}. \tag{5.50}$$

With Eqs.(5.26) and (5.27), the relative jerk is given by

$$J_\alpha^{(i)}(t) = -2\zeta_\alpha^{(i)}\ddot{z}_\alpha^{(i)} - (\omega_\alpha^{(i)})^2 \dot{z}_\alpha^{(i)} - Q_\alpha^{(i)}\Omega\sin\Omega t$$
$$- \dddot{x}_{\bar\alpha}^{(\bar i)} - 2\zeta_\alpha^{(i)}\ddot{x}_{\bar\alpha}^{(\bar i)} - (\omega_\alpha^{(i)})^2 \dot{x}_{\bar\alpha}^{(\bar i)}. \tag{5.51}$$

Therefore, for this case, the function $G_i^{(\alpha,1)}(\mathbf{z}_\alpha^{(i)}, \mathbf{x}_{\bar\alpha}^{(\bar i)}, t)$ is a relative jerk in domain $\Omega_\alpha^{(i)}$. The function $g_\alpha^{(i)}(\mathbf{z}_\alpha^{(i)}, \mathbf{x}_{\bar\alpha}^{(\bar i)}, t)$ is a relative acceleration or a relative force per unit mass.

With Eq.(5.49), the forming conditions for the stick motion in domain $\Omega_\alpha^{(i)}$ ($\alpha = 1, 3$) are

$$\left.\begin{array}{l}(-1)^i g_2^{(i)}(\mathbf{z}_2^{(i)}, \mathbf{x}_2^{(\bar i)}, t_{m-}) < 0, \\ (-1)^i g_1^{(i)}(\mathbf{z}_1^{(i)}, \mathbf{x}_3^{(\bar i)}, t_{m+}) < 0, \end{array}\right\} \text{ on } \partial\Omega_{21}^{(i)},$$
$$\left.\begin{array}{l}(-1)^i g_2^{(i)}(\mathbf{z}_2^{(i)}, \mathbf{x}_2^{(\bar i)}, t_{m-}) > 0, \\ (-1)^i g_3^{(i)}(\mathbf{z}_3^{(i)}, \mathbf{x}_1^{(\bar i)}, t_{m+}) > 0, \end{array}\right\} \text{ on } \partial\Omega_{23}^{(i)}. \tag{5.52}$$

The foregoing equation means that, for stick to occur for $i = 1$, the relative force per unit mass (or relative acceleration) in $\Omega_2^{(1)}$ and $\Omega_1^{(1)}$ must be positive on the boundary $\partial\Omega_{21}^{(1)}$, and the relative acceleration in $\Omega_2^{(1)}$ and $\Omega_3^{(1)}$ must be negative on the boundary $\partial\Omega_{23}^{(1)}$. The requirement to keep the stick motion in domain $\Omega_\alpha^{(i)}$ ($\alpha = 1,3$) is given by

$$\begin{aligned} (-1)^i g_1^{(i)}(\mathbf{z}_1^{(i)}, \mathbf{x}_3^{(\bar{i})}, t) < 0 & \quad \text{in } \Omega_1^{(i)}, \\ (-1)^i g_3^{(i)}(\mathbf{z}_3^{(i)}, \mathbf{x}_1^{(\bar{i})}, t) > 0 & \quad \text{in } \Omega_3^{(i)}. \end{aligned} \quad (5.53)$$

For example, to maintain stick motion if $i = 1$, the relative acceleration must be positive in $\Omega_1^{(1)}$ and negative in $\Omega_3^{(1)}$. In domain $\Omega_\alpha^{(i)}$ ($\alpha = 1,3$), the vanishing of the stick motion requires

$$\left.\begin{aligned} g_1^{(i)}(\mathbf{z}_1^{(i)}, \mathbf{x}_3^{(\bar{i})}, t_{m-}) &= 0, \\ (-1)^i G_i^{(1,1)}(\mathbf{z}_1^{(i)}, \mathbf{x}_3^{(\bar{i})}, t_{m+}) &> 0, \\ (-1)^i G_i^{(1,2)}(\mathbf{z}_2^{(i)}, \mathbf{x}_2^{(\bar{i})}, t_{m+}) &> 0, \end{aligned}\right\} \text{ on } \partial\Omega_{12}^{(i)},$$
$$\left.\begin{aligned} g_3^{(i)}(\mathbf{z}_3^{(i)}, \mathbf{x}_1^{(\bar{i})}, t_{m-}) &= 0, \\ (-1)^i G_i^{(1,3)}(\mathbf{z}_3^{(i)}, \mathbf{x}_1^{(\bar{i})}, t_{m+}) &< 0, \\ (-1)^i G_i^{(1,2)}(\mathbf{z}_2^{(i)}, \mathbf{x}_2^{(\bar{i})}, t_{m+}) &< 0, \end{aligned}\right\} \text{ on } \partial\Omega_{32}^{(i)}. \quad (5.54)$$

From the foregoing equation for $i = 1$, the relative acceleration must be zero and the relative jerks must be negative for stick vanishing on $\partial\Omega_{12}^{(1)}$. On $\partial\Omega_{32}^{(1)}$ the relative jerks must be positive for stick vanishing. The grazing of the stick motion requires

$$\left.\begin{aligned} g_1^{(i)}(\mathbf{z}_1^{(i)}, \mathbf{x}_3^{(\bar{i})}, t_{m\pm}) &= 0, \\ (-1)^i G_i^{(1,1)}(\mathbf{z}_1^{(i)}, \mathbf{x}_3^{(\bar{i})}, t_{m\pm}) &< 0, \\ (-1)^i G_i^{(1,2)}(\mathbf{z}_2^{(i)}, \mathbf{x}_2^{(\bar{i})}, t_{m+}) &< 0, \end{aligned}\right\} \text{ on } \partial\Omega_{12}^{(i)},$$
$$\left.\begin{aligned} g_3^{(i)}(\mathbf{z}_3^{(i)}, \mathbf{x}_1^{(\bar{i})}, t_{m\pm}) &= 0, \\ (-1)^i G_i^{(1,3)}(\mathbf{z}_3^{(i)}, \mathbf{x}_1^{(\bar{i})}, t_{m\pm}) &> 0, \\ (-1)^i G_i^{(1,2)}(\mathbf{z}_2^{(i)}, \mathbf{x}_2^{(\bar{i})}, t_{m+}) &> 0, \end{aligned}\right\} \text{ on } \partial\Omega_{32}^{(i)}. \quad (5.55)$$

For $i = 1$, the relative jerks on $\partial\Omega_{12}^{(1)}$ are positive, and on $\partial\Omega_{32}^{(1)}$ they are negative. This is opposite to the jerk conditions in Eq.(5.54), and such a condition keeps the stick motion continue. In domain $\Omega_\alpha^{(i)}$ with $\alpha = 1,3$ and $i = 1,2$, the following relations hold

$$\dot{z}_\alpha^{(i)} = (-1)^{i+1}\frac{d}{2}\,\text{sgn}\,(2-\alpha), \ddot{z}_\alpha^{(i)} = 0 \text{ and } \dddot{z}_\alpha^{(i)} = 0. \quad (5.56)$$

In other words, Eq. (5.56) is equivalent to

5.3 Mechanism of stick and grazing

$$\left.\begin{aligned} x_\alpha^{(i)} &= x_{\bar{\alpha}}^{(\bar{i})} - (-1)^i \frac{d}{2} \operatorname{sgn}(2-\alpha), \\ \dot{x}_\alpha^{(i)} &= \dot{x}_{\bar{\alpha}}^{(\bar{i})} \text{ and } \ddot{x}_\alpha^{(i)} = \ddot{x}_{\bar{\alpha}}^{(\bar{i})}. \end{aligned}\right\} \quad (5.57)$$

From the foregoing two equations, the relative force per unit mass is

$$g_\alpha^{(i)}(\mathbf{x}_\alpha^{(i)}, t) = b_\alpha^{(i)} + Q_\alpha^{(i)} \cos \Omega t - \ddot{x}_\alpha^{(i)} - 2\zeta_\alpha^{(i)} \dot{x}_\alpha^{(i)} - (\omega_\alpha^{(i)})^2 x_\alpha^{(i)}. \quad (5.58)$$

With Eq.(5.3), the relative force per unit mass with the internal force has the following relation:

$$g_\alpha^{(i)}(\mathbf{x}_\alpha^{(i)}, t) = f_\alpha^{(i)}/m_i \quad \text{for } \alpha = 1, 3. \quad (5.59)$$

So the function of $g_\alpha^{(i)}(\mathbf{x}_\alpha^{(i)}, t)$ $(\alpha = 1, 3)$ is the internal force per unit mass on the ith gear mass. The internal force $f_\alpha^{(i)} = 0$ implies that the two gears make contact but without interacting.

5.3.2 Physical interpretation

At the stick onset boundary, a motion from a free-flight state to stick possesses the following behaviors:

$$\left.\begin{aligned} x_\alpha^{(i)} &= x_\alpha^{(\bar{i})} - (-1)^i \frac{d}{2} \operatorname{sgn}(2-\alpha), \\ \dot{x}_\alpha^{(i)} &= \dot{x}_\alpha^{(\bar{i})} \text{ and } \ddot{x}_\alpha^{(i)} \ne \ddot{x}_\alpha^{(\bar{i})}. \end{aligned}\right\} \quad (5.60)$$

Before the two gears stick together, $g_\alpha^{(i)}(\mathbf{x}_\alpha^{(i)}, t)$ is the relative acceleration or relative force per unit mass. Just before the sticking of two gears occurs at the stick boundary, the relative velocity and acceleration are zero (i.e., $\dot{z}_\alpha^{(i)} = 0$ and $\ddot{z}_\alpha^{(i)} = 0$). However, the relative force per unit mass is $g_2^{(i)} \ne 0$, which cannot be thought as an internal force. Just after the stick begins, the relative force per unit mass is $g_\alpha^{(i)} \ne 0$ ($\alpha = 1, 3$) and $g_2^{(i)} \ne g_\alpha^{(i)}$, and the relative forces possess the property of $g_2^{(i)} \cdot g_\alpha^{(i)} > 0$ on the boundaries as in Eq.(5.52). Once the two gears stick together, the two accelerations should be the same (i.e., $\ddot{x}_\alpha^{(i)} = \ddot{x}_\alpha^{(\bar{i})}$). This means that the relative acceleration is zero. However, the relative force remains nonzero ($g_\alpha^{(i)} \ne 0$) for $\alpha = 1, 3$.

When the stick motion vanishes, the internal force and relative acceleration disappear (i.e., $g_\alpha^{(i)}(\mathbf{x}_\alpha^{(i)}, t) = 0$), which implies $\ddot{z}_\alpha^{(i)} = 0$ at the stick vanishing boundary. Also, the jerk of the ith gear should satisfy the condition in Eq.(5.54). With the initial conditions on $\partial\Omega_{12}^{(i)}$ (i.e., $(-1)^i z_2^{(i)} = -d/2$ and $\dot{z}_2^{(i)} = 0$) and $\ddot{z}_2^{(i)} = 0$, the jerk $(-1)^i J_\alpha^{(i)}(t) > 0$ for $t > t_{m+}$ leads to $(-1)^i \dddot{z}_2^{(i)} > 0$. From the acceleration, the relative velocity becomes $(-1)^i \dot{z}_2^{(i)} > 0$. Further, the relative displacement $(-1)^i z_2^{(i)} > -d/2$, which indicates that the ith and the $\bar{i}^{\text{th}}$ gear are in a state of free-flight motion in domain $\Omega_2^{(i)}$. Consider the initial conditions on the boundary $\partial \Omega_{23}^{(i)}$ (i.e., $(-1)^i z_2^{(i)} = d/2$ and $\dot{z}_2^{(i)} = 0$). The jerk $(-1)^i J_\alpha^{(i)}(t) < 0$ for $t > t_{m+}$ leads to $(-1)^i \ddot{z}_2^{(i)} < 0$, and the relative velocity is $(-1)^i \dot{z}_2^{(i)} < 0$. Furthermore, the relative displacement satisfies $(-1)^i z_2^{(i)} < d/2$, which indicates that the two gears are in the state of free-flight motion in domain $\Omega_2^{(i)}$.

The grazing condition of stick motion in Eq.(5.55) is also discussed herein. Because of $g_1^{(i)}(\mathbf{z}_1^{(i)}, t_{m\pm}) = 0$ on the stick boundary $\partial \Omega_{12}^{(i)}$, if $(-1)^i G_i^{(1,1)}(\mathbf{z}_1^{(i)}, t) < 0$ for $t > t_{m\pm}$, the internal force $(-1)^i g_1^{(i)}(\mathbf{z}_1^{(i)}, t) < 0$ can be obtained, which keeps the stick motion in domain $\Omega_1^{(i)}$. On the stick boundary $\partial \Omega_{23}^{(i)}$, the relative acceleration $g_3^{(i)}(\mathbf{z}_3^{(i)}, t_{m\pm}) = 0$ is obtained. Because of $(-1)^i G_i^{(3,1)}(\mathbf{z}_3^{(i)}, t) > 0$, the corresponding internal force is $(-1)^i g_3^{(i)}(\mathbf{z}_3^{(i)}, t) > 0$ for $t > t_{m\pm}$. Therefore, the stick motion still exists in domain $\Omega_3^{(i)}$.

5.4 Mapping structures and motions

To describe periodic and chaotic impacting chatter with and without stick in the two-oscillator system, the mapping structure will be developed from the separation boundaries. Before the mapping structure for a prescribed impacting chatter motion is developed, the switching planes will be defined first. From the switching planes, the basic mappings will be defined for a gear transmission system. Basic mappings will also be defined in the relative frame. A bifurcation scenario will be presented to illustrate complicated motions existing in the two-oscillator system. The mapping structure is a kind of symbolic dynamics description of chaotic motions.

5.4.1 Switching sets and basic mappings

From the discontinuous boundary, the switching planes based on the two impacting chatter boundaries are defined as

$$\left. \begin{aligned} {}^R\Sigma_{2\infty}^{(i)} &= \left\{ (t_k, x_k^{(i)}, \dot{x}_k^{(i)}, \dot{x}_k^{(\bar{i})}) \Big| x_k^{(i)} = x_k^{(\bar{i})} - \frac{d}{2}, \dot{x}_k^{(i)} \neq \dot{x}_k^{(\bar{i})} \right\}, \\ {}^L\Sigma_{2\infty}^{(i)} &= \left\{ (t_k, x_k^{(i)}, \dot{x}_k^{(i)}, \dot{x}_k^{(\bar{i})}) \Big| x_k^{(i)} = x_k^{(\bar{i})} + \frac{d}{2}, \dot{x}_k^{(i)} \neq \dot{x}_k^{(\bar{i})} \right\}. \end{aligned} \right\} \quad (5.61)$$

From now on, $x_k^{(i)} \equiv x^{(i)}(t_k)$ and $\dot{x}_k^{(i)} \equiv \dot{x}^{(i)}(t_k)$ on the separation boundary at time t_k are switching displacement and velocity in the absolute frame. The switching phase is defined by $\varphi_k = \mod(\Omega t_k, 2\pi)$. In the relative frame, the switching planes are expressed as

5.4 Mapping structures and motions

$$\left.\begin{aligned}{}^R\Sigma_{2\infty}^{(i)} &= \left\{(t_k,\dot{z}_k^{(i)},x_k^{(\bar{i})},\dot{x}_k^{(\bar{i})})\Big|z_k^{(i)}=\frac{d}{2},\dot{z}_k^{(i)}\neq 0\right\}, \\ {}^L\Sigma_{2\infty}^{(i)} &= \left\{(t_k,\dot{z}_k^{(i)},x_k^{(\bar{i})},\dot{x}_k^{(\bar{i})})\Big|z_k^{(i)}=-\frac{d}{2},\dot{z}_k^{(i)}\neq 0\right\}.\end{aligned}\right\} \quad (5.62)$$

The two switching sets are then decomposed as

$$ {}^R\Sigma_{2\infty}^{(i)} = {}^R_+\Sigma_{2\infty}^{(i)} \cup {}^R_-\Sigma_{2\infty}^{(i)} \text{ and } {}^L\Sigma_{2\infty}^{(i)} = {}^L_+\Sigma_{2\infty}^{(i)} \cup {}^L_-\Sigma_{2\infty}^{(i)}, \quad (5.63)$$

where

$$\left.\begin{aligned}{}^R_+\Sigma_{2\infty}^{(i)} &= \left\{(t_k,\dot{z}_k^{(i)},x_k^{(\bar{i})},\dot{x}_k^{(\bar{i})})\Big|z_k^{(i)}=\frac{d}{2},\dot{z}_k^{(i)}>0\right\}, \\ {}^R_-\Sigma_{2\infty}^{(i)} &= \left\{(t_k,\dot{z}_k^{(i)},x_k^{(\bar{i})},\dot{x}_k^{(\bar{i})})\Big|z_k^{(i)}=\frac{d}{2},\dot{z}_k^{(i)}<0\right\}, \\ {}^L_+\Sigma_{2\infty}^{(i)} &= \left\{(t_k,\dot{z}_k^{(i)},x_k^{(\bar{i})},\dot{x}_k^{(\bar{i})})\Big|z_k^{(i)}=-\frac{d}{2},\dot{z}_k^{(i)}>0\right\}, \\ {}^L_-\Sigma_{2\infty}^{(i)} &= \left\{(t_k,\dot{z}_k^{(i)},x_k^{(\bar{i})},\dot{x}_k^{(\bar{i})})\Big|z_k^{(i)}=-\frac{d}{2},\dot{z}_k^{(i)}<0\right\}.\end{aligned}\right\} \quad (5.64)$$

Based on the above definitions of switching planes, four mappings are defined in the absolute frame as

$$\left.\begin{aligned} P_2: {}^R\Sigma_{2\infty}^{(i)} &\to {}^R\Sigma_{2\infty}^{(i)}, \; P_3: {}^R\Sigma_{2\infty}^{(i)} \to {}^L\Sigma_{2\infty}^{(i)}, \\ P_5: {}^L\Sigma_{2\infty}^{(i)} &\to {}^L\Sigma_{2\infty}^{(i)}, \; P_6: {}^L\Sigma_{2\infty}^{(i)} \to {}^R\Sigma_{2\infty}^{(i)}. \end{aligned}\right\} \quad (5.65)$$

And in the relative frame, the mappings are also defined as

$$\left.\begin{aligned} P_2: {}^R_-\Sigma_{2\infty}^{(i)} &\to {}^R_+\Sigma_{2\infty}^{(i)}, \; P_3: {}^R_-\Sigma_{2\infty}^{(i)} \to {}^L_-\Sigma_{2\infty}^{(i)}, \\ P_5: {}^L_+\Sigma_{2\infty}^{(i)} &\to {}^L_-\Sigma_{2\infty}^{(i)}, \; P_6: {}^L_+\Sigma_{2\infty}^{(i)} \to {}^R_+\Sigma_{2\infty}^{(i)}. \end{aligned}\right\} \quad (5.66)$$

Among four basic mappings, the two mappings (P_2 and P_5) are local and the other two mappings (P_3 and P_6) are global. The local mappings will map the motion from a switching plane onto itself. On the other hand, the global mappings will map the motion from one switching plane to a different switching plane. Such mappings are sketched in Fig.5.7, where the corresponding switching planes are labeled by the dotted lines accordingly. The mappings for the absolute and relative frames are arranged in Fig.5.7(a) and (b), respectively. On the impacting chatter boundaries, impacts are expressed by thin straight lines with arrows. In the absolute frame, the switching plane changes with time. In other words, the impacting chatters do not occur at the same location. However, for the relative frame, the impact chatter takes place at the same location. In addition to mappings for impacting chatter, mappings for stick motions must be discussed. For stick motions in the two-oscillator system, the switching planes for stick are defined as

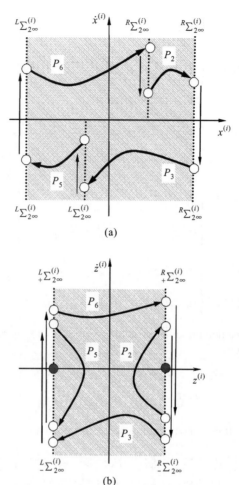

Fig. 5.7 Basic impacting chatter mappings: (a) absolute frame; (b) relative frame

$$\left.\begin{aligned}
\Sigma_{12}^{(i)} &= \left\{ (t_k, x_k^{(i)}, \dot{x}_k^{(i)}, \dot{x}_k^{(\bar{i})}) \,\Big|\, x_k^{(i)} = {}^R x_{2+}^{(\bar{i})} + \frac{d}{2},\ \dot{x}_k^{(i)} = {}^R \dot{x}_{2+}^{(\bar{i})} \right\}, \\
\Sigma_{21}^{(i)} &= \left\{ (t_k, x_k^{(i)}, \dot{x}_k^{(i)}, \dot{x}_k^{(\bar{i})}) \,\Big|\, x_k^{(i)} = {}^R x_{2-}^{(\bar{i})} + \frac{d}{2},\ \dot{x}_k^{(i)} = {}^R \dot{x}_{2-}^{(\bar{i})} \right\}, \\
\Sigma_{32}^{(i)} &= \left\{ (t_k, x_k^{(i)}, \dot{x}_k^{(i)}, \dot{x}_k^{(\bar{i})}) \,\Big|\, x_k^{(i)} = {}^L x_{2+}^{(\bar{i})} - \frac{d}{2},\ \dot{x}_k^{(i)} = {}^L \dot{x}_{2+}^{(\bar{i})} \right\}, \\
\Sigma_{23}^{(i)} &= \left\{ (t_k, x_k^{(i)}, \dot{x}_k^{(i)}, \dot{x}_k^{(\bar{i})}) \,\Big|\, x_k^{(i)} = {}^L x_{2-}^{(\bar{i})} - \frac{d}{2},\ \dot{x}_k^{(i)} = {}^L \dot{x}_{2-}^{(\bar{i})} \right\},
\end{aligned}\right\} \quad (5.67)$$

for the absolute frame and

5.4 Mapping structures and motions

$$\left.\begin{aligned}\Sigma_{12}^{(i)} &= \left\{(t_k,\dot{z}_k^{(i)},x_k^{(\bar{i})},\dot{x}_k^{(\bar{i})}) \,\Big|\, z_k^{(i)} = \frac{d}{2}, \dot{z}_k^{(i)} = 0_+ \right\}, \\ \Sigma_{21}^{(i)} &= \left\{(t_k,\dot{z}_k^{(i)},x_k^{(\bar{i})},\dot{x}_k^{(\bar{i})}) \,\Big|\, z_k^{(i)} = \frac{d}{2}, \dot{z}_k^{(i)} = 0_- \right\}, \\ \Sigma_{32}^{(i)} &= \left\{(t_k,\dot{z}_k^{(i)},x_k^{(\bar{i})},\dot{x}_k^{(\bar{i})}) \,\Big|\, z_k^{(i)} = -\frac{d}{2}, \dot{z}_k^{(i)} = 0_+ \right\}, \\ \Sigma_{23}^{(i)} &= \left\{(t_k,\dot{z}_k^{(i)},x_k^{(\bar{i})},\dot{x}_k^{(\bar{i})}) \,\Big|\, z_k^{(i)} = -\frac{d}{2}, \dot{z}_k^{(i)} = 0_- \right\},\end{aligned}\right\} \quad (5.68)$$

for the relative frame. The switching planes in Eqs.(5.64) and (5.68) for the relative frame are almost the same. In Eq.(5.64), the relative velocity is nonzero (i.e., $\dot{z}_k^{(i)} \neq 0$) but in Eq.(5.68), the relative velocity is zero (i.e., $\dot{z}_k^{(i)} = 0$). From the stick switching planes, the mappings are defined as

$$\left.\begin{aligned} P_1 &: \Sigma_{21}^{(i)} \to \Sigma_{12}^{(i)},\; P_2 : \Sigma_{12}^{(i)} \to \Sigma_{21}^{(i)},\; P_3 : \Sigma_{12}^{(i)} \to \Sigma_{23}^{(i)}, \\ P_4 &: \Sigma_{23}^{(i)} \to \Sigma_{32}^{(i)},\; P_5 : \Sigma_{23}^{(i)} \to \Sigma_{32}^{(i)},\; P_6 : \Sigma_{32}^{(i)} \to \Sigma_{21}^{(i)}. \end{aligned}\right\} \quad (5.69)$$

The corresponding switching planes for the mappings relative to impacting chatter with and without stick can be treated as the same. Accordingly, the mappings in Eq.(5.69) are the same as in Eqs.(5.65) and (5.66) except for the addition of P_1 and P_4. The mappings related to the stick switching planes are sketched in Fig.5.8. In Fig.5.8(a), the two stick mappings P_1 and P_4 are new, and the other four mappings are the same as in Fig.5.7. Similarly, mappings based on the relative switching planes can be defined, and within the relative frame, the switching planes are points as described in Fig.5.8(b).

With mixed switching planes for chatter with and without stick, four mappings are defined as

$$\left.\begin{aligned} P_2 &: \Sigma_{12}^{(i)} \to {}^R\Sigma_{2\infty}^{(i)} \text{ and } P_2 : {}^R\Sigma_{2\infty}^{(i)} \to \Sigma_{21}^{(i)}, \\ P_3 &: \Sigma_{12}^{(i)} \to {}^L\Sigma_{2\infty}^{(i)} \text{ and } P_3 : {}^R\Sigma_{2\infty}^{(i)} \to \Sigma_{23}^{(i)}, \\ P_5 &: \Sigma_{23}^{(i)} \to {}^L\Sigma_{2\infty}^{(i)} \text{ and } P_5 : {}^L\Sigma_{2\infty}^{(i)} \to \Sigma_{32}^{(i)}, \\ P_6 &: \Sigma_{32}^{(i)} \to {}^R\Sigma_{2\infty}^{(i)} \text{ and } P_6 : {}^L\Sigma_{2\infty}^{(i)} \to \Sigma_{21}^{(i)}. \end{aligned}\right\} \quad (5.70)$$

The stick mapping is difficult to illustrate, but the possible mappings based on the stick and impact switching planes are presented in Fig.5.9(a) and (b).

5.4.2 Mapping equations

For mappings in the absolute and relative frames, set the vectors as

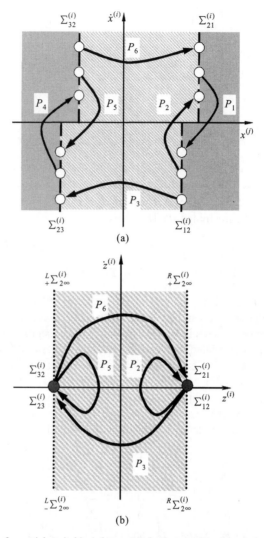

Fig. 5.8 Mappings from stick switching planes: (a) absolute motion; (b) relative motion

$$\left.\begin{array}{l}\mathbf{y}_k \equiv \left(t_k, x_k^{(i)}, \dot{x}_k^{(i)}, \dot{x}_k^{(\bar{i})}\right)^{\mathrm{T}}, \\ \mathbf{w}_k \equiv \left(t_k, \dot{z}_k^{(i)}, x_k^{(i)}, \dot{x}_k^{(\bar{i})}\right)^{\mathrm{T}}.\end{array}\right\} \quad (5.71)$$

For impacting maps $P_\sigma (\sigma = 2, 3, 5, 6)$ in the absolute frame, $\mathbf{y}_{k+1} = P_\sigma \mathbf{y}_k$ can be expressed by

$$P_\sigma : (t_k, x_k^{(i)}, \dot{x}_k^{(i)}, \dot{x}_k^{(\bar{i})}) \to (t_{k+1}, x_{k+1}^{(i)}, \dot{x}_{k+1}^{(i)}, \dot{x}_{k+1}^{(\bar{i})}). \quad (5.72)$$

5.4 Mapping structures and motions

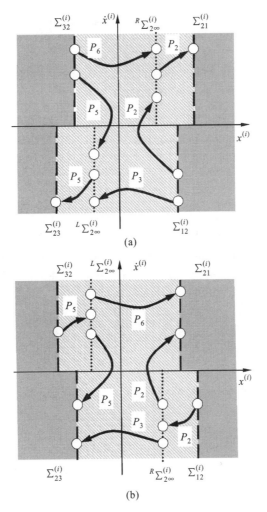

Fig. 5.9 Mappings between the switching planes of stick and impact

From Appendix, the absolute displacement and velocity for two oscillators are obtained with initial conditions $(t_k, x_k^{(i)}, \dot{x}_k^{(i)})$ and $(t_k, x_k^{(\bar{i})}, \dot{x}_k^{(\bar{i})})$. The switching planes require $x_\gamma^{(\bar{i})} = x_\gamma^{(i)} \pm d/2$ ($\gamma = k, k+1$), so the final state for time t_{k+1} can be given. The four equations of displacement and velocity for two masses give a set of four algebraic equations, i.e.,

$$\mathbf{f}^{(\sigma)}(\mathbf{y}_k, \mathbf{y}_{k+1}) = 0, \qquad (5.73)$$

where

$$\mathbf{f}^{(\sigma)} = (f_1^{(\sigma)}, f_2^{(\sigma)}, f_3^{(\sigma)}, f_4^{(\sigma)})^\mathrm{T}. \qquad (5.74)$$

For the stick motion, the displacement and velocity of the ith gear will be adopted. In addition, the stick vanishing conditions in Eq.(5.54) will be used.

$$\left.\begin{array}{l} \dot{x}_{k+1}^{(i)} = \dot{x}_{k+1}^{(i)}, \\ g_\alpha^{(i)}(x_{k+1}^{(i)}, \dot{x}_{k+1}^{(i)}, \ddot{x}_{k+1}^{(i)}, t_{k+1}) = 0. \end{array}\right\} \quad (5.75)$$

With the condition $x_\gamma^{(i)} = x_\gamma^{(i)} \pm d/2$ ($\gamma = k, k+1$), the algebraic equations in Eq.(5.73) can be obtained. If a mapping starts or ends at the stick boundary, the corresponding displacement plus the following equation can be employed to obtain Eq.(5.73).

$$\dot{x}_\gamma^{(i)} = \dot{x}_\gamma^{(i)}, \text{ for } \gamma \in \{k, k+1\}. \quad (5.76)$$

Based on the relative frame of $(z^{(i)}, \dot{z}^{(i)})$, the relative displacement and velocity can be used with the initial condition $(t_k, z_k^{(i)}, \dot{z}_k^{(i)})$ on the switching boundary. The displacement and velocity with an initial condition $(t_k, x_k^{(i)}, \dot{x}_k^{(i)})$ can be given. With conditions $z_\gamma^{(i)} = \pm d/2$ ($\gamma = k, k+1$), the relative and absolute displacements and velocities generate a set of four algebraic equations as

$$\mathbf{h}^{(\sigma)}(\mathbf{w}_k, \mathbf{w}_{k+1}) = 0, \quad (5.77)$$

where

$$\mathbf{h}^{(\sigma)} = (h_1^{(\sigma)}, h_2^{(\sigma)}, h_3^{(\sigma)} h_4^{(\sigma)})^{\text{T}}. \quad (5.78)$$

In a similar fashion, for stick motion, the stick vanishing condition in Eq.(5.54) gives

$$\left.\begin{array}{l} \dot{z}_{k+1}^{(i)} = \dot{z}_k^{(i)} = 0, \\ g_\alpha^{(i)}(x_{k+1}^{(i)}, \dot{x}_{k+1}^{(i)}, \ddot{x}_{k+1}^{(i)}, t_{k+1}) = 0. \end{array}\right\} \quad (5.79)$$

If a mapping starts or ends at the stick boundary, the relative displacement plus the following equation can be used to obtain Eq.(5.78).

$$\dot{z}_\gamma^{(i)} = 0, \text{ for } \gamma \in \{k, k+1\}. \quad (5.80)$$

Finally, the impact mapping on the impact boundaries is defined as

$$P_0 : {}^R\Sigma_{2\infty}^{(i)} \to {}^R\Sigma_{2\infty}^{(i)} \text{ and } P_0 : {}^L\Sigma_{2\infty}^{(i)} \to {}^L\Sigma_{2\infty}^{(i)}, \quad (5.81)$$

in the absolute frame, and in the relative frame

$$P_0 : {}_+^R\Sigma_{2\infty}^{(i)} \to {}_-^R\Sigma_{2\infty}^{(i)} \text{ and } P_0 : {}_-^L\Sigma_{2\infty}^{(i)} \to {}_+^L\Sigma_{2\infty}^{(i)}. \quad (5.82)$$

The corresponding functions in Eq.(5.73) and (5.77), respectively, are

5.4 Mapping structures and motions

$$\left.\begin{aligned} f_1^{(0)} &= t_{k+1} - t_k, \\ f_2^{(0)} &= x_{k+1}^{(i)} - x_k^{(i)}, \\ f_3^{(0)} &= \dot{x}_{k+1}^{(i)} - I_1^{(i)} \dot{x}_k^{(i)} - I_2^{(i)} \dot{x}_k^{(\bar{i})}, \\ f_4^{(0)} &= \dot{x}_{k+1}^{(\bar{i})} - I_1^{(\bar{i})} \dot{x}_k^{(\bar{i})} - I_2^{(\bar{i})} \dot{x}_k^{(i)}, \end{aligned}\right\} \quad (5.83)$$

$$\left.\begin{aligned} h_1^{(0)} &= t_{k+1} - t_k, \\ h_2^{(0)} &= \dot{z}_{k+1}^{(i)} + e \dot{z}_k^{(i)}, \\ h_3^{(0)} &= x_{k+1}^{(i)} - x_k^{(i)}, \\ h_4^{(0)} &= \dot{x}_{k+1}^{(i)} - \dot{x}_k^{(i)} - \frac{m_i}{m_1 + m_2}(\dot{z}_{k+1}^{(i)} - \dot{z}_k^{(i)}). \end{aligned}\right\} \quad (5.84)$$

For simplicity of mapping structures of periodic motions, the impact mapping will be dropped from now on, but the impact relation will be embedded.

5.4.3 Mapping structures

For periodic and chaotic motions in such a gear transmission system, the notation for mapping actions of basic mappings is introduced as in Luo (2006)

$$P_{n_k \cdots n_2 n_1} \equiv P_{n_k} \circ \cdots \circ P_{n_2} \circ P_{n_1}, \quad (5.85)$$

where the mapping $P_{n_j}(n_j \in \{1, 2, \cdots, 6\}, j = 1, 2, \ldots, k)$ is defined in the previous section.

Consider a generalized mapping structure as

$$P_{\underbrace{(65^{k_{s4}}4^{k_{s3}}3 1^{k_{s2}}2^{k_{s1}})\cdots(65^{k_{14}}4^{k_{13}}3 1^{k_{12}}2^{k_{11}})}_{s\text{-terms}}}$$
$$= P_{\underbrace{(65^{k_{s4}}4^{k_{s3}}3 1^{k_{s2}}2^{k_{s1}}) \circ \cdots \circ P_{(65^{k_{14}}4^{k_{13}}3 1^{k_{12}}2^{k_{11}})}}_{s\text{-terms}}}, \quad (5.86)$$

where ($k_{\mu\nu} \in \{0, \mathbb{N}\}, \mu = 1, 2, \ldots, s, \nu = 1, 2, 3, 4$). From the generalized mapping structure, consider a simple mapping structure of periodic motions for impacting chatter. For instance, the mapping structure is

$$P_{65^n 3 2^m} = P_6 \circ P_{5^n} \circ P_3 \circ P_{2^m}, \quad (5.87)$$

where $m, n \in \{0, \mathbb{N}\}$. Such a mapping structure gives m-impacts on the boundary $^R\partial\Omega_{2\infty}^{(i)}$ and n-impacts on the boundary $^L\partial\Omega_{2\infty}^{(i)}$, which are described by mappings P_2 and P_5, respectively. Through the global mappings P_3 and P_6, the impacting chatters on the two boundaries are connected together. Consider a periodic motion of $P_{65^n 3 2^m}$ with period $T_1 = k_1 T$ ($k_1 \in \mathbb{N}$). If the mapping structure copies itself, the new mapping structure is

$$P_{(65^n32^m)^2} = P_{65^n32^m} \circ P_{65^n32^m} \tag{5.88}$$

The periodic motion of $P_{(65^n32^m)^2}$ is obtained during a period of $2T_1$. In a similar fashion, such an action of mapping structure continues to copy itself with period-$2^l T_1$.

$$P_{(65^n32^m)^{2^l}} = P_{(65^n32^m)^{2^{l-1}}} \circ P_{(65^n32^m)^{2^{l-1}}}. \tag{5.89}$$

As $l \to \infty$, a chaotic motion of $P_{65^n32^m}$ is formed. The prescribed chaos is generated by period-doubling. However, if grazing bifurcation occurs, such a mapping structure may not be copied by itself. The new mapping structures are combined by the two different mapping structures, i.e.,

$$\left.\begin{array}{l} P_{65^{n_2}32^{m_2}65^{n_1}32^{m_1}} = P_{65^{n_2}32^{m_2}} \circ P_{65^{n_1}32^{m_1}}, \\ \quad\vdots \\ P_{65^{n_l}32^{m_l}\ldots 65^{n_1}32^{m_1}} = \underbrace{P_{65^{n_l}32^{m_l}} \circ \cdots \circ P_{65^{n_1}32^{m_1}}}_{l\text{-terms}}. \end{array}\right\} \tag{5.90}$$

Such a gazing bifurcation will cause the discontinuity of periodic motions, and chaotic motions may exist between periodic motions of $P_{65^{n_l}32^{m_l}\ldots 65^{n_1}32^{m_1}}$ and $P_{65^{n_l-1}32^{m_l-1}\ldots 65^{n_1}32^{m_1}}$.

For low excitation frequency, the impacting chatter accompanying stick motion exists in the gear transmission system. Consider a simple chatter with stick motion by the following mapping structure

$$P_{645^n312^m} = P_6 \circ P_4 \circ P_{5^n} \circ P_3 \circ P_1 \circ P_{2^m}. \tag{5.91}$$

From the above mapping structure, there are m-impacts on the boundary $^R\partial\Omega_{2\infty}^{(i)}$ and n-impacts on the boundary $^L\partial\Omega_{2\infty}^{(i)}$, which are described by mappings P_2 and P_5, respectively. In addition, the mth mapping of P_2 and the nth mapping of P_5 map the impacting boundary to the stick boundary, and the corresponding stick mappings are P_1 and P_4, respectively. The two global mappings P_3 and P_6 connect the impact and stick boundaries. For period-doubling, the mapping structures are given by

$$\left.\begin{array}{l} P_{(645^n312^m)^2} = P_{645^n312^m} \circ P_{645^n312^m}, \\ \quad\vdots \\ P_{(645^n312^m)^{2^l}} = P_{(645^n312^m)^{2^{l-1}}} \circ P_{(645^n312^m)^{2^{l-1}}}. \end{array}\right\} \tag{5.92}$$

Due to grazing bifurcation, the mapping structures are

$$\left.\begin{array}{l} P_{645^{n_2}32^{m_2}65^{n_1}312^{m_1}} = P_{645^{n_2}312^{m_2}} \circ P_{645^{n_1}312^{m_1}}, \\ \quad\vdots \\ P_{645^{n_l}312^m\ldots 645^{n_1}312^{m_1}} = \underbrace{P_{645^{n_l}312^{m_l}} \circ \cdots \circ P_{645^{n_1}312^{m_1}}}_{l\text{-terms}}. \end{array}\right\} \tag{5.93}$$

5.4 Mapping structures and motions

To understand the two types of mapping structures, two simple mapping structures are shown in Fig.5.10(a) and (b) for the impacting chatter with and without stick. Similarly, the other mapping structures can be discussed through the generalized mapping structure in Eq.(5.86). Furthermore, the periodic and chaotic motions relative to a certain mapping structure can be determined.

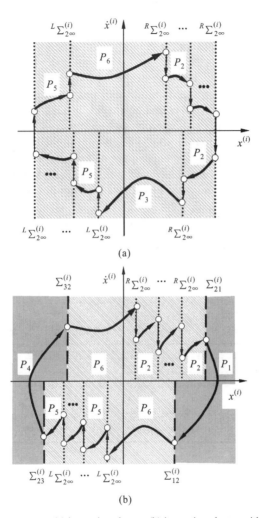

Fig. 5.10 Mapping structures: (a) impacting chatter; (b) impacting chatter with stick

5.4.4 Bifurcation scenario

Before discussing impacting chatter in the gear transmission system, a bifurcation scenario is presented through the switching displacements, velocities and phases of two gear oscillators. All the numerical computations are completed from closed-form solutions in Appendix. The parameters ($m_1 = 2, m_2 = 1, r_1 = r_2 = 0.6, k_1 = 30$, $k_2 = 20, Q_0 = 35, d = 1.0$, $e = 0.7$) are considered. The impacting chatter with and without stick varying with excitation frequency are presented in Figs.5.11. The "C-chatter" and "S-chatter" represent the complicated and simple impacting chatters, respectively. The "Chatter w/stick" denotes the impacting chatter with stick for periodic motions. For high excitation frequency, it is observed that the impacting chatter motion possesses a simple mapping structure. The excitation frequency lies in the interval of excitation frequency $\Omega \in (4.159, 16.433)$ for the simplest impacting-chatter periodic motion relative to P_{63}. At $\Omega \approx 16.433$, the motion disappears and for $\Omega > 16.433$ the two gear oscillators will not contact. However, the complex impacting-chatter motion exists for $\Omega \in (1.398, 4.159)$. For $\Omega \in (0, 1.398)$, the impacting chatter with stick exists. The details of mapping structures for impacting chatter with and without stick are tabulated in Table 5.1 and Table 5.2, respectively.

Table 5.1 The summary of excitation frequency for impacting chatter with stick ($m_1 = 2$, $m_2 = 1$, $r_1 = r_2 = 0.6$, $k_1 = 30$, $k_2 = 20$, $Q_0 = 35$, $d = 1.0$ and $e = 0.7$)

Mapping structure		Excitation frequency
$P_{6^{226}13^{526}4}$	$P(T)$	(1.382, 1.397)
$P_{6^{227}13^{527}4}$	$P(T)$	(1.264, 1.382)
$P_{6^{228}13^{528}4}$	$P(T)$	(0.908, 1.264)
$P_{6^{227}13^{527}4}$	$P(T)$	(0.680, 0.908)
$P_{6^{226}13^{526}4}$	$P(T)$	(0.635, 0.680)
$P_{6^{225}13^{525}4}$	$P(T)$	(0.613, 0.635)
$P_{6^{224}13^{524}4}$	$P(T)$	(0.455, 0.613)
$P_{6^{225}13^{525}4}$	$P(T)$	(0.346, 0.455)
$P_{6^{224}13^{524}4}$	$P(T)$	(0.317, 0.346)
$P_{6^{223}13^{523}4}$	$P(T)$	(0.216, 0.317)
$P_{6^{2k}13^{5k}4}$	$P(T)$	(0.000, 0.216)

5.5 Periodic motion prediction

In this section, the analytical prediction of periodic motions will be completed through mapping structures, and the corresponding local stability and bifurcation analysis will be carried out through eigenvalue analysis. The generalized methodology for analytical prediction and stability analysis of periodic motions will be presented first. The two periodic motions with and without stick will be discussed.

5.5 Periodic motion prediction

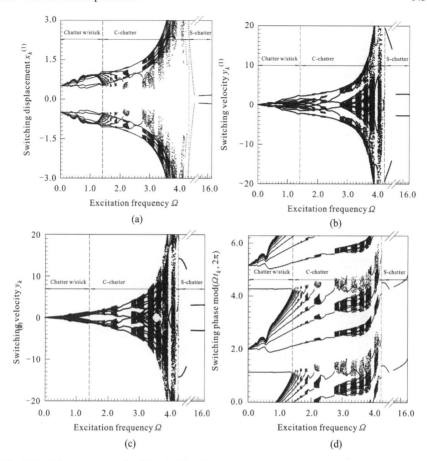

Fig. 5.11 Bifurcation scenario: (a) switching displacement (b) switching velocity of the first mass (c) switching velocity of the second mass and (d) switching phase ($m_1 = 2, m_2 = 1, r_1 = r_2 = 0.6, k_1 = 30, k_2 = 20, Q_0 = 35, d = 1.0$ and $e = 0.7$).

The switching displacement, velocity, and phase varying with excitation frequency will be given. Finally, the stability and bifurcation analysis will be presented through the eigenvalues of the prescribed periodic motions.

5.5.1 Approach

From mapping structures of periodic motions, the switching sets for any specific periodic motion can be determined through solving a set of nonlinear algebraic equations. Consider a periodic motion of mapping structure $P_{(65^{k_{s4}}4^{k_{s3}}31^{k_{s2}}2^{k_{s1}})\cdots(65^{k_{14}}4^{k_{13}}31^{k_{12}}2^{k_{11}})}$ and the following relation holds.

Table 5.2 The summary of excitation frequency for impacting chatter ($m_1 = 2$, $m_2 = 1$, $r_1 = r_2 = 0.6$, $k_1 = 30$, $k_2 = 20$, $Q_0 = 35$, $d = 1.0$ and $e = 0.7$)

Mapping structure		Excitation frequency
P_{63}	$P(T)$	(4.159,16.433)
P_{6532}	$P(T)$	(4.155,4.252)
$P_{(6532)^2}$	$P(2T)$	(4.150,4.155)
	chaos	(4.10,4.150)
$P_{(6553265)(3226532)}$	$P(3T)$	(4.095,4.100), (3.683,3.724)
$P_{((6553265)(3226532))^2}$	$P(6T)$	(4.094, 4.095), (3.725,3.727)
$P_{((6553265)(3226232))^4}$	$P(12T)$	(3.727,3.728)
P_{655322} or P_{322655}	$P(T)$	(3.909,4.022)
	chaos	(3.728,3.909)
$P_{((65532653265)(32265326532))^4}$	$P(20T)$	(3.528,3.724)
$P_{((65532653265)(32265326532))^2}$	$P(10T)$	(3.524,3.528)
$P_{(65532653265)(32265326532)}$	$P(5T)$	(3.455,3.524)
	chaos	(3.267,3.455)
$P_{(65532653265)(32265326532)}$	$P(5T)$	(3.150,3.267)
	chaos	(3.020,3.150)
$P_{(65532)(65322)(6532)(6532)}$	$P(4T)$	(3.014,3.020)
	chaos	(2.925,3.014)
$P_{(6553265)(3226532)}$	$P(3T)$	(2.889,2.925)
	chaos	(2.748,2.889)
$P_{65^2 32^2}$	$P(T)$	(2.309-2.748)
$P_{65^m 32^m}$ ($m = 1, 2, \ldots, 10$) and chaos	$P(T)$	(1.378,2.309)

$$P_{(65^{k_{s4}} 4^{k_{s3}} 3 1^{k_{s2}} 2^{k_{s1}}) \cdots (65^{k_{14}} 4^{k_{13}} 3 1^{k_{12}} 2^{k_{11}})} \mathbf{y}_k = \mathbf{y}_{k+2s+\sum_{m=1}^{s}\sum_{j=1}^{4} k_{mj}}, \quad (5.94)$$

where $\mathbf{y}_k = (t_k, x_k^{(i)}, \dot{x}_k^{(i)}, \dot{x}_k^{(\bar{i})})^{\mathrm{T}}$. A set of vector equations are

$$\left.\begin{array}{l} \mathbf{f}^{(2)}(\mathbf{y}_{k+1}, \mathbf{y}_k) = 0, \\ \mathbf{f}^{(2)}(\mathbf{y}_{k+2}, \mathbf{y}_{k+1}) = 0, \\ \quad \vdots \\ \mathbf{f}^{(6)}\left(\mathbf{y}_{k+2s+\sum_{m=1}^{s}\sum_{j=1}^{4} k_{mj}}, \mathbf{y}_{k+2s+\sum_{m=1}^{s}\sum_{j=1}^{4} k_{mj}-1}\right) = 0, \end{array}\right\} \quad (5.95)$$

where $\mathbf{f}^{(\sigma)} = (f_1^{(\sigma)}, f_2^{(\sigma)}, f_3^{(\sigma)}, f_4^{(\sigma)})^{\mathrm{T}}$ is relative to governing equations of mapping P_σ ($\sigma \in \{1, 2, \ldots, 6\}$). The periodicity of the period-1 motion per N-periods requires

$$\mathbf{y}_{k+2s+\sum_{m=1}^{s}\sum_{j=1}^{4} k_{mj}} = \mathbf{y}_k, \quad (5.96)$$

5.5 Periodic motion prediction

$$\left.\begin{array}{c} x^{(i)}_{k+2s+\sum\limits_{m=1}^{s}\sum\limits_{j=1}^{4}k_{mj}} = x^{(i)}_k, \dot{x}^{(i)}_{k+2s+\sum\limits_{m=1}^{s}\sum\limits_{j=1}^{4}k_{mj}} = \dot{x}^{(i)}_k, \ \dot{x}^{(\bar{i})}_{k+2s+\sum\limits_{m=1}^{s}\sum\limits_{j=1}^{4}k_{mj}} = \dot{x}^{(\bar{i})}_k, \\ \Omega t_{k+2s+\sum\limits_{m=1}^{s}\sum\limits_{j=1}^{4}k_{mj}} \equiv \Omega t_k + 2N\pi. \end{array}\right\} \quad (5.97)$$

Solving Eqs.(5.95) and (5.97) generates the switching sets for periodic motions. Once the analytical prediction of any periodic motion is obtained, the corresponding stability and bifurcation analysis can be completed. The local stability and bifurcation for such a period-1 motion is determined through the corresponding Jacobian matrix of the Poincare mapping. From Eq.(5.94), the Jacobian matrix is computed by the chain rule

$$DP = DP_{(65^{k_{n4}}4^{k_{n3}}31^{k_{n2}}2^{k_{n1}})\cdots(65^{k_{14}}4^{k_{13}}31^{k_{12}}2^{k_{11}})}$$
$$= \prod_{m=1}^{n}\left(DP_6 \cdot DP_5^{k_{m4}} \cdot DP_4^{k_{m3}} \cdot DP_3 \cdot DP_1^{k_{m2}} \cdot DP_2^{k_{m1}}\right) \quad (5.98)$$

where

$$DP_\lambda = \left[\frac{\partial(t_{v+1},x^{(i)}_{v+1},y^{(i)}_{v+1},y^{(\bar{i})}_{v+1})}{\partial(t_v,x^{(i)}_v,y^{(i)}_v,y^{(\bar{i})}_v)}\right], \quad (5.99)$$

for $v = k, k+1, \ldots, k+2s+\sum\limits_{m=1}^{s}\sum\limits_{j=1}^{4}k_{mj}-1$, and all the Jacobian matrix components can be computed by Eq.(5.95). The variational equation for a set of switching points $\{\mathbf{y}^*_k,\ldots,\mathbf{y}^*_{k+2s+\sum\limits_{m=1}^{s}\sum\limits_{j=1}^{4}k_{mj}-1}\}$ is

$$\Delta\mathbf{y}_{k+2s+\sum\limits_{m=1}^{s}\sum\limits_{j=1}^{4}k_{mj}} = DP(\mathbf{y}^*_k)\Delta\mathbf{y}_k. \quad (5.100)$$

The eigenvalues are computed by

$$|DP - \lambda\mathbf{I}| = 0. \quad (5.101)$$

The Jacobian matrix DP is 4×4, and the eigenvalues can be determined. Because DP is a 4×4 matrix, there are four eigenvalues. If the four eigenvalues lie inside the unit circle, then the period-1 motion is stable. If one of them lies outside the unit circle, the periodic motion is unstable. The detailed local stability and bifurcation conditions can be refered to Appendix.

5.5.2 Impacting chatter

Using the mapping structure in Eq.(5.86), all the periodic motions for the entire range of excitation frequency can be determined analytically. The mapping struc-

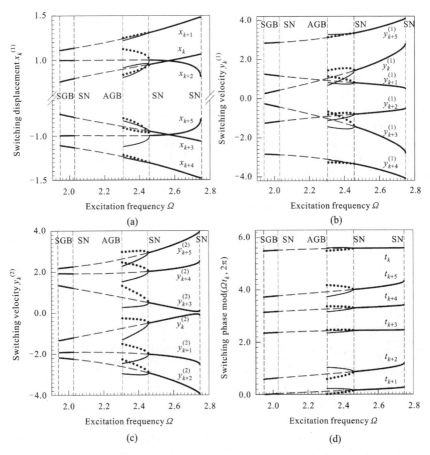

Fig. 5.12 Analytical prediction of periodic motion $P_{6 5^2 3 2^2}$: (a) switching displacement and (b) switching velocity of the first mass (c) switching velocity of the second mass and (d) switching phase ($m_1 = 2$, $m_2 = 1$, $r_1 = r_2 = 0.6$, $k_1 = 30$, $k_2 = 20$, $Q_0 = 35$, $d = 1.0$ and $e = 0.7$)

ture gives the nonlinear algebraic equations similar to Eqs.(5.95) and (5.97), which can be solved by the Newton-Raphson method. Once the first solution is obtained, the rest of the solutions with varying parameters can be determined through the corresponding mapping structure. The parameters ($m_1 = 2$, $m_2 = 1$, $r_1 = r_2 = 0.6$, $k_1 = 30$, $k_2 = 20$, $Q_0 = 35$, $d = 1.0$ and $e = 0.7$) are also used for the analytical prediction of periodic motions. The analytical prediction of the periodic motion of $P_{6 4^2 3 2^2}$ is shown in Figs.5.12 The switching displacement and velocity of the first mass are plotted in Fig.5.12(a) and (b), and for the second mass, only the switching velocity is presented in Fig.5.12(c). At switching points, the switching displacement of the second gear can be obtained from the switching displacement of the first gear, i.e., $x_k^{(2)} = x_k^{(1)} \pm d/2$. The switching phase is presented in Fig.5.12(d). Also, the switching sets are recorded before impact rather than after impact. The corre-

5.5 Periodic motion prediction

sponding eigenvalues of symmetric and asymmetric periodic motion are presented in Figs.5.13, respectively. From local stability analysis, the excitation frequency range for stable and unstable periodic motion is given. The range for stable, symmetric motion of $P_{6 4^2 3 2^2}$ is $\Omega \in \{(2.457, 2.748), (1.946, 2.028)\}$, depicted by the thick curves. An unstable symmetric periodic motion for such a mapping structure lies in $\Omega \in (2.028, 2.457)$, depicted by dashed curves. For the same mapping structure, the asymmetric, periodic motion exists in the range of $\Omega \in (2.380, 2.457)$, and there are two branches of solutions. One of the two solutions is presented through the dotted curves, and the other is shown with the thin curves. The saddle-node bifurcation of the symmetric, periodic motion occurs at $\Omega \approx 2.028, 2.457$ and 2.748, while the saddle-node bifurcation of the asymmetric periodic motion occurs at $\Omega \approx 2.457$. Such a saddle-node bifurcation causes the symmetry break of impacting chatter mo-

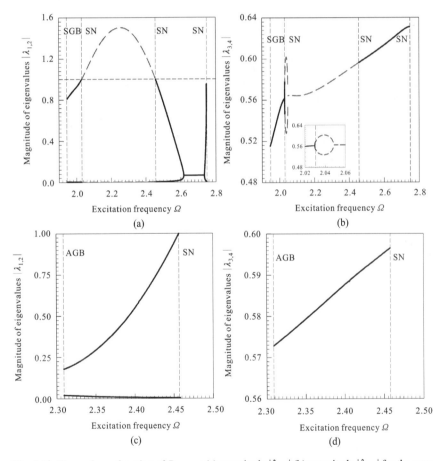

Fig. 5.13 Eigenvalues of motion of $P_{6 5^2 3 2^2}$: (a) magnitude $|\lambda_{1,2}|$ (b) magnitude $|\lambda_{3,4}|$ for the symmetric motion; (c) magnitude $|\lambda_{1,2}|$ and (d) magnitude $|\lambda_{3,4}|$ for asymmetric motion ($m_1 = 2$, $m_2 = 1$, $r_1 = r_2 = 0.6$, $k_1 = 30$, $k_2 = 20$, $Q_0 = 35$, $d = 1.0$ and $e = 0.7$)

tion. The grazing bifurcations for the symmetric and asymmetric periodic motions are $\Omega \approx 2.380$ and 1.946. For the grazing periodic motion, the eigenvalues show that the two periodic motions are stable, i.e., the eigenvalues of the grazing periodic motion lie in the unit circle. However, grazing causes the mapping structure to change. Thus, the motion relative to the old mapping structure disappears.

5.5.3 Impacting chatter with stick

The analytical prediction of periodic motions with stick is also given herein. Using parameters ($m_1 = 2$, $m_2 = 1$, $r_1 = r_2 = 0.6$, $k_1 = 30$, $k_2 = 20$, $Q_0 = 35$, $d = 1.0$ and $e = 0.7$), the corresponding switching sets varying with excitation frequency are

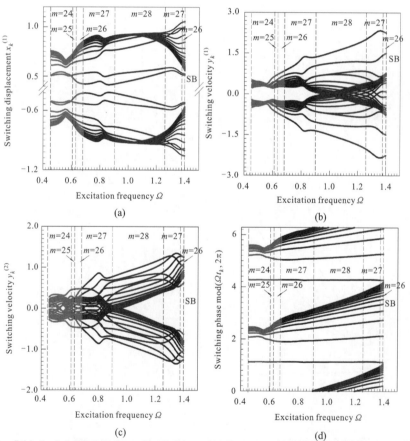

Fig. 5.14 Analytical prediction of periodic motion $P_{6 4 5^m 3 1 2^m}$: (a) switching displacement (b) switching velocity of the first mass (c) switching velocity of the second mass and (d) switching phase ($m_1 = 2$, $m_2 = 1$, $r_1 = r_2 = 0.6$, $k_1 = 30$, $k_2 = 20$, $Q_0 = 35$, $d = 1.0$ and $e = 0.7$)

5.5 Periodic motion prediction

presented in Figs.5.14 for the mapping structure of $P_{645^m312^m}$ ($m = 24, 25, \cdots, 28$). Because it is very time-consuming, the impacting chatters with stick are predicted only in the range of excitation frequency $\Omega \in (0.455, 1.39859)$. The eigenvalues of impacting chatter with stick motion are presented in Fig.5.15. Note that eigenvalues of $\lambda_{3,4}$ are zero for the entire range of stick motion. The stick bifurcation (SB) occurs at $\Omega \approx 1.39859$. Once the stick motion appears, the impacting chatter with stick for $P_{645^{26}31^{2^{26}}}$ lies in the range of $\Omega \in (1.328, 1.39859)$ and $(0.635, 0.6789)$. The stick motion of $P_{654^{27}31^{2^{27}}}$ lies in the two regions of $\Omega \in (1.264, 1.328)$ and $(0.680, 0.9068)$. The stick motion of $P_{645^{28}31^{2^{28}}}$ exists within $\Omega \in (0.9068, 1.264)$. The stick motions of $P_{645^{25}31^{2^{25}}}$ and $P_{645^{24}31^{2^{24}}}$ are in $\Omega \in (0.612, 0.6789)$ and $(0.455, 0.612)$, respectively. In analytical computation, $\dot{x}_k^{(1)} = \dot{x}_k^{(2)}$ is set directly.

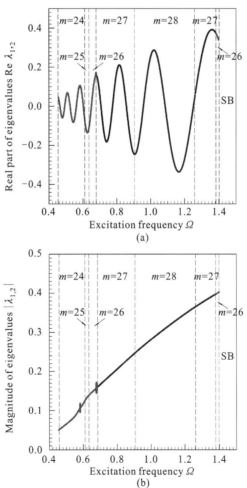

Fig. 5.15 Eigenvalues for the chatter with stick $P_{645^m312^m}$: (a) real part Re $\lambda_{1,2}$ (b) magnitude $|\lambda_{1,2}|$ ($m_1 = 2$, $m_2 = 1$, $r_1 = r_2 = 0.6$, $k_1 = 30$, $k_2 = 20$, $Q_0 = 35$, $d = 1.0$ and $e = 0.7$)

However, the computation criterion for stick motion (i.e. $|\dot{x}_k^{(1)} - \dot{x}_k^{(2)}| < 10^{-6}$) was embedded in numerical simulation. If the tolerance is set too small, both the analytical and the numerical predictions may not match. Such a problem may be caused by the computational error. In other stick motions, the analytical prediction can be carried out.

5.5.4 Parameter maps

The parameter maps for excitation frequency Ω versus restitution e are shown in Fig.5.16 for parameters, ($m_1 = 2$, $m_2 = 1$, $r_1 = r_2 = 0.6$, $k_1 = 30$, $k_2 = 20$, $Q_0 = 50$, $d = 1.0$). In Fig.5.16(a), the entire range of excitation frequency for two masses

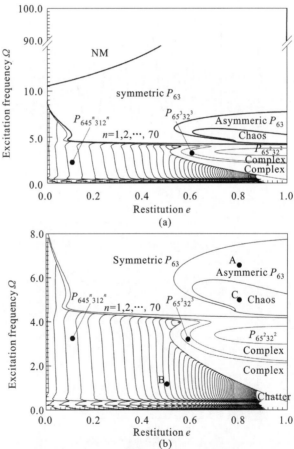

Fig. 5.16 Parameter map for excitation frequency versus restitution: (a) overview and (b) zoomed view ($m_1 = 2, m_2 = 1, r_1 = r_2 = 0.6, k_1 = 30, k_2 = 20, Q_0 = 50.0$ and $d = 1.0$)

experiencing interaction is presented. The zoomed view of the parameter map is given in Fig.5.16(b) for $\Omega \in [0,8]$. The chatter with stick possesses a mapping structure of $P_{645^n312^n}$ for $n = 1,2,\ldots,70$. The number of impacting chatters increases with the increasing e. The region labeled by "Chatter" represents chatter with stick where the chatter impacts number approaches infinity as $e \to 1$. The region just above the chatters with stick has complex mapping structure. Within the "complex motion" region, chaotic and periodic motions of impacting chatter without stick exist, and the corresponding mapping structures are relative to $P_{65^2 32^2}$ and $P_{65^3 32^3}$. With increasing excitation frequency, symmetric and asymmetric periodic motions with the mapping structure of P_{63} are presented. The larger region for the motion of P_{63} is symmetric, while the smaller region for the motion of P_{63} is asymmetric. For higher excitation frequency, the two masses will not contact each other, and such a region is labeled by "NM". It means the two masses do not transfer any energy.

5.6 Numerical illustrations

In this section, numerical illustrations of periodic and chaotic motions will help one better understand the motion mechanisms of such a two-oscillator system. From the analytical prediction, periodic and chaotic motions will be plotted for chatter with and without stick. In addition, the analytical condition for grazing and stick motions will be illustrated through the force and jerk responses. The Poincare mapping section based on the switching points will be presented for illustration of complexity in such a dynamical system on time-varying domains.

5.6.1 Impacting chatter

For parameters ($m_1 = 2$, $m_2 = 1$, $r_1 = r_2 = 0.6$, $k_1 = 30$, $k_2 = 20$, $Q_0 = 35$, $d = 1.0$ and $e = 0.7$), consider the excitation frequency $\Omega = 2.72$. The analytical prediction gives the initial condition (i.e., $t_0 \approx 2.0559169$, $x_0^{(1)} \approx 1.0505790$, $x_0^{(2)} \approx 0.5505790$, $y_0^{(1)} \approx 1.68706528$, $y_0^{(2)} \approx 4.5219068$) for the stable periodic impacting chatter of $P_{64^2 32^2}$. With displacement and velocity time-histories, trajectories of the first and second gears in phase plane are presented in Figs.5.17. All the responses of the first and second masses are represented by the thick and thin curves, respectively. The left and right sides of the second mass are presented by the intermediate thick curves. The switching points before and after impacts are represented by circles. The hollow and gray-filled circles are for the first and second masses, respectively. The switching points for stick are represented by the large filled circles.

The displacement responses of the two masses are presented in Fig.5.17(a). The motion starts at the right side (R-side) of the second gear. The first gear has two impacts with the second gear. Afterwards, the two gears have a free-flying motion. Once the first gear arrives to the left side (L-side) of the second gear, the next two

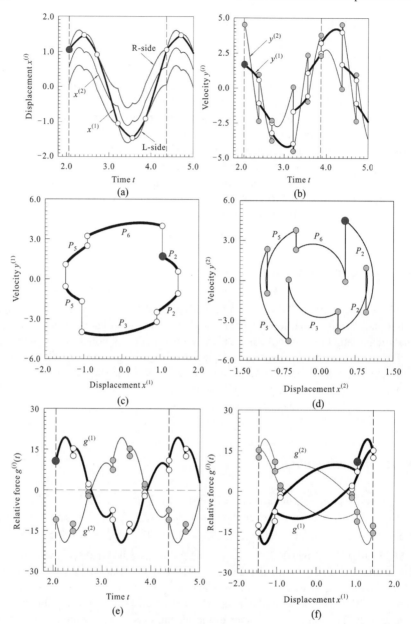

Fig. 5.17 Periodic motion of impacting chatter for $P_{6 5^2 3 2^2}$ ($\Omega = 2.72$): (a) displacement (b) velocity (c) phase plane of the first mass (d) phase plane of the second mass (e) relative force time-history and (f) relative force distribution along the displacement of the first mass. Initial condition ($t_0 \approx 2.055\,916\,9$, $x_0^{(1)} \approx 1.050\,5790$, $x_0^{(2)} \approx 0.550\,5790$, $y_0^{(1)} \approx 1.687\,065\,28$ and $y_0^{(2)} \approx 4.521\,906\,8$) ($m_1 = 2$, $m_2 = 1$, $r_1 = r_2 = 0.6$, $k_1 = 30$, $k_2 = 20$, $Q_0 = 35$, $d = 1.0$ and $e = 0.7$)

5.6 Numerical illustrations

impacts between the two gears occur. After these two impacts, the two gears have another free-flying motion, and the two gears meet at the right side of the second gear. Such a motion process will repeat and form a periodic motion. Due to impacts between two gears at the left and right sides, the impacting velocities become discontinuous. Thus, the time-histories of velocity for two masses are presented in Fig.5.17(b). Such discontinuity is observed at the impact locations. After the right side impact, the first mass velocity is less than the second mass velocity (i.e., $y_k^{(1)} < y_k^{(2)}$), and vice versa for the left side. Before and after impacts, the motions for two gears are separated because $y_k^{(1)} \neq y_k^{(2)}$. To observe the periodic behaviors of the two gears, the trajectories of such a periodic motion in phase plane for the first and second masses are presented in Fig.5.17(c) and (d), accordingly. The mapping structure of $P_{64^2 3 2^2}$ is observed via a periodic loop. In Fig.5.22(d), the absolute displacement of the central location of the second masses is employed for illustration, instead of the left and right sides of the second mass. However, the left and right tooth locations of the second mass can be obtained by adding the two offsets.

To understand such impacting chatter in the two-oscillator system, the relative force responses should be discussed. After the first right-side impact, to make the repeated impact occur, the relative force should be greater than zero (i.e., $g_\alpha^{(1)}(t_k) > 0$). Since the absolute velocity relation ($y_k^{(1)} < y_k^{(2)}$) holds just after impact, the relative velocity is less than zero ($\dot{z}_k^{(1)} = y_k^{(1)} - y_k^{(2)} < 0$). Thus, just after the impact, the relative displacement is less than half the gap (i.e., $z_k^{(1)} = x_k^{(1)} - x_k^{(2)} < d/2$). To allow the first gear to catch up with the second one, the relative force should be greater than zero (i.e., $g_\alpha^{(1)}(t_k) > 0$) before the next impact. Otherwise, the next impact at the right-side will not occur. In a similar fashion, just after the first impact at the left-side impact, the relative force should be less than zero (i.e., $g_\alpha^{(1)}(t_k) < 0$) to achieve the second impact. From Fig.5.17(e) and (f), such a relation is observed. Based on the perspective of the second gear, the relative force criterion of the periodic motion is described by $z^{(2)} = x^{(2)} - x^{(1)} = -z^{(1)}$, and the relative velocity and force for the second gear is opposite to that of the first gear as shown in Fig.5.17(e) and (f).

5.6.2 Impacting chatter with stick

The impacting chatter with stick in the gear transmission system will be discussed by the motion of $P_{5^{24} 3 1 2^{24} 6 4}$. The impacting chatter with stick relative to $P_{5^{24} 3 1 2^{24} 6 4}$ is equivalent to the one of $P_{6 4 5^{24} 3 1 2^{24}}$ except that the starting points are from different switching points. For $\Omega = 0.5$, the same parameters as mentioned before ($m_1 = 2$, $m_2 = 1$, $r_1 = r_2 = 0.6$, $k_1 = 30$, $k_2 = 20$, $Q_0 = 35$, $d = 1.0$ and $e = 0.7$)) will be adopted for this illustration. From the analytical prediction, the initial condition is selected as $t_0 \approx 4.8888960$, $x_0^{(1)} \approx -0.7387947$, $x_0^{(2)} \approx -0.2387947$ and $y_0^{(1)} = y_0^{(2)} \approx -0.2754161$. From such an initial condition, the stick motion will appear on the left-side of the second gear. After the stick vanishes, the first gear moves to the

156   5 Two Oscillators with Impacts and Stick

right-side of the second gear through the free-flying motion. Because of different velocities after impact, the two gears experience impacting chatter until $y_k^{(1)} = y_k^{(2)}$ and the stick motion is observed. In Fig.5.18(a) and (b), displacement and velocity time-histories are presented.

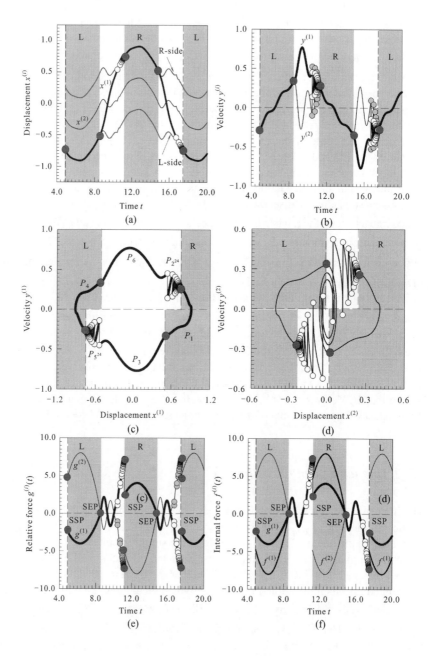

5.6 Numerical illustrations

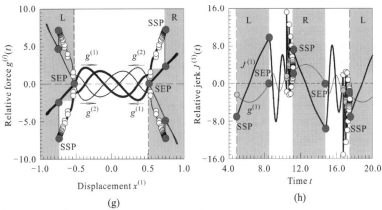

Fig. 5.18 Displacement and velocity time-histories for impacting chatter with stick for $P_{6_4 5^{24} 3 1^{24}}(\Omega = 0.5)$: (a) displacement (b) velocity (c) phase plane of the first mass (d) phase plane of the second mass (e) relative force time–history (f) internal force time-history (g) relative force distribution along the displacement of the first mass (h) jerk time-history. Initial condition ($t_0 \approx 4.8888960$, $x_0^{(1)} \approx -0.7387947$, $x_0^{(2)} \approx -0.2387947$ and $y_0^{(1)} = y_0^{(2)} \approx -0.2754161$) ($m_1 = 2, m_2 = 1$, $r_1 = r_2 = 0.6$, $k_1 = 30$, $k_2 = 20, Q_0 = 35, d = 1.0$, and $e = 0.7$)

The shaded area is used for stick motion, and the letters "L" and "R" represent the left-side and right-side of the second gear, respectively. In Fig.5.18(a), the displacement responses of the two gears are presented. Before stick motion is observed, 24 impacts occur on the left and right sides of the second gear. To further confirm the impacts, velocity responses for the two gears are presented in Fig.5.18(b). The phase plane trajectories of the first and second gears are plotted in Fig.5.18(c) and (d), respectively. For a further understanding of impacting chatter with stick, the relative force and jerk for the two gears are illustrated in Fig.5.18(e)∼(h). The impacting motion between the two gears has been discussed in Fig.5.17. Herein, the discussion focuses on the portion of the stick motion. On the switching boundary with $y_k^{(1)} = y_k^{(2)}$, the stick motion requires $g_\alpha^{(1)}(t_k) > 0$ (or $g_\alpha^{(2)}(t_k) < 0$) on the right-side for $\alpha = 1, 2$ and $g_\alpha^{(1)}(t_k) < 0$ (or $g_\alpha^{(2)}(t_k) > 0$) on the left-side for $\alpha = 2, 3$. Because the vector fields in different domains are different, the relative forces and jerks in the different domains are completely different (i.e., $g_\alpha^{(i)}(t_k) \neq g_\beta^{(i)}(t_k)$, $\alpha, \beta \in \{1, 2, 3\}$ and $\alpha \neq \beta$). The onset of the stick motion requires the two forces $g_\alpha^{(i)}(t_k)$ and $g_\beta^{(i)}(t_k)$ have the same sign. To obtain a stick motion in domain Ω_1 requires $g_1^{(1)}(t_k) > 0$ and $g_2^{(1)}(t_k) > 0$, and to obtain the stick motion in Ω_3 requires $g_2^{(1)}(t_k) < 0$ and $g_3^{(1)}(t_k) < 0$. Such a relative force characteristic is presented in Fig.5.18(e). In Fig.5.18(f), the internal forces in the stick domains for two gears are presented because the internal forces between the two gears only exist during stick. When the stick motion disappears, the internal forces vanish. In other words, the internal forces are zero at the vanishing of the stick motion. The vanishing condition for stick is $g_\alpha^{(i)}(t_k) = 0$, which is observed in Fig.5.18(e) and (f).

To correspond with the trajectory in phase plane, the relative force distribution along the displacement of the first gear is shown in Fig.5.18(g). The aforementioned conditions for the onset and vanishing of stick motion are observed. However, the condition of $g_\alpha^{(i)}(\mathbf{x}_\alpha^{(i)},t) = 0$ is a necessary condition. The sufficient condition for the vanishing of the stick motion requires $J_\alpha^{(1)}(t_k) < 0$ (or $J_\alpha^{(2)}(t_k) > 0$) at the right-side and $J_\alpha^{(1)}(t_k) > 0$ (or $J_\alpha^{(2)}(t_k) < 0$) at the left-side. Therefore, in Fig.5.18(h), the jerk $J_\alpha^{(1)}(t)$ of the first gear is presented, and such a sufficient condition is observed. When the stick vanishes at $z_k^{(1)} = x_k^{(1)} - x_k^{(2)} = d/2$, the relative velocity and acceleration are zero (i.e., $y_k^{(1)} = 0$ and $\ddot{z}_k^{(1)} = 0$). If $J_\alpha^{(1)}(t) < 0$ for $t > t_k + \varepsilon$, the corresponding relative velocity and acceleration are less than zero, which leads to the relative displacement satisfying $z^{(1)} = x^{(1)} - x^{(2)} < d/2$. In other words, the two gears lie in the state of free-flying motion, and the stick disappears. In a similar manner, the stick motion disappearance at the left-side can be discussed. The jerk condition of $J_\alpha^{(1)}(t_k) > 0$ leads to $z^{(1)} = x^{(1)} - x^{(2)} > -d/2$. Therefore, the first gear just begins to separate from the left-side of the second gear, and the stick motion disappears.

5.6.3 Further illustrations

To demonstrate motions with specific mapping structures in the parameter map, three sets of excitation frequency and restitution are used, and they are labeled via points A, B and C in Fig.5.20(b). The parameters ($m_1 = 2$, $m_2 = 1$, $r_1 = r_2 = 0.6$, $k_1 = 30$, $k_2 = 20$, $Q_0 = 50$, $d = 1.0$, $e = 0.7$) are adopted. At the point "A", $\Omega = 6.6$ and $e = 0.8$ are selected. The initial conditions are $t_0 \approx 0.8852$, $x_0^{(1)} \approx -0.4179$, $y_0^{(1)} \approx -0.0770$ and $x_0^{(2)} \approx -0.9179$, $y_0^{(2)} \approx -4.0045$. The corresponding phase plane is plotted in Fig.5.19(a). The motion starts with just after the driving gear impacts at the right hand side of the driven gear. The next impact takes place at the left hand side of the second mass and then returns back to the right-side again. The asymmetric motion is relative to mapping P_6 and P_3, its twin asymmetric motion will not presented and the detailed discussion can be referred to Luo (2006). For parameters (i.e., $\Omega = 1.0$ and $e = 0.5$) labeled "B" in Fig.5.16(b), the periodic motion of impacting chatter with stick $P_{5^{15}3^{12}1^{5}64}$ is plotted in Fig.5.19(b) with initial conditions ($t_0 \approx 2.6000$, $x_0^{(1)} \approx -1.1201$, $y_0^{(1)} \approx -0.5273$, $x_0^{(2)} \approx -0.6201$, $y_0^{(2)} \approx -0.5271$). The first mass begins at the onset of stick motion relative to P_4 on the left hand side of the driven gear. Crossing the gap from the left to right side of the second mass is the mapping of P_6. The two masses impact fifteen times (i.e., P_2^{15}) before a new stick is formed on the right-side, and the stick motion is described by the mapping of P_1. The second half of the periodic motion can be described in a similar fashion. Finally, chaotic motions are demonstrated via Poincare mapping sections at point C (i.e., $\Omega = 5.0$ and $e = 0.8$). The initial condition is $t_0 \approx 0.0641$, $x_0^{(1)} \approx -1.5161$, $y_0^{(1)} \approx 6.0031$, $x_0^{(2)} \approx -2.0161$, $y_0^{(2)} \approx 3.5209$. The

5.6 Numerical illustrations

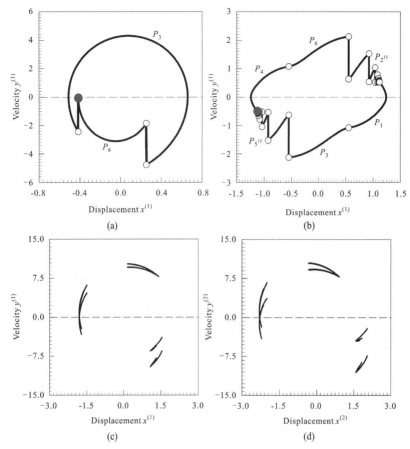

Fig. 5.19 Phase planes: (a) asymmetric impacts P_{63} ($\Omega = 6.6$ and $e = 0.8$; $t_0 \approx 0.8852$, $x_0^{(1)} \approx -0.4179$, $y_0^{(1)} \approx -0.0770$ and $x_0^{(2)} \approx -0.9179$, $y_0^{(2)} \approx -4.0045$) (b) impact chatter with stick $P_{5^{15}3^{12}1^{5}64}$ ($\Omega = 1.0$ and $e = 0.5$; $t_0 \approx 2.6000$, $x_0^{(1)} \approx -1.1201$, $y_0^{(1)} \approx -0.5273$ and $x_0^{(2)} \approx -0.6201$, $y_0^{(2)} \approx -0.5271$); poincare mapping section: (c) the first mass and (d) the second mass ($\Omega = 5.0$ and $e = 0.8$; $t_0 \approx 0.0641$, $x_0^{(1)} \approx -1.5161$, $y_0^{(1)} \approx 6.0031$ and $x_0^{(2)} \approx -2.0161$, $y_0^{(2)} \approx 3.5209$) ($m_1 = 2$, $m_2 = 1$, $r_1 = r_2 = 0.6$, $k_1 = 30$, $k_2 = 20$, $Q_0 = 50$, $d = 1.0$ and $e = 0.7$)

switching points are plotted in Fig.5.19 for ten thousand periods ($10^4 T$) of the excitation forcing. The Poincare mapping sections of switching points for the first and second gears are given in Fig.5.19(c) and (d), respectively. The switching points describe the displacement and velocity of the two masses upon contact. The switching points form a strange attractor of chaotic motions for such two-oscillator system. In a similar fashion, the other periodic and chaotic motions can be illustrated.

References

Han, R. P. S., Luo, A. C. J. and Deng, W. (1995), Chaotic motion of a horizontal impact pair, *Journal of Sound and Vibration*, **181**, pp.231-250.

Luo, A. C. J. (1995), *Analytical Modeling of Bifurcations, Chaos and Fractals in Nonlinear Dynamics*, Ph.D. Dissertation, University of Manitoba, Winnipeg, Canada.

Luo, A. C. J. (2005), A theory for non-smooth dynamic systems on the connectable domains, *Communications in Nonlinear Science and Numerical Simulation*, 10, pp.1-55.

Luo, A. C. J. (2006), *Singularity and Dynamics on Discontinuous Vector Fields*, Amsterdam: Elsevier.

Luo, A. C. J. (2007), Differential geometry of flows in nonlinear dynamical systems, *Proceedings of IDECT'07*, 2007 ASME International Design Engineering Technical Conferences, September 4-7, 2007, Las Vegas, Nevada, USA. DETC2007-84754.

Luo, A. C. J. (2008a), A theory for flow switchability in discontinuous dynamical systems, *Nonlinear Analysis: Hybrid Systems*, **2**, pp.1030-1061.

Luo, A. C. J. (2008b), *Global Transversality, Resonance and Chaotic dynamics*, Singapore: World Scientific.

Luo, A. C. J. (2008c), On the differential geometry of flows in nonlinear dynamic systems, ASME *Journal of Computational and Nonlinear Dynamics*, 021104-1-10.

Luo, A. C. J. and O'Connor, D. (2007a), Nonlinear dynamics of a gear transmission system, Part I: mechanism of impacting chatter with stick, *Proceedings of IDETC'07*, 2007 ASME International Design Engineering Conferences and Exposition, September 4-7, 2007, Las Vegas, Nevada. IDETC2007-34881.

Luo, A. C. J. and O'Connor, D. (2007b), Nonlinear dynamics of a gear transmission system, Part II: periodic impacting chatter and stick, *Proceedings of IMECE'07*, 2007 ASME *International Mechanical Engineering Congress and Exposition*, November 10-16, 2007, Seattle, Washington. IMECE2007-43192.

Luo, A. C. J. and Guo, Y.(2008), Switching bifurcation and chaos in an extended Fermi-acceleration oscillator, *Proceedings of IMECE'08*, 2008 ASME *International Mechanical Engineering Congress and Exposition*, October 31-November 6, 2008, Boston, Massachusetts. IMECE2008-68003.

Chapter 6
Dynamical Systems with Frictions

In this chapter, two dynamical systems connected with frictions will be presented, and such two discontinuous dynamical systems possess a moving boundary. For such a problem, the analytical conditions for motion switching and sliding on the time-varying boundary will be developed. Based on the time-varying boundary, the basic mappings are introduced and the mapping structures for periodic motions will be developed. From a certain mapping structure, the corresponding periodic motions will be predicted analytically and the corresponding local stability and bifurcation analysis will be completed. To understand the singularity and switchability of motions to the time-varying boundary in such a two-mass system, illustrations of periodic motions will be given from analytical predictions. The relative forces will be presented for illustration of the analytical conditions.

6.1 Problem statement

As in Luo and Thapa (2007, 2009), consider two oscillators contacting each other on a surface with a dry friction, as shown in Fig.6.1. The first oscillator includes the primary mass m_1, a spring of stiffness k_1 and a damper of coefficient r_1. The primary mass is connected with the spring and damper, and both of them are fixed on the wall. The second oscillator consists of a second mass m_2, a spring of stiffness k_2 and a damper of coefficient r_2. The spring and damper connect both the second mass and the fixed wall. An external excitation $A_0^{(1)} \cos \Omega t$ exerts on the second mass, where $A_0^{(1)}$ and Ω are excitation amplitude and frequency, respectively. The displacements of the primary and second masses are measured from their equilibrium positions through the coordinates $x^{(1)}$ and $x^{(2)}$, accordingly.

Because the two masses contact each other, they can slide or stick upon each other and then move together. Without loss generality, consider a dry-friction force for the ith mass ($i = 1, 2$) as

$$\bar{F}_{\mathrm{f}}^{(i)} \begin{cases} = \mu_k F_{\mathrm{N}}, & \text{for } \dot{x}^{(i)} > \dot{x}^{(\bar{i})} \\ \in [-\mu_k F_{\mathrm{N}}, \mu_k F_{\mathrm{N}}], & \text{for } \dot{x}^{(i)} = \dot{x}^{(\bar{i})} \\ = -\mu_k F_{\mathrm{N}}, & \text{for } \dot{x}^{(i)} < \dot{x}^{(\bar{i})} \end{cases} \tag{6.1}$$

where F_{N} is a normal force acting on the two masses, as sketched in Fig. 6.2. $\dot{x}^{(i)}$ is the velocity of the ith mass, and $\dot{x}^{(\bar{i})}$ is the velocity of the $\bar{i}^{\mathrm{th}}$ mass, where $\bar{i} = 2$ if $i = 1$ or $\bar{i} = 1$ if $i = 2$. μ_k is the dynamic coefficient of friction.

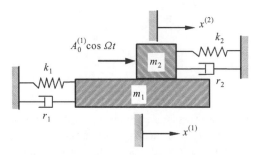

Fig. 6.1 Mechanical model of a system of two oscillators connected with friction

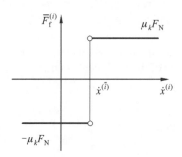

Fig. 6.2 Friction force between the two masses

For the ith-mass m_i, the non-friction forces acting on the mass m_i in the $x^{(i)}$-direction per unit mass is given by

$$F_{\mathrm{nf}}^{(i)} = -2d^{(i)}\dot{x}^{(i)} - (\omega^{(i)})^2 x^{(i)} + Q_0^{(i)} \cos \Omega t, \tag{6.2}$$

where

$$\left. \begin{aligned} 2d^{(i)} &= \frac{r_i}{m_i}, \ \omega^{(i)} = \sqrt{\frac{k_i}{m_i}}, \ F_{\mathrm{f}}^{(i)} = \frac{\mu F_{\mathrm{N}}}{m_i}, \ \text{for } i = 1, 2, \\ Q_0^{(2)} &= \frac{A_0^{(2)}}{m_2}, Q_0^{(1)} = 0. \end{aligned} \right\} \tag{6.3}$$

6.1 Problem statement

If $|F_{nf}^{(i)}| \geq F_f^{(i)}$, the non-friction force can overcome the friction force, the two masses will not stick together to move. So, the total force per unit mass becomes

$$F^{(i)} = -2d^{(i)}\dot{x}^{(i)} - (\omega^{(i)})^2 x^{(i)} + Q_0^{(i)} \cos \Omega t - F_f^{(i)} \text{sgn}(\dot{x}^{(i)} - \dot{x}^{(\bar{i})}), \quad (6.4)$$

where

$$\text{sgn}(\dot{x}^{(i)} - \dot{x}^{(\bar{i})}) = \begin{cases} 1, & \text{if } \dot{x}^{(i)} - \dot{x}^{(\bar{i})} > 0, \\ -1, & \text{if } \dot{x}^{(i)} - \dot{x}^{(\bar{i})} < 0. \end{cases} \quad (6.5)$$

The equation of non-stick motion for $i = 1, 2$ becomes

$$\ddot{x}^{(i)} + 2d^{(i)}\dot{x}^{(i)} + (\omega^{(i)})^2 x^{(i)} = Q_0^{(i)} \cos \Omega t - F_f^{(i)} \text{sgn}(\dot{x}^{(i)} - \dot{x}^{(\bar{i})}). \quad (6.6)$$

If $|F_{nf}^{(i)}| < F_f^{(i)}$, the non-friction force cannot overcome the static friction force. The two masses will stick together. So the two masses have the same velocity (i.e., $\dot{x}^{(i)} = \dot{x}^{(\bar{i})}$). The equation of stick motion for the ith- mass is

$$\ddot{x}^{(i)} + 2d^{(0)}\dot{x}^{(i)} + (\omega^{(0)})^2 x^{(i)} = Q_0^{(0)} \cos \Omega t - K^{(i)}, \quad (6.7)$$

where

$$\left. \begin{array}{c} 2d^{(0)} = \dfrac{r_1 + r_2}{m_1 + m_2}, \; \omega^{(0)} = \sqrt{\dfrac{k_1 + k_2}{m_1 + m_2}}, \; Q_0^{(0)} = \dfrac{A_0^{(2)}}{m_1 + m_2}, \\ K^{(i)} = -\dfrac{k_i}{m_1 + m_2}({}^0 x^{(i)} - {}^0 x^{(\bar{i})}), \end{array} \right\} \quad (6.8)$$

and ${}^0 x^{(i)}$ and ${}^0 x^{(\bar{i})}$ are two displacements for the starting of stick motion on the velocity boundary. $K^{(i)}$ is constant before vanishing of stick motion on the boundary. However, the interaction forces per unit mass between the two masses are computed by

$$F_{\text{int}}^{(i)} = -\ddot{x}^{(i)} - 2d^{(i)}\dot{x}^{(i)} - (\omega^{(i)})^2 x^{(i)} + Q_0^{(i)} \cos \Omega t, \text{ for } i = 1, 2, \quad (6.9)$$

and from the Newton's third law, we have

$$m_1 F_{\text{int}}^{(1)} + m_2 F_{\text{int}}^{(2)} = 0. \quad (6.10)$$

If $|F_{\text{int}}^{(i)}| \geq F_f^{(i)}$, the stick motion will disappear.

Introduce the relative displacement, velocity and accelerations as

$$z^{(i)} = x^{(i)} - x^{(\bar{i})}, \dot{z}^{(i)} = \dot{x}^{(i)} - \dot{x}^{(\bar{i})}, \ddot{z}^{(i)} = \ddot{x}^{(i)} - \ddot{x}^{(\bar{i})}. \quad (6.11)$$

The equation of non-stick motion for $i = 1, 2$ becomes

$$\left. \begin{array}{c} \ddot{z}^{(i)} + 2d^{(i)}\dot{z}^{(i)} + (\omega^{(i)})^2 z^{(i)} = Q_0^{(i)} \cos \Omega t - F_f^{(i)} \text{sgn}(\dot{z}^{(i)}) \\ \qquad -\ddot{x}^{(\bar{i})} - 2d^{(i)}\dot{x}^{(\bar{i})} - (\omega^{(i)})^2 x^{(\bar{i})}, \\ \ddot{x}^{(\bar{i})} + 2d^{(\bar{i})}\dot{x}^{(\bar{i})} + (\omega^{(\bar{i})})^2 x^{(\bar{i})} = Q_0^{(\bar{i})} \cos \Omega t - F_f^{(\bar{i})} \text{sgn}(\dot{z}^{(\bar{i})}). \end{array} \right\} \quad (6.12)$$

for a stick motion, we have

$$\ddot{z}^{(i)} = 0 \text{ and } \dot{z}^{(i)} = 0. \tag{6.13}$$

from which $z^{(i)} = C$ is constant. The corresponding absolute motion is determined by Eq.(6.7).

6.2 Switching and stick motions

In this section, the phase spaces of the two systems connected with friction will be partitioned into domains and boundary, and the corresponding equations of motion will be described on each sub-domain and boundary through vector fields. Furthermore, the analytical conditions for stick and non-stick motions on the time-varying boundary will be developed.

6.2.1 Equations of motion

The direction of friction force is dependent on the direction of relative velocities between two masses. Once the direction of the relative velocity between the two masses switches, the friction force will change the direction, the total forces of the mass will be discontinuous. To investigate such a problem, the phase plane for the ith-mass ($i = 1, 2$) is partitioned into two domains, as shown in Fig.6.3. For each mass, there are two domains expressed by $\Omega_\alpha^{(i)}$ ($\alpha = 1, 2$). The boundary is expressed by $\partial \Omega_{\alpha\beta}^{(i)}$ ($\alpha, \beta = 1, 2$ and $\alpha \neq \beta$), which means the boundary for a flow of the ith-mass from domain $\Omega_\alpha^{(i)}$ to $\Omega_\beta^{(i)}$. For simplicity, let $y^{(i)} = \dot{x}^{(i)}$ and the corresponding domain and boundary for the ith-mass are defined as

$$\left.\begin{aligned}
\Omega_1^{(i)} &= \left\{ (x^{(i)}, y^{(i)}) | y^{(i)} \in (y^{(\bar{i})}(t), \infty), t \in [0, \infty) \right\}, \\
\Omega_2^{(i)} &= \left\{ (x^{(i)}, y^{(i)}) | y^{(i)} \in (-\infty, y^{(\bar{i})}(t)), t \in [0, \infty) \right\}, \\
\partial \Omega_{\alpha\beta}^{(i)} &= \left\{ (x^{(i)}, y^{(i)}) | \varphi_{\alpha\beta}(x, y) = y^{(i)} - y^{(\bar{i})} = 0 \right\}, \\
&\text{for } \alpha, \beta \in \{1, 2\}, \alpha \neq \beta.
\end{aligned}\right\} \tag{6.14}$$

From the defined sub-domains and boundary, the equations of motion in Eqs.(6.6) and (6.7) are expressed by the state vector as

$$\dot{\mathbf{x}}^{(i)} = \mathbf{F}_\lambda^{(i)}(\mathbf{x}^{(i)}, t), i \in \{1, 2\}, \lambda \in \{0, \alpha\}, \tag{6.15}$$

where

$$\mathbf{x}^{(i)} = (x^{(i)}, y^{(i)})^{\text{T}}, \tag{6.16}$$

6.2 Switching and stick motions

$$\left. \begin{aligned} &\mathbf{F}_\alpha^{(i)}(\mathbf{x}^{(i)},t) = (y^{(i)}, F_\alpha^{(i)}(\mathbf{x}^{(i)}, \Omega t))^{\mathrm{T}} \text{ in } \Omega_\alpha^{(i)}, \text{ for } \alpha \in \{1,2\}, \\ &\mathbf{F}_0^{(i)}(\mathbf{x}^{(i)},t) \in [\mathbf{F}_\alpha^{(i)}(\mathbf{x}^{(i)},t), \mathbf{F}_\beta^{(i)}(\mathbf{x}^{(i)},t)], \text{ on } \partial\Omega_{\alpha\beta}^{(i)} \text{ for non-stick,} \\ &\mathbf{F}_0^{(i)}(\mathbf{x}^{(i)},t) = (y^{(0)}, F_0^{(i)}(\mathbf{x}^{(i)}, \Omega t))^{\mathrm{T}}, \text{ on } \partial\Omega_{\alpha\beta}^{(i)} \text{ for stick,} \\ &F_\alpha^{(i)}\left(\mathbf{x}^{(i)}, \Omega t\right) = -2d_\alpha^{(i)} \dot{x}^{(i)} - (\omega_\alpha^{(i)})^2 x^{(i)} + Q_0^{(i)} \cos \Omega t + b_\alpha^{(i)}, \\ &F_0^{(i)}(\mathbf{x}^{(i)}, \Omega t) = -2d^{(0)} \dot{x}^{(i)} - (\omega^{(0)})^2 x^{(i)} + Q_0^{(0)} \cos \Omega t - K^{(i)}. \end{aligned} \right\} \quad (6.17)$$

Note that $b_1^{(i)} = (-1)^i F_f^{(i)}$ and $b_2^{(i)} = (-1)^{i+1} F_f^{(i)}$. For the mechanical model in Fig.6.1, we have $d_1^{(i)} = d_2^{(i)} = d^{(i)}$, $\omega_1^{(i)} = \omega_2^{(i)} = \omega^{(i)}$.

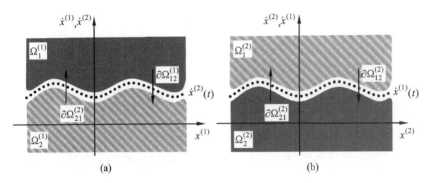

Fig. 6.3 Phase plane partitions in the absolute frame for the two masses : (a) the first mass and (b) the second mass

For the absolute frame, the friction forces switching of the ith-mass is dependent on the velocity of the $\bar{i}^{\mathrm{th}}$-mass. Such a velocity changes with time and displacement. It is very difficult to handle the switching and sliding motion on the boundary. Therefore, the relative frame is introduced to investigate such switching and sliding motions. In the relative frame, equation (6.11) will be adopted. Let $v^{(i)} = \dot{z}^{(i)}$ and the corresponding domain and boundary for the ith-mass are defined as

$$\left. \begin{aligned} &\Omega_1^{(i)} = \left\{ (z^{(i)}, v^{(i)}) | v^{(i)} \in (0, \infty), t \in [0, \infty) \right\}, \\ &\Omega_2^{(i)} = \left\{ (z^{(i)}, v^{(i)}) | v^{(i)} \in (-\infty, 0), t \in [0, \infty) \right\}, \\ &\partial\Omega_{\alpha\beta}^{(i)} = \left\{ (z^{(i)}, v^{(i)}) | \varphi_{\alpha\beta}^{(i)}(z,v) = v^{(i)} = 0 \right\} \\ &\text{for } \alpha, \beta \in \{1,2\}, \alpha \neq \beta. \end{aligned} \right\} \quad (6.18)$$

The foregoing domains and boundary are shown in Fig.6.4. The boundary becomes a straight line on the z-axis. The equation of motion for the ith-mass becomes

$$\left. \begin{aligned} &\dot{\mathbf{z}}^{(i)} = \mathbf{g}_\lambda^{(i)}(\mathbf{z}^{(i)}, \mathbf{x}^{(\bar{i})}, t), i \in \{1,2\}, \lambda \in \{0, \alpha\}, \\ &\dot{\mathbf{x}}^{(\bar{i})} = \mathbf{F}_{\bar{\lambda}}^{(\bar{i})}(\mathbf{x}^{(\bar{i})}, t), \bar{i} \in \{1,2\}, \bar{\lambda} \in \{0, \bar{\alpha}\}, \end{aligned} \right\} \quad (6.19)$$

where $\bar{\alpha} = 2$ if $\alpha = 1; \bar{\alpha} = 1$ if $\alpha = 2$, and

$$\mathbf{z}^{(i)} = (z^{(i)}, v^{(i)})^{\mathrm{T}}, \tag{6.20}$$

$$\left.\begin{array}{l}\mathbf{g}_\alpha^{(i)}(\mathbf{z}^{(i)},\mathbf{x}^{(\bar{i})},t) = (v^{(i)}, g_\alpha^{(i)}(\mathbf{z}^{(i)},\mathbf{x}^{(\bar{i})},\Omega t))^{\mathrm{T}}, \text{ in } \Omega_\alpha^{(i)} \text{ for } \alpha \in \{1,2\}, \\ \mathbf{g}_0^{(i)}(\mathbf{z}^{(i)},\mathbf{x}^{(\bar{i})},t) \in [\mathbf{g}_\alpha^{(i)}(\mathbf{z}^{(i)},\mathbf{x}^{(\bar{i})},t), \mathbf{g}_\beta^{(i)}(\mathbf{z}^{(i)},\mathbf{x}^{(\bar{i})},t)], \\ \qquad \text{on } \partial \Omega_{\alpha\beta}^{(i)} \text{ for non-stick,} \\ \mathbf{g}_0^{(i)}(\mathbf{z}^{(i)},\mathbf{x}^{(\bar{i})},t) = (0,0)^{\mathrm{T}} \text{ on } \partial \Omega_{\alpha\beta}^{(i)} \text{ for stick,} \\ g_\alpha^{(i)}(\mathbf{z}^{(i)},\mathbf{x}^{(\bar{i})},\Omega t) = -2d_\alpha^{(i)} \dot{z}^{(i)} - (\omega_\alpha^{(i)})^2 z^{(i)} + Q_0^{(i)} \cos \Omega t - b_\alpha^{(i)} \\ \qquad - \ddot{x}_{\bar{\alpha}}^{(\bar{i})} - 2d_\alpha^{(i)} \dot{x}_{\bar{\alpha}}^{(\bar{i})} - (\omega_\alpha^{(i)})^2 x_{\bar{\alpha}}^{(\bar{i})}. \end{array}\right\} \tag{6.21}$$

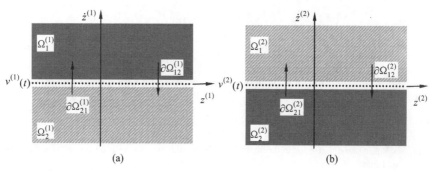

Fig. 6.4 Phase plane partitions in the absolute for the two masses: (a) the first mass and (b) the second mass

6.2.2 Analytical conditions

Since the discontinuous boundary is dependent on the time in the absolute frame, it is very difficult that the corresponding analytical conditions are obtained. Herein, in the relative frame, the analytical conditions will be developed for a flow switching from one domain to another one. Before presenting analytical conditions, the following G-functions are introduced from Chapter 2 (also see Luo (2007a,b, 2008a,b)), i.e.,

$$G_i^{(0,\alpha)}(\mathbf{z}_\alpha^{(i)},\mathbf{x}_{\bar{\alpha}}^{(\bar{i})},t_{m\pm}) = \mathbf{n}_{\partial\Omega_{\alpha\beta}^{(i)}}^{\mathrm{T}} \cdot [\mathbf{g}_\alpha^{(i)}(\mathbf{z}_\alpha^{(i)},\mathbf{x}_{\bar{\alpha}}^{(\bar{i})},t_{m\pm}) - \mathbf{g}_{\alpha\beta}^{(0)}(\mathbf{z}_{\alpha\beta}^{(0)},t_{m\pm})], \tag{6.22}$$

$$G_i^{(1,\alpha)}(\mathbf{z}_\alpha^{(i)},\mathbf{x}_{\bar{\alpha}}^{(\bar{i})},t_{m\pm}) = 2D\mathbf{n}_{\partial\Omega_{\alpha\beta}^{(i)}}^{\mathrm{T}} \cdot [\mathbf{g}_\alpha^{(i)}(\mathbf{z}_\alpha^{(i)},\mathbf{x}_{\bar{\alpha}}^{(\bar{i})},t_{m\pm}) - \mathbf{g}_{\alpha\beta}^{(0)}(\mathbf{z}_{\alpha\beta}^{(0)},t_{m\pm})]$$
$$+ \mathbf{n}_{\partial\Omega_{\alpha\beta}^{(i)}}^{\mathrm{T}} \cdot [D\mathbf{g}_\alpha^{(i)}(\mathbf{z}_\alpha^{(i)},\mathbf{x}_{\bar{\alpha}}^{(\bar{i})},t_{m\pm}) - D\mathbf{g}_{\alpha\beta}^{(0)}(\mathbf{z}_{\alpha\beta}^{(0)},t_{m\pm})], \tag{6.23}$$

6.2 Switching and stick motions

where $D = \dot{z}\partial(\ )/\partial z + \dot{v}\partial(\ )/\partial v + \partial(\ )/\partial t$. The switching time t_m is the time for motion on the boundary, and $t_{m\pm} = t_m \pm 0$ reflects time for responses in the domains instead of on the boundary. The vector field $\mathbf{g}_{\alpha\beta}^{(0)}(\mathbf{z}_{\alpha\beta}^{(i)}, t)$ is for a flow on the boundary. The normal vector $\mathbf{n}_{\partial\Omega_{\alpha\beta}^{(i)}}$ of the boundary $\partial\Omega_{\alpha\beta}^{(i)}$ is computed by

$$\mathbf{n}_{\partial\Omega_{\alpha\beta}^{(i)}} = \nabla\varphi_{\alpha\beta}^{(i)} = \left(\frac{\partial\varphi_{\alpha\beta}^{(i)}}{\partial z^{(i)}}, \frac{\partial\varphi_{\alpha\beta}^{(i)}}{\partial v^{(i)}}\right)^T, \quad (6.24)$$

where $\nabla = (\partial/\partial z, \partial/\partial v)^T$ is the Hamilton operator.

Equation (6.22) gives the normal components of the difference between the system vector field and the boundary vector field in the normal direction of the separation boundary. Equation (6.23) gives the time-change rate of the normal component of such a vector field difference. Because the boundary $\partial\Omega_{\alpha\beta}^{(i)}$ in the relative frame is a straight line independent of time t, $D\mathbf{n}_{\partial\Omega_{\alpha\beta}^{(i)}}^T = 0$ exists. Due to $\mathbf{n}_{\partial\Omega_{\alpha\beta}^{(i)}}^T \cdot \mathbf{g}_{\alpha\beta}^{(0)}(\mathbf{z}, t_{m\pm}) = 0$, we have

$$D\mathbf{n}_{\partial\Omega_{\alpha\beta}^{(i)}}^T \cdot \mathbf{g}_{\alpha\beta}^{(0)}(\mathbf{z}, t_{m\pm}) + \mathbf{n}_{\partial\Omega_{\alpha\beta}^{(i)}}^T \cdot D\mathbf{g}_{\alpha\beta}^{(0)}(\mathbf{z}, t_{m\pm}) = 0. \quad (6.25)$$

Therefore

$$\mathbf{n}_{\partial\Omega_{\alpha\beta}^{(i)}}^T \cdot D\mathbf{g}_{\alpha\beta}^{(0)}(\mathbf{z}, t_{m\pm}) = 0. \quad (6.26)$$

Equations (6.22) and (6.23) reduce to

$$\left.\begin{aligned}G_i^{(0,\alpha)}(\mathbf{z}_\alpha^{(i)}, t_{m\pm}) &= \mathbf{n}_{\partial\Omega_{\alpha\beta}^{(i)}}^T \cdot \mathbf{g}_\alpha^{(i)}(\mathbf{z}_\alpha^{(i)}, \mathbf{x}_{\bar{\alpha}}^{(\bar{i})}, t_{m\pm}), \\ G_i^{(1,\alpha)}(\mathbf{z}_\alpha^{(i)}, t_{m\pm}) &= \mathbf{n}_{\partial\Omega_{\alpha\beta}^{(i)}}^T \cdot D\mathbf{g}_\alpha^{(i)}(\mathbf{z}_\alpha^{(i)}, \mathbf{x}_{\bar{\alpha}}^{(\bar{i})}, t_{m\pm}).\end{aligned}\right\} \quad (6.27)$$

For a general case, equations (6.22) and (6.23) instead of Eq.(6.27) will be used. Notice that $\mathbf{g}_{\alpha\beta}^{(0)}(\mathbf{z}, t) = (0,0)^T$. From Luo (2005a,b), the grazing motion to the boundary $\partial\Omega_{\alpha\beta}^{(i)}$ in domain $\Omega_{\alpha\beta}^{(i)}$ is guaranteed by

$$\left.\begin{aligned}G_i^{(0,\alpha)}(\mathbf{z}_\alpha^{(i)}, t_{m\pm}) &= \mathbf{n}_{\partial\Omega_{\alpha\beta}^{(i)}}^T \cdot \mathbf{g}_\alpha^{(i)}(\mathbf{z}_\alpha^{(i)}, \mathbf{x}_{\bar{\alpha}}^{(\bar{i})}, t_{m\pm}) = 0 \\ &\text{for } \alpha,\beta \in \{1,2\}, \\ G_i^{(1,\alpha)}(\mathbf{z}_\alpha^{(i)}, t_{m\pm}) &= \mathbf{n}_{\partial\Omega_{\alpha\beta}^{(i)}}^T \cdot D\mathbf{g}_\alpha^{(i)}(\mathbf{z}_\alpha^{(i)}, \mathbf{x}_{\bar{\alpha}}^{(\bar{i})}, t_{m\pm}) > 0 \\ &\text{for } \mathbf{n}_{\partial\Omega_{\alpha\beta}^{(i)}} \to \Omega_\beta^{(i)} \text{ or} \\ G_i^{(1,\alpha)}(\mathbf{z}_\alpha^{(i)}, t_{m\pm}) &= \mathbf{n}_{\partial\Omega_{\alpha\beta}^{(i)}}^T \cdot D\mathbf{g}_\alpha^{(i)}(\mathbf{z}_\alpha^{(i)}, \mathbf{x}_{\bar{\alpha}}^{(\bar{i})}, t_{m\pm}) < 0 \\ &\text{for } \mathbf{n}_{\partial\Omega_{\alpha\beta}^{(i)}} \to \Omega_\alpha^{(i)},\end{aligned}\right\} \quad (6.28)$$

where $\mathbf{n}_{\partial\Omega_{\alpha\beta}^{(i)}} \to \Omega_\beta^{(i)}$ means that the normal vector of the boundary $\partial\Omega_{\alpha\beta}^{(i)}$ (i.e., $\mathbf{n}_{\partial\Omega_{\alpha\beta}^{(i)}}$) points to the domain $\Omega_\beta^{(i)}$, and

$$D\mathbf{g}_\alpha^{(i)}(\mathbf{z}_\alpha^{(i)}, \mathbf{x}_{\tilde{\alpha}}^{(\tilde{i})}, t)$$
$$= \left[g_\alpha^{(i)}(\mathbf{z}_\alpha^{(i)}, \mathbf{x}_{\tilde{\alpha}}^{(\tilde{i})}, t), \nabla g_\alpha^{(i)}(\mathbf{z}_\alpha^{(i)}, \mathbf{x}_{\tilde{\alpha}}^{(\tilde{i})}, t) \cdot \mathbf{g}_\alpha^{(i)}(\mathbf{z}_\alpha^{(i)}, \mathbf{x}_{\tilde{\alpha}}^{(\tilde{i})}, t) + \frac{\partial g_\alpha^{(i)}(\mathbf{z}_\alpha^{(i)}, \mathbf{x}_{\tilde{\alpha}}^{(\tilde{i})}, t)}{\partial t} \right]^{\mathrm{T}}. \quad (6.29)$$

From Chapter 2 (or Luo (2006, 2007a,b)), a motion passable to the boundary $\partial\Omega_{\alpha\beta}^{(i)}$ is guaranteed by

$$_iL_{\alpha\beta}^{(0,0)}(t_m) = G_i^{(0,\alpha)}(\mathbf{z}_\alpha^{(i)}, t_{m-}) \times G_i^{(0,\beta)}(\mathbf{z}_\beta^{(i)}, t_{m+}) > 0. \quad (6.30)$$

In other words, the foregoing conditions can be expressed by

$$\left. \begin{array}{l} G_i^{(0,\alpha)}(\mathbf{z}_\alpha^{(i)}, t_{m-}) < 0, G_i^{(0,\beta)}(\mathbf{z}_\beta^{(i)}, t_{m+}) < 0, \text{ for } \mathbf{n}_{\partial\Omega_{\alpha\beta}^{(i)}} \to \Omega_\alpha^{(i)}, \\ G_i^{(0,\alpha)}(\mathbf{z}_\alpha^{(i)}, t_{m-}) > 0, G_i^{(0,\beta)}(\mathbf{z}_\beta^{(i)}, t_{m+}) > 0, \text{ for } \mathbf{n}_{\partial\Omega_{\alpha\beta}^{(i)}} \to \Omega_\beta^{(i)}. \end{array} \right\} \quad (6.31)$$

From Chapter 2 (or Luo (2005a, 2006, 2007b)), a sliding motion on the boundary $\partial\Omega_{\alpha\beta}^{(i)}$ is guaranteed by

$$_iL_{\alpha\beta}^{(0,0)}(t_m) = G_i^{(0,\alpha)}(\mathbf{z}_\alpha^{(i)}, t_{m-}) \times G_i^{(0,\beta)}(\mathbf{z}_\beta^{(i)}, t_{m-}) < 0. \quad (6.32)$$

The foregoing conditions can also be expressed by

$$\left. \begin{array}{l} G_i^{(0,\alpha)}(\mathbf{z}_2^{(i)}, t_{m-}) < 0, G_i^{(0,\beta)}(\mathbf{z}_1^{(i)}, t_{m-}) > 0, \text{ for } \mathbf{n}_{\partial\Omega_{\alpha\beta}^{(i)}} \to \Omega_\alpha^{(i)}, \\ G_i^{(0,\alpha)}(\mathbf{z}_2^{(i)}, t_{m-}) > 0, G_i^{(0,\beta)}(\mathbf{z}_1^{(i)}, t_{m-}) < 0, \text{ for } \mathbf{n}_{\partial\Omega_{\alpha\beta}^{(i)}} \to \Omega_\beta^{(i)}. \end{array} \right\} \quad (6.33)$$

For the discontinuous boundaries $\partial\Omega_{\alpha\beta}^{(i)}$ in Eq.(6.18), the normal vector is

$$\mathbf{n}_{\partial\Omega_{23}^{(i)}} = \mathbf{n}_{\partial\Omega_{12}^{(i)}} = (0,1)^{\mathrm{T}}. \quad (6.34)$$

The corresponding G-functions are

$$\left. \begin{array}{l} G_i^{(0,\alpha)}(\mathbf{z}_\alpha^{(i)}, t) = \mathbf{n}_{\partial\Omega_{\alpha\beta}^{(i)}}^{\mathrm{T}} \cdot \mathbf{g}_\alpha^{(i)}(\mathbf{z}_\alpha^{(i)}, \mathbf{x}_{\tilde{\alpha}}^{(\tilde{i})}, t) = g_\alpha^{(i)}(\mathbf{z}_\alpha^{(i)}, \mathbf{x}_{\tilde{\alpha}}^{(\tilde{i})}, t), \\ G_i^{(1,\alpha)}(\mathbf{z}_\alpha^{(i)}, t) = \mathbf{n}_{\partial\Omega_{\alpha\beta}^{(i)}}^{\mathrm{T}} \cdot D\mathbf{g}_\alpha^{(i)}(\mathbf{z}_\alpha^{(i)}, \mathbf{x}_{\tilde{\alpha}}^{(\tilde{i})}, t) \\ \qquad = \nabla g_\alpha^{(i)}(\mathbf{z}_\alpha^{(i)}, \mathbf{x}_{\tilde{\alpha}}^{(\tilde{i})}, t) \cdot \mathbf{g}_\alpha^{(i)}(\mathbf{z}_\alpha^{(i)}, \mathbf{x}_{\tilde{\alpha}}^{(\tilde{i})}, t) + \frac{\partial g_\alpha^{(i)}(\mathbf{z}_\alpha^{(i)}, \mathbf{x}_{\tilde{\alpha}}^{(\tilde{i})}, t)}{\partial t}. \end{array} \right\} \quad (6.35)$$

6.2 Switching and stick motions

$$G_i^{(1,\alpha)}(\mathbf{z}^{(i)},t) = -2d_\alpha^{(i)}\ddot{z}_\alpha^{(i)} - (\omega_\alpha^{(i)})^2\dot{z}_\alpha^{(i)} - Q_\alpha^{(i)}\Omega\sin\Omega t$$
$$- \dddot{x}_{\bar{\alpha}}^{(\bar{i})} - 2d_\alpha^{(i)}\ddot{x}_{\bar{\alpha}}^{(\bar{i})} - (\omega_\alpha^{(i)})^2\dot{x}_{\bar{\alpha}}^{(\bar{i})}. \quad (6.36)$$

From Eqs.(6.19) and (6.21), the relative jerk is given by

$$J_\alpha^{(i)}(t) = -2d_\alpha^{(i)}\ddot{z}_\alpha^{(i)} - (\omega_\alpha^{(i)})^2\dot{z}_\alpha^{(i)} - Q_\alpha^{(i)}\Omega\sin\Omega t$$
$$- \dddot{x}_{\bar{\alpha}}^{(\bar{i})} - 2d_\alpha^{(i)}\ddot{x}_{\bar{\alpha}}^{(\bar{i})} - (\omega_\alpha^{(i)})^2\dot{x}_{\bar{\alpha}}^{(\bar{i})}. \quad (6.37)$$

Therefore, for this case, the function $G_i^{(\alpha,1)}(\mathbf{z}_\alpha^{(i)},t)$ is a relative jerk in domain $\Omega_\alpha^{(i)}$. The function $g_\alpha^{(i)}(\mathbf{z}_\alpha^{(i)},t)$ is a relative acceleration or a relative force per unit mass. Both of the relative acceleration and jerk can be expressed by two absolute state variables as

$$\left.\begin{aligned}
g_\alpha^{(i)}(\mathbf{z}_\alpha^{(i)},\mathbf{x}_{\bar{\alpha}}^{(\bar{i})},t) &= \Gamma_\alpha^{(i)}(\mathbf{x}_\alpha^{(i)},\mathbf{x}_{\bar{\alpha}}^{(\bar{i})},t) \\
&= -2d_\alpha^{(i)}\dot{x}_\alpha^{(i)} - (\omega_\alpha^{(i)})^2 x_\alpha^{(i)} + Q_0^{(i)}\cos\Omega t - b_\alpha^{(i)} \\
&\quad + 2d_{\bar{\alpha}}^{(\bar{i})}\dot{x}_{\bar{\alpha}}^{(\bar{i})} + (\omega_{\bar{\alpha}}^{(\bar{i})})^2 x_{\bar{\alpha}}^{(\bar{i})} - Q_0^{(i)}\cos\Omega t - b_{\bar{\alpha}}^{(\bar{i})}, \\
G_i^{(1,\alpha)}(\mathbf{z}_\alpha^{(i)},\mathbf{x}_{\bar{\alpha}}^{(\bar{i})},t) &= \Gamma_\alpha^{(\alpha,i)}(\mathbf{x}_\alpha^{(i)},\mathbf{x}_{\bar{\alpha}}^{(\bar{i})},t) \\
&= -2d_\alpha^{(i)}\ddot{x}_\alpha^{(i)} - (\omega_\alpha^{(i)})^2 \dot{x}_\alpha^{(i)} - Q_0^{(i)}\Omega\sin\Omega t \\
&\quad + 2d_{\bar{\alpha}}^{(\bar{i})}\ddot{x}_{\bar{\alpha}}^{(\bar{i})} + (\omega_{\bar{\alpha}}^{(\bar{i})})^2 \dot{x}_{\bar{\alpha}}^{(\bar{i})} + Q_0^{(i)}\Omega\sin\Omega t.
\end{aligned}\right\} \quad (6.38)$$

With Eqs.(6.31) and (6.35), the conditions for passable motion from domain $\Omega_\alpha^{(i)}$ ($\alpha = 1,2$) to $\Omega_\beta^{(i)}$ ($\beta = 2,1$) are

$$\left.\begin{aligned}
g_1^{(i)}(\mathbf{z}_1^{(i)},\mathbf{x}_2^{(\bar{i})},t_{m-}) < 0,\ g_2^{(i)}(\mathbf{z}_2^{(i)},\mathbf{x}_1^{(\bar{i})},t_{m+}) < 0 \text{ on } \partial\Omega_{12}^{(i)}, \\
g_2^{(i)}(\mathbf{z}_2^{(i)},\mathbf{x}_1^{(\bar{i})},t_{m-}) > 0,\ g_1^{(i)}(\mathbf{z}_1^{(i)},\mathbf{x}_2^{(\bar{i})},t_{m+}) > 0 \text{ on } \partial\Omega_{21}^{(i)}.
\end{aligned}\right\} \quad (6.39)$$

For the sliding motion along the discontinuous boundary $\partial\Omega_{\alpha\beta}^{(i)}$ ($\alpha,\beta = 1,2$ and $\alpha \neq \beta$), one obtains

$$g_1^{(i)}(\mathbf{z}_1^{(i)},\mathbf{x}_2^{(\bar{i})},t_{m-}) < 0,\ g_2^{(i)}(\mathbf{z}_2^{(i)},\mathbf{x}_1^{(\bar{i})},t_{m-}) > 0 \text{ on } \partial\Omega_{\alpha\beta}^{(i)}. \quad (6.40)$$

The sliding motion vanishing from the boundary $\partial\Omega_{\alpha\beta}^{(i)}$ and getting into domain $\Omega_\alpha^{(i)}$ ($\alpha = 1,2$) requires

$$\left.\begin{aligned}
g_\alpha^{(i)}(\mathbf{z}_\alpha^{(i)},\mathbf{x}_{\bar{\alpha}}^{(\bar{i})},t_{m\pm}) &= 0,\ g_\beta^{(i)}(\mathbf{z}_\beta^{(i)},\mathbf{x}_{\bar{\beta}}^{(\bar{i})},t_{m-}) > 0, \\
G_i^{(1,\alpha)}(\mathbf{z}_\alpha^{(i)},t_{m\pm}) &> 0 \text{ if } \mathbf{n}_{\partial\Omega_{\alpha\beta}^{(i)}} \to \Omega_\alpha^{(i)}, \\
G_i^{(1,\alpha)}(\mathbf{z}_\alpha^{(i)},t_{m\pm}) &< 0 \text{ if } \mathbf{n}_{\partial\Omega_{\alpha\beta}^{(i)}} \to \Omega_\beta^{(i)},
\end{aligned}\right\} \text{ on } \partial\Omega_{\alpha\beta}^{(i)}. \quad (6.41)$$

The condition in the foregoing equation can be expressed by the product of the normal components of the vector fields

$$\left.\begin{aligned}
{}_iL^{(0,0)}_{\alpha\beta}(t_m) &= G^{(\alpha,0)}_i(\mathbf{z}^{(i)}_\alpha, t_{m\pm}) \times G^{(\beta,0)}_i(\mathbf{z}^{(i)}_\beta, t_{m-}) = 0, \\
G^{(1,\alpha)}_i(\mathbf{z}^{(i)}_\alpha, t_{m\pm}) &> 0 \text{ if } \mathbf{n}_{\partial\Omega^{(i)}_{\alpha\beta}} \to \Omega^{(i)}_\alpha, \\
G^{(1,\alpha)}_i(\mathbf{z}^{(i)}_\alpha, t_{m\pm}) &< 0 \text{ if } \mathbf{n}_{\partial\Omega^{(i)}_{\alpha\beta}} \to \Omega^{(i)}_\beta.
\end{aligned}\right\} \qquad (6.42)$$

From Eq.(6.28), the conditions for the grazing of a flow to the boundary $\partial\Omega_{\alpha\beta}$ in domain Ω_α requires

$$\left.\begin{aligned}
g^{(i)}_\alpha(\mathbf{z}^{(i)}_\alpha, \mathbf{x}^{(\bar{i})}_{\bar{\alpha}}, t_{m\pm}) &= 0, \\
G^{(1,\alpha)}_i(\mathbf{z}^{(i)}_\alpha, t_{m\pm}) &> 0 \text{ if } \mathbf{n}_{\partial\Omega^{(i)}_{\alpha\beta}} \to \Omega^{(i)}_\alpha, \\
G^{(1,\alpha)}_i(\mathbf{z}^{(i)}_\alpha, t_{m\pm}) &< 0 \text{ if } \mathbf{n}_{\partial\Omega^{(i)}_{\alpha\beta}} \to \Omega^{(i)}_\beta,
\end{aligned}\right\} \text{ on } \partial\Omega^{(i)}_{\alpha\beta}. \qquad (6.43)$$

From the derivation, the analytical conditions are based on the normal component of the vector field on the normal direction of the boundary. In Fig.6.5, vector fields and flows in absolute phase space are illustrated, including the sliding motion on $\partial\Omega^{(i)}_{12}$ and passable motion for $\Omega^{(i)}_1$ to $\Omega^{(i)}_2$, and grazing motion in $\Omega^{(i)}_1$ and $\Omega^{(i)}_2$. The corresponding conditions are also presented in Fig.6.6 through the relative phase space.

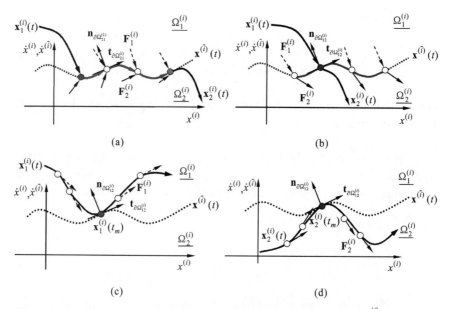

Fig. 6.5 Vector fields and flows in absolute phase space: (a) sliding motion on $\partial\Omega^{(i)}_{12}$, (b) passable motion for $\Omega^{(i)}_1$ to $\Omega^{(i)}_2$, (c) grazing motion in $\Omega^{(i)}_1$ and (d) grazing motion in $\Omega^{(i)}_2$

6.2 Switching and stick motions

The dotted curve is a boundary, labeled by $\mathbf{x}^{(\bar{i})}$. The thick solid curves are the flow of motion in two domains ($\Omega_1^{(i)}$ and $\Omega_2^{(i)}$), and the corresponding vector fields are presented by the solid and dashed arrows. The coordinates for the ith and $\bar{i}^{\text{th}}$ masses may not have the same origin. The sliding motion on the boundary requires both of the normal components of vector fields to the boundary have opposite signs, as sketched in Fig.6.5(a). The passable motion requires that the normal vector field components have the same signs, as depicted in Fig.6.5(b). The grazing motion requires the normal vector field components of the first order is zero and the time change rate of the normal vector field component is non-zero. Such a condition of grazing motion to the boundary is presented in Fig.6.5(c) and (d). The boundary varies with time, which implies that the normal vector of the boundary always changes with time. However, in the relative frame, the normal vector of the boundary is invariant. The conditions can be easily developed, and the corresponding illustrations are presented in Fig.6.6. In Fig.6.6(a), the sliding motion in the relative frame is a point, but the vector field will change with time until the sliding motion disappears. The switchable motion on the boundary is sketched in Fig.6.6(b). The relative velocity is zero with different relative displacements. In Fig.6.6 (c) and (d), the grazing motion is tangential to the axis of relative displacement, which can be determined very easily.

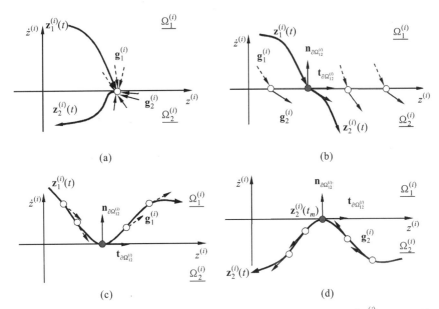

Fig. 6.6 Vector fields and flows in relative phase space: (a) sliding motion on $\partial\Omega_{12}^{(i)}$, (b) passable motion for $\Omega_1^{(i)}$ to $\Omega_2^{(i)}$, (c) grazing motion in $\Omega_1^{(i)}$ and (d) grazing motion in $\Omega_2^{(i)}$

6.3 Periodic motions

In this section, the analytical prediction of periodic motions in such a two-mass system will be presented. First, the switching sets and the corresponding mappings will be defined. From the basic mappings, mapping structures for periodic motions will be developed. From such mapping structures, the periodic motions will be determined through nonlinear algebraic equations, and the corresponding local stability and bifurcation analysis will be carried out by the eigenvalue analysis.

6.3.1 Switching planes and mappings

To investigate periodic motions of a two-mass system in Eq.(6.15), the mapping concept will be adopted from the discontinuous boundary. Before generic mappings are defined, the switching planes will be introduced first. From the generic mappings, the mapping structure of a periodic motion will be developed. From Eq.(6.14), switching planes are defined as

$$\Sigma^0 = {}^{(1)}\Sigma^0 \otimes {}^{(2)}\Sigma^0, \Sigma^+ = {}^{(1)}\Sigma^+ \otimes {}^{(2)}\Sigma^-, \Sigma^- = {}^{(1)}\Sigma^- \otimes {}^{(2)}\Sigma^+, \qquad (6.44)$$

where for $i = 1, 2$ and $\bar{i} = 2, 1$

$$\begin{aligned}
{}^{(i)}\Sigma^0 &= \left\{ (x_k^{(i)}, \dot{x}_k^{(i)}, \Omega t_k) \,\middle|\, \dot{x}_k^{(i)} = \dot{x}_k^{(\bar{i})} \right\}, \\
{}^{(i)}\Sigma^+ &= \left\{ (x_k^{(i)}, \dot{x}_k^{(i)}, \Omega t_k) \,\middle|\, \dot{x}_k^{(i)} = \dot{x}_{k+}^{(\bar{i})} \right\}, \\
{}^{(i)}\Sigma^- &= \left\{ (x_k^{(i)}, \dot{x}_k^{(i)}, \Omega t_k) \,\middle|\, \dot{x}_k^{(i)} = \dot{x}_{k-}^{(\bar{i})} \right\},
\end{aligned} \qquad (6.45)$$

and

$$\dot{x}_{k\pm}^{(i)} = \lim_{\delta \to 0} (\dot{x}_k^{(i)} \pm \delta). \qquad (6.46)$$

From the foregoing equations, we have $\dot{x}_{k\pm}^{(1)} = \dot{x}_{k\mp}^{(2)}$. Using the basic switching planes, three basic mappings are defined as

$$P_0 : \Sigma^0 \to \Sigma^0, P_1 : \Sigma^+ \to \Sigma^+, P_2 : \Sigma^- \to \Sigma^-, \qquad (6.47)$$

where the generic mapping have two components

$$P_0 = ({}^{(1)}P_0, {}^{(2)}P_0)^{\mathrm{T}}, P_1 = ({}^{(1)}P_1, {}^{(2)}P_1)^{\mathrm{T}}, P_2 = ({}^{(1)}P_2, {}^{(2)}P_2)^{\mathrm{T}}, \qquad (6.48)$$

and

$$\begin{aligned}
{}^{(i)}P_0 &: {}^{(i)}\Sigma^0 \to {}^{(i)}\Sigma^0, \text{ for } i = 1,2, \\
{}^{(1)}P_1 &: {}^{(1)}\Sigma^+ \to {}^{(1)}\Sigma^+, {}^{(2)}P_1 : {}^{(2)}\Sigma^- \to {}^{(2)}\Sigma^-, \\
{}^{(1)}P_2 &: {}^{(1)}\Sigma^- \to {}^{(1)}\Sigma^-, {}^{(2)}P_2 : {}^{(2)}\Sigma^+ \to {}^{(2)}\Sigma^+.
\end{aligned} \qquad (6.49)$$

6.3 Periodic motions

From the previously defined generic mappings, any periodic motion can be determined through a certain mapping structure and be labeled through such basic mappings. The generic mappings can be illustrated in Fig.6.7. In Fig.6.7(a), the sub-mappings relative to the first mass are sketched. For a motion in domain $\Omega_1^{(1)}$ (i.e., $\dot{x}^{(1)} > \dot{x}^{(2)}$), the mapping is depicted by the generic mapping $^{(1)}P_1$. For a motion in domain $\Omega_2^{(1)}$ (i.e., $\dot{x}^{(1)} < \dot{x}^{(2)}$), the mapping is depicted by the generic mapping $^{(1)}P_2$. For a motion on the boundary $\partial \Omega_{12}^{(1)}$ (i.e., $\dot{x}^{(1)} = \dot{x}^{(2)}$), the generic mapping is given by $^{(1)}P_0$. Similarly, for a motion in domain $\Omega_1^{(2)}$ (i.e., $\dot{x}^{(2)} > \dot{x}^{(1)}$), compared to the motion of the first mass, the mapping is depicted by the generic mapping of $^{(2)}P_2$. For a motion in domain $\Omega_2^{(2)}$, we have $\dot{x}^{(2)} < \dot{x}^{(1)}$, and the generic mapping should be $^{(2)}P_1$ corresponding to $^{(1)}P_1$. For a motion on the boundary (i.e., $\dot{x}^{(1)} = \dot{x}^{(2)}$), the generic mapping is given by $^{(2)}P_0$. For a better understanding of mappings, consider the generic mappings in the same phase space, as shown in Fig.6.8. In Fig.6.8(a), the two non-sliding mappings are presented for the two masses. The sliding mappings for the two masses are sketched in Fig.6.8(b). Although the two masses have the same velocity on the boundary, the corresponding displacements may be different. Therefore, a flow of the two masses cannot be overlapping each other for the sliding motion, as shown in Fig.6.8(b).

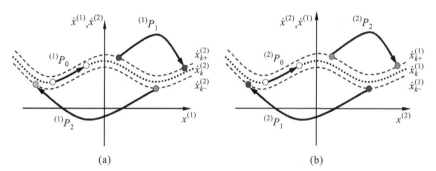

Fig. 6.7 Generic mappings in phase space: (a) sub-mappings for the first mass and (b) sub-mappings for the second mass

The total switching plane can be expressed by

$$\begin{aligned}
\Sigma^0 &= \left\{ (x_k^{(1)}, \dot{x}_k^{(1)}, x_k^{(2)}, \dot{x}_k^{(2)}, \Omega t_k) \,\big|\, \dot{x}_k^{(1)} = \dot{x}_k^{(2)} \text{ and } x_k^{(1)} - x_k^{(2)} = C_k \right\}, \\
\Sigma^+ &= \left\{ (x_k^{(1)}, \dot{x}_k^{(1)}, x_k^{(2)}, \dot{x}_k^{(2)}, \Omega t_k) \,\big|\, \dot{x}_k^{(1)} = \dot{x}_{k+}^{(2)} \text{ or } \dot{x}_k^{(2)} = \dot{x}_{k-}^{(1)} \right\}, \\
\Sigma^- &= \left\{ (x_k^{(1)}, \dot{x}_k^{(1)}, x_k^{(2)}, \dot{x}_k^{(2)}, \Omega t_k) \,\big|\, \dot{x}_k^{(1)} = \dot{x}_{k-}^{(2)} \text{ or } \dot{x}_k^{(2)} = \dot{x}_{k+}^{(1)} \right\},
\end{aligned} \qquad (6.50)$$

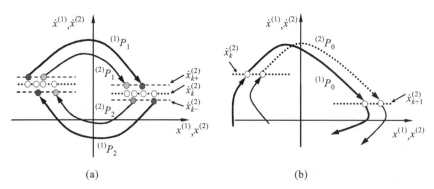

Fig. 6.8 Generic mappings in phase space: (a) non-sliding, sub-mappings and (b) sliding sub-mappings

$$\left.\begin{aligned}
P_0 &: (x_k^{(1)}, \dot{x}_k^{(1)}, x_k^{(2)}, \dot{x}_k^{(2)}, \Omega t_k) \to (x_{k+1}^{(1)}, \dot{x}_{k+1}^{(1)}, x_{k+1}^{(2)}, \dot{x}_{k+1}^{(2)}, \Omega t_{k+1}) \\
&\quad \text{with } \dot{x}_\lambda^{(1)} = \dot{x}_\lambda^{(2)} \text{ and } x_\lambda^{(1)} - x_\lambda^{(2)} = C_k, \lambda = k, k+1, \\
P_1 &: (x_k^{(1)}, \dot{x}_k^{(1)}, x_k^{(2)}, \dot{x}_k^{(2)}, \Omega t_k) \to (x_{k+1}^{(1)}, \dot{x}_{k+1}^{(1)}, x_{k+1}^{(2)}, \dot{x}_{k+1}^{(2)}, \Omega t_{k+1}) \\
&\quad \text{with } \dot{x}_\lambda^{(1)} = \dot{x}_{\lambda+}^{(2)} \text{ or } \dot{x}_\lambda^{(2)} = \dot{x}_{\lambda-}^{(1)}, \ \lambda = k, k+1, \\
P_2 &: (x_k^{(1)}, \dot{x}_k^{(1)}, x_k^{(2)}, \dot{x}_k^{(2)}, \Omega t_k) \to (x_{k+1}^{(1)}, \dot{x}_{k+1}^{(1)}, x_{k+1}^{(2)}, \dot{x}_{k+1}^{(2)}, \Omega t_{k+1}) \\
&\quad \text{with } \dot{x}_\lambda^{(1)} = \dot{x}_{\lambda-}^{(2)} \text{ or } \dot{x}_\lambda^{(2)} = \dot{x}_{\lambda+}^{(1)}, \ \lambda = k, k+1.
\end{aligned}\right\} \quad (6.51)$$

The sliding mapping P_0 can be determined by

$$\left.\begin{aligned}
&{}^{(i)}f_1^{(0)}\left(x_k^{(i)}, \dot{x}_k^{(i)}, \Omega t_k, x_{k+1}^{(i)}, \dot{x}_{k+1}^{(i)}, \Omega t_{k+1}\right) = 0, \\
&{}^{(i)}f_2^{(0)}\left(x_k^{(i)}, \dot{x}_k^{(i)}, \Omega t_k, x_{k+1}^{(i)}, \dot{x}_{k+1}^{(i)}, \Omega t_{k+1}\right) = 0, \\
&\dot{x}_k^{(\bar{i})} = \dot{x}_k^{(i)}, \dot{x}_{k+1}^{(\bar{i})} = \dot{x}_{k+1}^{(i)}, \\
&x_k^{(i)} - x_k^{(\bar{i})} = C_k, x_{k+1}^{(i)} - x_{k+1}^{(\bar{i})} = C_k, \\
&\Gamma_\alpha^{(i)}\left(x_{k+1}^{(i)}, \dot{x}_{k+1}^{(i)}, \Omega t_{k+1}\right) = 0.
\end{aligned}\right\} \quad (6.52)$$

The mapping P_α in domain $\Omega_\alpha = \Omega_\alpha^{(i)} \otimes \Omega_{\bar{\alpha}}^{(\bar{i})}$ can be determined by

6.3 Periodic motions

$$\left.\begin{array}{l}{}^{(1)}f_1^{(\alpha)}\left(x_k^{(1)},\dot{x}_k^{(1)},\Omega t_k,x_{k+1}^{(1)},\dot{x}_{k+1}^{(1)},\Omega t_{k+1}\right)=0,\\ {}^{(1)}f_2^{(\alpha)}\left(x_k^{(1)},\dot{x}_k^{(1)},\Omega t_k,x_{k+1}^{(1)},\dot{x}_{k+1}^{(1)},\Omega t_{k+1}\right)=0,\\ {}^{(2)}f_1^{(\bar{\alpha})}\left(x_k^{(2)},\dot{x}_k^{(2)},\Omega t_k,x_{k+1}^{(2)},\dot{x}_{k+1}^{(2)},\Omega t_{k+1}\right)=0,\\ {}^{(2)}f_2^{(\bar{\alpha})}\left(x_k^{(2)},\dot{x}_k^{(2)},\Omega t_k,x_{k+1}^{(2)},\dot{x}_{k+1}^{(2)},\Omega t_{k+1}\right)=0,\\ \dot{x}_\lambda^{(1)}=\dot{x}_{\lambda+}^{(2)}\ \text{or}\ \dot{x}_\lambda^{(2)}=\dot{x}_{\lambda-}^{(1)}\ \text{for}\ \alpha=1, \lambda=k,k+1,\\ \dot{x}_\lambda^{(1)}=\dot{x}_{\lambda-}^{(2)}\ \text{or}\ \dot{x}_\lambda^{(2)}=\dot{x}_{\lambda+}^{(1)}\ \text{for}\ \alpha=2, \lambda=k,k+1.\end{array}\right\} \quad (6.53)$$

The first two equations of Eq.(6.52) are given from equation of motion in Eq.(6.7), the third and fourth equations are based on the sliding motion requirement because $\dot{x}^{(i)}(t_m)=\dot{x}^{(i)}(t_m)$ and $x^{(i)}(t_m)-x^{(i)}(t_m)=C_k$ for $t_m \in [t_k,t_{k+1}]$. With Eq.(6.38), the fifth equation of Eq.(6.52) is given by the vanishing condition of the sliding motion in Eq.(6.41). During the sliding motion, equation (6.40) should be satisfied for $t_m \in [t_k,t_{k+1})$. It can be found that the sliding motion possesses many constraint conditions. For the non-sliding mapping P_α, the first two equations in Eq.(6.53) are determined by the displacement and velocity solution of the first mass from Appendix, the third and fourth equations are given by the displacement and velocity of the second mass. The fifth equation in Eq.(6.53) is the condition for a passable motion on the boundary.

6.3.2 Mapping structures and motions

Consider a simple periodic motion without stick, the mapping structure shown in Fig.6.9 is expressed by

$$P = P_2 \circ P_1 : \Sigma^+ \to \Sigma^-. \quad (6.54)$$

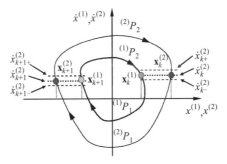

Fig. 6.9 A mapping structure $P = P_2 \circ P_1$ for a periodic motion

Without loss of generality, introduce the following vectors

$$\left.\begin{aligned}
\mathbf{X}_k &= (x_k^{(1)}, \dot{x}_k^{(1)}, x_k^{(2)})^{\mathrm{T}} = (x_k^{(1)}, y_k^{(1)}, x_k^{(2)})^{\mathrm{T}}, \\
\mathbf{Y}_k &= (x_k^{(1)}, \dot{x}_k^{(1)}, x_k^{(2)}, \Omega t_k)^{\mathrm{T}} = (x_k^{(1)}, y_k^{(1)}, x_k^{(2)}, \varphi_k)^{\mathrm{T}}, \\
\mathbf{h}^{(\alpha)} &= (h_1^{(\alpha)}, h_2^{(\alpha)}, h_3^{(\alpha)}, h_4^{(\alpha)})^{\mathrm{T}},
\end{aligned}\right\} \quad (6.55)$$

where $\varphi_k = \Omega t_k$. Therefore, from Eqs.(6.52) and (6.53), we have

$$\left.\begin{aligned}
h_1^{(0)}(\mathbf{Y}_k, \mathbf{Y}_{k+1}) &= {}^{(i)}f_1^{(0)}\left(x_k^{(i)}, \dot{x}_k^{(i)}, \Omega t_k, x_{k+1}^{(i)}, \dot{x}_{k+1}^{(i)}, \Omega t_{k+1}\right), \\
h_2^{(0)}(\mathbf{Y}_k, \mathbf{Y}_{k+1}) &= {}^{(i)}f_2^{(0)}\left(x_k^{(i)}, \dot{x}_k^{(i)}, \Omega t_k, x_{k+1}^{(i)}, \dot{x}_{k+1}^{(i)}, \Omega t_{k+1}\right), \\
h_3^{(0)}(\mathbf{Y}_k, \mathbf{Y}_{k+1}) &= x_k^{(i)} - x_k^{(\bar{i})} - C_k, \\
h_4^{(0)}(\mathbf{Y}_k, \mathbf{Y}_{k+1}) &= \Gamma_\alpha^{(i)}\left(x_{k+1}^{(i)}, \dot{x}_{k+1}^{(i)}, \Omega t_{k+1}\right),
\end{aligned}\right\} \quad (6.56)$$

and

$$\left.\begin{aligned}
h_1^{(\alpha)}(\mathbf{Y}_k, \mathbf{Y}_{k+1}) &= {}^{(1)}f_1^{(\alpha)}\left(x_k^{(1)}, \dot{x}_k^{(1)}, \Omega t_k, x_{k+1}^{(1)}, \dot{x}_{k+1}^{(1)}, \Omega t_{k+1}\right), \\
h_2^{(\alpha)}(\mathbf{Y}_k, \mathbf{Y}_{k+1}) &= {}^{(1)}f_2^{(\alpha)}\left(x_k^{(1)}, \dot{x}_k^{(1)}, \Omega t_k, x_{k+1}^{(1)}, \dot{x}_{k+1}^{(1)}, \Omega t_{k+1}\right), \\
h_3^{(\alpha)}(\mathbf{Y}_k, \mathbf{Y}_{k+1}) &= {}^{(2)}f_1^{(\bar{\alpha})}\left(x_k^{(2)}, \dot{x}_k^{(2)}, \Omega t_k, x_{k+1}^{(2)}, \dot{x}_{k+1}^{(2)}, \Omega t_{k+1}\right), \\
h_4^{(\alpha)}(\mathbf{Y}_k, \mathbf{Y}_{k+1}) &= {}^{(2)}f_2^{(\bar{\alpha})}\left(x_k^{(2)}, \dot{x}_k^{(2)}, \Omega t_k, x_{k+1}^{(2)}, \dot{x}_{k+1}^{(2)}, \Omega t_{k+1}\right).
\end{aligned}\right\} \quad (6.57)$$

From the mapping structure in Eq.(6.54), we have

$$\left.\begin{aligned}
\mathbf{h}^{(1)}(\mathbf{Y}_k, \mathbf{Y}_{k+1}) &= 0, \\
\mathbf{h}^{(2)}(\mathbf{Y}_{k+1}, \mathbf{Y}_{k+2}) &= 0.
\end{aligned}\right\} \quad (6.58)$$

For a periodic motion of $\mathbf{Y}_{k+2} = P\mathbf{Y}_k$, the periodicity of this periodic motion requires

$$\mathbf{Y}_{k+2} = \mathbf{Y}_k. \quad (6.59)$$

More strictly speaking, we have

$$\mathbf{X}_{k+2} = \mathbf{X}_k, \Omega t_{k+2} = \Omega t_k + 2N\pi. \quad (6.60)$$

From Eqs.(6.58) and (6.60), the switching points of periodic motion on the boundary can be obtained by solving a set of nonlinear algebraic equations.

Consider a periodic motion including the sliding motion with a mapping structure $P = P_2 \circ P_1 \circ P_0$, as shown in Fig.6.10. Such a mapping is

$$P = P_2 \circ P_1 \circ P_0 : \Sigma^0 \to \Sigma^-. \quad (6.61)$$

The corresponding relations are

6.3 Periodic motions

$$\left.\begin{array}{l}\mathbf{h}^{(0)}(\mathbf{Y}_k, \mathbf{Y}_{k+1}) = 0, \\ \mathbf{h}^{(1)}(\mathbf{Y}_{k+1}, \mathbf{Y}_{k+2}) = 0, \\ \mathbf{h}^{(2)}(\mathbf{Y}_{k+2}, \mathbf{Y}_{k+3}) = 0.\end{array}\right\} \quad (6.62)$$

With the periodicity of the periodic motion similar to Eq.(6.60), the switching points of such a periodic motion on the boundary can be determined.

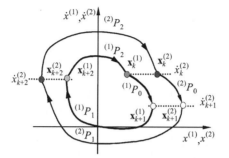

Fig. 6.10 A mapping structure $P = P_2 \circ P_1 \circ P_0$ for a periodic motion with sliding

To describe a generalized periodic motion, consider a generalized mapping structure as

$$P = \underbrace{P_2^{(k_{n2})} \circ P_1^{(k_{n1})} \circ P_0^{(k_{n0})} \circ \cdots \circ P_2^{(k_{12})} \circ P_1^{(k_{11})} \circ P_0^{(k_{10})}}_{n\text{-terms}}, \quad (6.63)$$

where $k_{l\lambda} \in \{0,1\}$ for $l = 1, 2, \ldots, n$ and $\lambda = 0, 1, 2$. $P_\lambda^{(k)} = P_\lambda \circ P_\lambda^{(k-1)}$ and $P_\lambda^{(0)} = 1$. As stated before, the clockwise or counter-clockwise rotation of the order of mappings in Eq.(6.63) will not change the profile of the periodic motion. However, the corresponding initial conditions for such a periodic motion are different. For simplicity, the following notation can be adopted:

$$P = \underbrace{P_{2\ 1\ 0}^{(k_{n2})(k_1)(k_{n0})} \cdots {}^{(k_{12})(k_{11})(k_{10})}_{2\ 1\ 0}}_{n\text{-terms}}. \quad (6.64)$$

Based on the generalized mapping structure, the periodic motion of such a mapping structure is

$$\mathbf{Y}_{k+\sum_{l=1}^{n} k_{l2}+k_{l1}+k_{l0}} = P\mathbf{Y}_k, \quad (6.65)$$

which can be determined. Further, the local stability of such a periodic motion can be determined by the eigenvalue analysis of the Jacobian matrix of the periodic motion. The corresponding matrix is computed by

$$DP = \left.\frac{\partial \mathbf{Y}_{k+\Sigma_{l=1}^{n}(k_{l2}+k_{l1}+k_{l0})}}{\partial \mathbf{Y}_k}\right|_{\mathbf{Y}_k^*}$$

$$= \left[\frac{\partial \left(\mathbf{X}_{k+\Sigma_{l=1}^{m}(k_{l2}+k_{l1}+k_{l0})}, \varphi_{k+\Sigma_{l=1}^{m}(k_{l2}+k_{l1}+k_{l0})}\right)}{\partial (\mathbf{X}_k, \Omega t_k,)}\right]_{(\mathbf{X}_k^*, \varphi_k)}$$

$$= \underbrace{(DP_2^{(k_{n2})} \cdot DP_1^{(k_{n1})} \cdot DP_0^{(k_{n0})}) \cdot \, \cdots \, \cdot (DP_2^{(k_{12})} \cdot DP_1^{(k_{11})} \cdot DP_0^{(k_{10})})}_{n-terms}, \quad (6.66)$$

where

$$DP_\lambda = \begin{bmatrix} \frac{\partial x_{v+1}^{(1)}}{\partial x_v^{(1)}} & \frac{\partial x_{v+1}^{(1)}}{\partial y_v^{(1)}} & \frac{\partial x_{v+1}^{(1)}}{\partial x_v^{(2)}} & \frac{\partial x_{v+1}^{(1)}}{\partial \varphi_v} \\ \frac{\partial y_{v+1}^{(1)}}{\partial x_v^{(1)}} & \frac{\partial y_{v+1}^{(1)}}{\partial y_v^{(1)}} & \frac{\partial y_{v+1}^{(1)}}{\partial x_v^{(2)}} & \frac{\partial y_{v+1}^{(1)}}{\partial \varphi_v} \\ \frac{\partial x_{v+1}^{(2)}}{\partial x_v^{(1)}} & \frac{\partial x_{v+1}^{(2)}}{\partial y_v^{(1)}} & \frac{\partial x_{v+1}^{(2)}}{\partial x_v^{(2)}} & \frac{\partial x_{v+1}^{(2)}}{\partial \varphi_v} \\ \frac{\partial \varphi_{v+1}}{\partial x_v^{(1)}} & \frac{\partial \varphi_{v+1}}{\partial y_v^{(1)}} & \frac{\partial \varphi_{v+1}}{\partial x_v^{(2)}} & \frac{\partial \varphi_{v+1}}{\partial \varphi_v} \end{bmatrix}, \quad (6.67)$$

for $\lambda = 0, 1, 2$ and $v = k, k+1, \ldots, k+\Sigma_{l=1}^{n}(k_{12}+k_{11}+k_{10})-1$, and the Jacobian matrix components for each mapping can be computed through Eqs.(6.56) and (6.57), and all the Jacobian matrix can be obtained through Eq.(6.66). The variational equation for a set of switching points $\{\mathbf{Y}_k^*, \ldots, \mathbf{Y}_{k+\Sigma_{l=1}^{n}(k_{12}+k_{11}+k_{10})-1}^*\}$ is

$$\Delta \mathbf{Y}_{k+\Sigma_{l=1}^{n}(k_{12}+k_{11}+k_{10})} = DP(\mathbf{Y}_k^*)\Delta \mathbf{Y}_k. \quad (6.68)$$

The eigenvalues are computed by

$$|DP - \lambda \mathbf{I}| = 0. \quad (6.69)$$

Because *DP* is a 4×4 matrix, there are four eigenvalues. If the four eigenvalues lie inside the unit circle, then the period-1 motion is stable. If one of them lies outside the unit circle, the periodic motion is unstable.

6.3.3 Bifurcation scenario

To understand the dynamic behaviors of the two-mass system, consider a set of parameters for bifurcation scenario.

$$m_1 = 2, k_1 = 70, r_1 = 1, m_2 = 1, k_2 = 100,$$
$$r_2 = 1, F_N = 7, \mu_k = 0.3, A_0^{(2)} = 10, A_0^{(1)} = 0.$$

6.3 Periodic motions

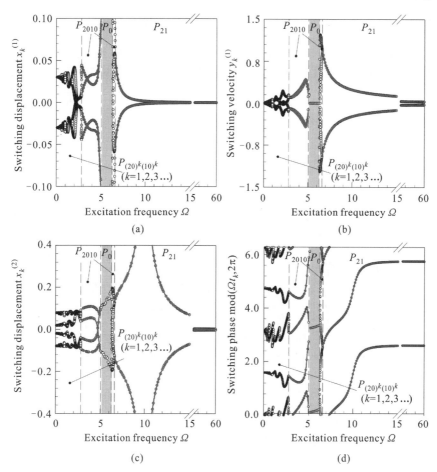

Fig. 6.11 Numerical and analytical prediction of switching points for motions of the two masses connected with friction: (a) switching displacement of the first mass, (b) switching velocity of the first mass, (c) switching displacement of the second mass and (d) switching phase. ($m_1 = 2, k_1 = 70, r_1 = 1, m_2 = 1, k_2 = 100, r_2 = 1, F_N = 7, \mu_k = 0.3$, $A_0^{(2)} = 10$, $A_0^{(1)} = 0$)

Because the closed-form solutions for the mechanical model can be obtained, such closed-form solutions are listed in Appendix and will be used in numerical computation. For a dynamical system without any closed-form solution, the numerical integration scheme should be adopted such as the Runge-Kutta method or the symplectic integration scheme. The switching values versus excitation frequency for motions of two masses in a two-mass system can be computed and are presented in Fig.6.11. The circular symbols represent numerical predictions. The solid curves represent the analytical predictions. The vertical, dashed lines are the boundaries for motion switching. For the analytical prediction, the eigenvalues of periodic motions are computed, as shown in Fig.6.12. Before the motion switching, the motion is always stable. The switching values of motions are presented through the switching displacement and velocity of the primary mass, the switching displacement of the second mass, and switching phase. Because the switching velocity of the second mass is the same as the one of the primary mass, the switching velocity of the second mass will not be presented. The corresponding ranges of excitation frequency are listed in Table 6.1. It is observed that the motion of P_{21} exists in a large range of excitation frequency. Especially, only the stick motion of P_0 exists in the range of $\Omega \in (5.115, 6.264)$.

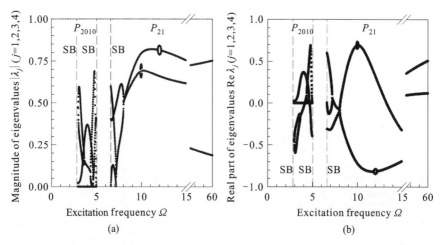

Fig. 6.12 Local stability for periodic motions of two masses connected with friction: (a) magnitude of eigenvalues, (b) real part of eigenvalues ($m_1 = 2, k_1 = 70, r_1 = 1, m_2 = 1, k_2 = 100, r_2 = 1, F_N = 7, \mu_k = 0.3, A_0^{(2)} = 10, A_0^{(1)} = 0$)

Table 6.1 Excitation frequency range for periodic motions($m_1 = 2$, $k_1 = 70$, $r_1 = 1$, $m_2 = 1$, $k_2 = 100$, $r_2 = 1$, $F_N = 7$, $\mu_k = 0.3$, $A_0^{(2)} = 10$, $A_0^{(1)} = 0$)

Mapping structure		Excitation frequency
P_{21}	P(T)	$(6.6, \infty)$
P_{0201}	P(T)	$(0.1, 0.194), (0, 233, 0.267), (0.277, 0.311), (0.341, 0.382),$ $(0.440, 0.588), (0.642, 0.938), (1.110, 1.771), (1.806, 2.758),$ $(2.803, 5.042), (6.291, 6.599)$
$P_{(02)^2(01)^2}$	P(T)	$(0.195, 0.232), (0.268, 0.276), (0.312, 0.340), (0.383, 0.439),$ $(0.589, 0.641), (0.939, 1.109), (1.772, 1.805), (2.759, 2.802)$
P_{20}	P(T)	$(5.043, 5.114)$
P_{01}	P(T)	$(6.265, 6.290)$
P_0	P(T)	$(5.115, 6.264)$

6.4 Numerical illustrations

From the analytical prediction of periodic motions, one of the switching points will be selected as an initial condition, and the numerical computation of such a periodic motion will be carried out. Phase plane, force distribution along displacement, velocity and force time-history will be illustrated in this section. The corresponding analytical conditions of motion on the discontinuous boundary can be illustrated to understand the dynamics of the two-mass system. Again, the closed-form solutions in Appendix will be used in numerical computation. The same system parameters will be used as before. The thin and thick curves represent the responses of the first and second mass, respectively. The circular points give the corresponding switching points on the discontinuous boundary. The shaded areas are the portion of the stick motion. For these analytical conditions, the solid and dashed curves are for the domain $\Omega_1^{(1)}$ (or $\Omega_2^{(2)}$) and $\Omega_2^{(1)}$ (or $\Omega_1^{(2)}$), respectively.

6.4.1 Periodic motion without stick

Consider a simple periodic motion of $P_{12}(\Omega = 6.6)$ with initial condition ($t_0 = 0.8631$, $x_0^{(1)} = 0.05245934$, $x_0^{(2)} = 0.128546350$ and $\dot{x}_0^{(1)} = \dot{x}_0^{(2)} = 0.94732666$). The trajectory in phase plane, displacement and velocity responses of the periodic motion relative to P_{12} are illustrated in Fig.6.13(a)~(c), respectively. In Fig.6.13(a), the trajectory of the first mass is inside the trajectory of the second mass. The two horizontal dashed lines give the switching points with the same velocity. At switching points, the discontinuity of the two trajectories is observed. In domain $\Omega_1^{(1)}$ (or $\Omega_2^{(2)}$), the velocity of the first mass in domain is greater than the one of the second mass in domain (i.e., $y^{(1)} > y^{(2)}$). However, in domain $\Omega_2^{(1)}$ (or $\Omega_1^{(2)}$), the velocity of the first mass is less than the one of the second mass (i.e., $y^{(1)} < y^{(2)}$). Such a characteristic is observed in Fig.6.13(c). In Fig.6.13(b), the time-histories of displacement for the two masses are plotted. The switching points are labeled by the circular symbols. The switching displacements are very smooth because the switching velocities are continuous. The discontinuity can be observed from the time-history of velocity in Fig.6.13(c) because the force is discontinuous.

For a better understanding of the singularity and dynamics of the two-mass system, the corresponding $g_\alpha^{(i)}$ ($i = 1, 2, \alpha = 1, 2$) will be presented in Fig.6.14. The dotted curves are the velocity boundaries. In phase planes of the the masses, the velocity boundary based on the velocity of the other masses is clearly presented. Once the two velocities of the masses match each other, the switchability and singularity of the periodic motion on the velocity boundary will occur, as shown in Fig.6.14(a) and (e) for the first and second masses, respectively. Thus the corresponding relative force distribution along the displacement are presented in Fig.6.14(b) and (f). At the starting point, the switchability condition in Eq.(6.39) cannot be satisfied

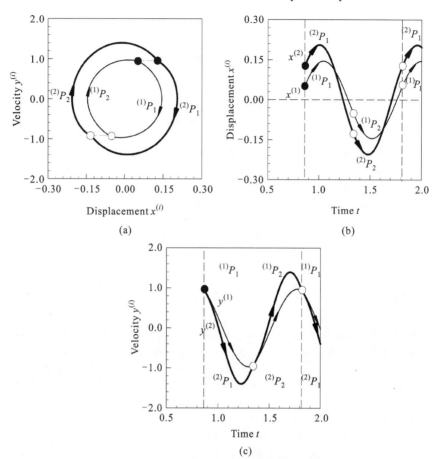

Fig. 6.13 Periodic motion relative to P_{21} ($\Omega = 6.6$): (a) trajectory in phase plane, (b) displacement-time history, and (c) velocity-time history. The thin and thick curves represent the responses of the first and second mass, respectively. The initial condition is $t_0 = 0.863\,1, x_0^{(1)} = 0.052\,459\,34, x_0^{(2)} = 0.128\,546\,35$ and $\dot{x}_0^{(1)} = \dot{x}_0^{(2)} = 0.947\,326\,66$. ($m_1 = 2$, $k_1 = 70$, $r_1 = 1$, $m_2 = 1$, $k_2 = 100$, $r_2 = 1$, $F_N = 7$, $\mu_k = 0.3$, $A_0^{(2)} = 10$, $A_0^{(1)} = 0$)

because $g_1^{(1)} = 0$ and $g_2^{(1)} > 0$ for the first mass and $g_1^{(2)} < 0$ and $g_2^{(2)} = 0$ for the second mass. However, from the relative force time-history, we have $g_1^{(1)}(t_{m-}) < 0$ and $g_2^{(1)}(t_{m+}) > 0$ for the first mass (i.e., $\frac{d}{dt}g_1^{(1)}(t_{m\pm}) = J_1^{(1)}(t_{m\pm}) > 0$). In other words, we have $G_1^{(1,1)}(\mathbf{z}_1^{(1)}, t_{m\pm}) > 0$. From Luo(2006a), the periodic flow for the first mass will pass over the boundary from $\Omega_2^{(1)}$ to $\Omega_1^{(1)}$ because $\mathbf{n}_{\partial\Omega_{12}^{(i)}} = (0,1)^T$ always points to $\Omega_1^{(1)}$ and $\Omega_2^{(2)}$ with $y^{(1)} - y^{(2)} > 0$. Similarly, for the second mass, $G_2^{(1,2)}(\mathbf{z}_2^{(2)}, t_{m+}) < 0$. Thus, the periodic flow for the second mass will pass

6.4 Numerical illustrations

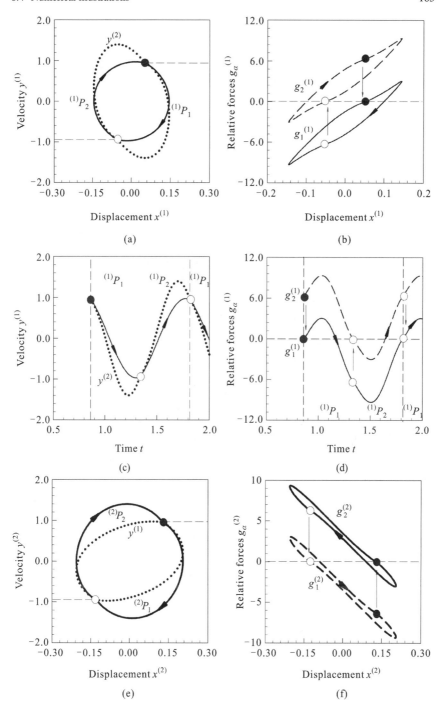

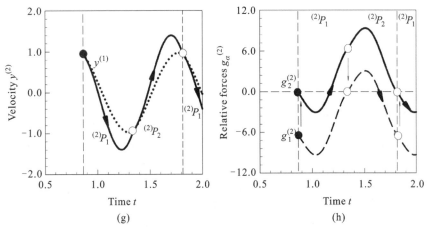

Fig. 6.14 Periodic motion relative to P_{21} ($\Omega = 6.6$): (a) and (e) trajectory with velocity boundary in phase plane, (b) and (f) relative force versus displacement, (c) and (g) velocity-time history with velocity, and (d) and (h) the relative force time-history. The initial condition is $t_0 = 0.8631, x_0^{(1)} = 0.052\,459\,34, x_0^{(2)} = 0.128\,546\,35$ and $\dot{x}_0^{(1)} = \dot{x}_0^{(2)} = 0.947\,326\,66$ ($m_1 = 2$, $k_1 = 70$, $r_1 = 1$, $m_2 = 1$, $k_2 = 100$, $r_2 = 1$, $F_N = 7$, $\mu_k = 0.3$, $A_0^{(2)} = 10$, $A_0^{(1)} = 0$)

over the boundary from $\Omega_1^{(2)}$ to $\Omega_2^{(2)}$. Following the similar fashion, the switchability for the next switching point on the boundary can be discussed. From illustrations in Fig.6.14(d) and (h), for the first mass at the next switching point, we have $G_1^{(1,2)}(\mathbf{z}_2^{(1)}, t_{m\pm}) < 0$ from which the flow will pass the boundary from $\Omega_2^{(1)}$ to $\Omega_1^{(1)}$. However, for the second mass at the next switching point, the condition $G_2^{(1,1)}(\mathbf{z}_1^{(2)}, t_{m\pm}) > 0$ exists, which implies that the flow will pass the boundary from $\Omega_2^{(2)}$ to $\Omega_1^{(2)}$. The relations between the velocity boundary and velocity response for the two masses are presented in Fig.6.14(c) and (g), and the corresponding analytical conditions on the discontinuous boundary are illustrated through the relative velocity in Fig.6.14(d) and (h). The single end arrow points to the motion from one domain switching to another. The curves with and without the arrows are corresponding to the real and imaginary vector fields in the domains.

6.4.2 Periodic motion with stick

Consider a simple periodic motion of a mapping structure P_{2010} with the initial condition (i.e., $t_0 = 0.3068, x_0^{(1)} = -0.032\,907\,56$, $x_0^{(2)} = 0.054\,864\,1$ and $\dot{x}_0^{(1)} = \dot{x}_0^{(2)} = -0.241\,354\,06$) for $\Omega = 4$. The shaded areas represent the stick portion of the periodic motion. The trajectories, displacement and velocity for the two masses in the two-mass system are illustrated. In Fig.6.15(a), the two horizontal narrow zones in phase plane are for two masses sticking together. For such zones, the velocities of

6.4 Numerical illustrations

the two masses are of the same, but the constant difference of the two displacements exists, which can be observed in Fig.6.15(b) and (c). Such a phenomenon is called the stick motion (or synchronization). From Fig.6.15(a) and (c), in domain $\Omega_1^{(1)}$ (or $\Omega_2^{(2)}$), it is also observed that the velocity of the first mass is greater than the one of the second mass in domain (i.e., $y^{(1)} > y^{(2)}$). However, in domain $\Omega_2^{(1)}$ (or $\Omega_1^{(2)}$), the velocity of the first mass is less than the one of the second mass (i.e., $y^{(1)} < y^{(2)}$).

To understand the switching and sliding of motion on the boundary, the corresponding trajectories in phase plane, velocity response and the corresponding rel-

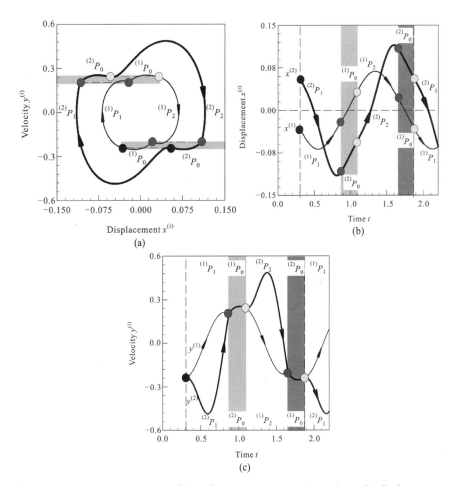

Fig. 6.15 Periodic motion of $P_{2010}(\Omega = 4)$: (a) trajectory in phase plane, (b) displacement-time history, and (c) velocity-time history. The thin and thick curves represent the first and second mass, respectively. The initial condition is $t_0 = 0.3068$, $x_0^{(1)} = -0.03290756$, $x_0^{(2)} = 0.0548641$ and $\dot{x}_0^{(1)} = \dot{x}_0^{(2)} = -0.24135406$ ($m_1 = 2$, $k_1 = 70$, $r_1 = 1, m_2 = 1$, $k_2 = 100$, $r_2 = 1$, $F_N = 7$, $\mu_k = 0.3$, $A_0^{(2)} = 10$, $A_0^{(1)} = 0$)

ative forces are illustrated in Fig.6.16. The single end arrow points to the motion from one domain switching to another, but the two end arrows represent the starting of the stick motion. The dotted curves are the velocity boundaries. In Fig.6.16(a) and (e), the velocity boundary based on the velocity of the other masses is plotted. The stick motion possesses the same velocity of the two masses. For the portion of non-stick in the periodic motion, the velocity relation of the two masses is clearly presented. From the analytical conditions in Eq.(6.40), the stick motion formation requires that the signs of the relative forces in the two domains be opposite, which is observed in Fig.6.16(b) and (f) and in Fig.6.16(d) and (h). In other words, once the periodic flow from domains $\Omega_1^{(1)}$ and $\Omega_2^{(2)}$ arrives to the boundary with $y^{(1)} = y^{(2)}$, $g_1^{(i)} \times g_2^{(i)} < 0$ ($i = 1, 2$) can be observed for stick motion. If such a condition keeps for a certain time interval, the stick motion on the boundary with $y^{(1)} = y^{(2)}$ will exist. Once $g_1^{(i)} \times g_2^{(i)} = 0$ ($i = 1, 2$) appears, equation (6.41) may exist, and then the stick motion may disappear. For the stick motion vanishing and entering into domain $\Omega_1^{(1)}$ and $\Omega_2^{(2)}$, we have $g_1^{(1)} = 0$ and $g_2^{(1)} > 0$ for the first mass

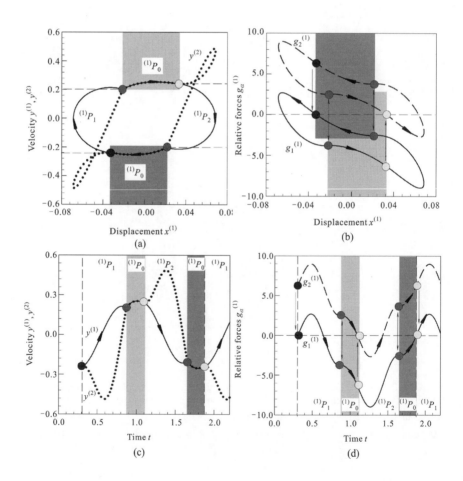

6.4 Numerical illustrations

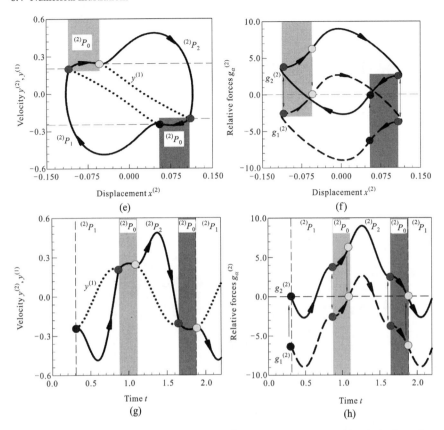

Fig. 6.16 Periodic motion relative to P_{2010} ($\Omega = 4$): (a) and (e) trajectory with velocity boundary in phase plane, (b) and (f) relative force versus displacement, (c) and (g) velocity-time history with velocity boundary, and (d) and (h) the relative force time-history response. The initial condition is $t_0 = 0.3068$, $x_0^{(1)} = -0.03290756$, $x_0^{(2)} = 0.0548641$ and $\dot{x}_0^{(1)} = \dot{x}_0^{(2)} = -0.24135406$ ($m_1 = 2$, $k_1 = 70$, $r_1 = 1$, $m_2 = 1$, $k_2 = 100$, $r_2 = 1$, $F_N = 7$, $\mu_k = 0.3$, $A_0^{(2)} = 10$, $A_0^{(1)} = 0$)

and $g_1^{(2)} < 0$ and $g_2^{(2)} = 0$ for the second mass. However, from the relative force time-history, we have $g_1^{(1)}(t_{m-}) < 0$ and $g_1^{(1)}(t_{m+}) > 0$ for the first mass, which implies $\frac{d}{dt}g_1^{(1)}(t_{m\pm}) = J_1^{(1)}(t_{m\pm}) > 0$. That is, $G_1^{(1,1)}(\mathbf{z}_1^{(1)}, t_{m\pm}) > 0$. Similarly, for the second mass, one obtains $G_2^{(2,1)}(\mathbf{z}_2^{(2)}, t_{m\pm}) < 0$. From Eq.(6.41), the stick motion on the boundary disappears and enters the domain $\Omega_1^{(1)}$ and $\Omega_2^{(2)}$, because $\mathbf{n}_{\partial\Omega_{12}^{(i)}} = (0,1)^T$ always points to domain $\Omega_1^{(1)}$ and $\Omega_2^{(2)}$, with $y^{(1)} - y^{(2)} > 0$. For the stick motion vanishing and entering into domain $\Omega_2^{(1)}$ and $\Omega_1^{(2)}$, one obtains $G_1^{(2,1)}(\mathbf{z}_2^{(1)}, t_{m\pm}) < 0$ for the first mass and $G_2^{(1,1)}(\mathbf{z}_1^{(2)}, t_{m\pm}) > 0$ for the second mass. Therefore, from Eq.(6.41), the stick motion on the boundary disappears and enters the domain $\Omega_2^{(1)}$

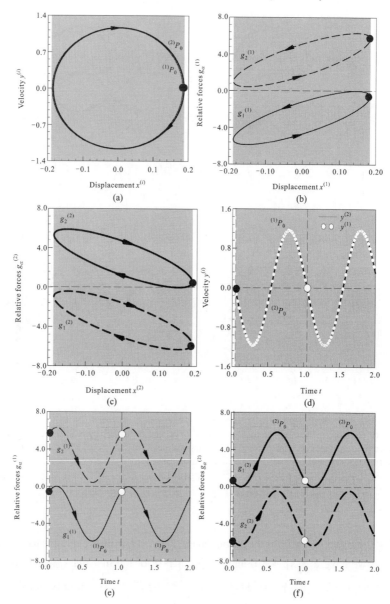

Fig. 6.17 Periodic motion relative to P_0 ($\Omega = 6.264$): (a) trajectory in phase plane, (b) and (c) relative force along displacement, (d) velocity-time history, and (e) and (f) relative force time-history responses. The initial condition is $t_0 = 0.8631$, $x_0^{(1)} = 0.17898521$, $x_0^{(2)} = 0.18238112$, $\dot{x}_0^{(1)} = \dot{x}_0^{(2)} = 0.26846663$ ($m_1 = 2$, $k_1 = 70$, $r_1 = 1$, $m_2 = 1$, $k_2 = 100$, $r_2 = 1$, $F_N = 7$, $\mu_k = 0.3$, $A_0^{(2)} = 10$, $A_0^{(1)} = 0$)

and $\Omega_1^{(2)}$. The velocity boundaries and velocity responses for the two masses are also presented in Fig.6.16(c) and (g), and the corresponding analytical conditions on the discontinuous boundary are illustrated by the relative velocity in Fig.6.16(d) and (h).

6.4.3 Periodic motion with stick only

The periodic motion with stick only will be presented for the same parameters with an initial condition ($t_0 = 0.8631$, $x_0^{(1)} = 0.17898521$, $x_0^{(2)} = 0.18238112$ and $\dot{x}_0^{(1)} = \dot{x}_0^{(2)} = 0.26846663$) for $\Omega = 6.264$. The trajectories of periodic motion in phase plane, the relative force distribution along two displacements, velocity responses and relative force responses are presented in Fig.6.17. In Fig.6.17(a), the two velocities are always of the same, but the difference of two displacement is a non-zero constant. In Fig.6.17(b) and (c), the relative force distributions along displacements give $g_1^{(1)} \leqslant 0$ and $g_2^{(1)} > 0$ for the first mass and $g_2^{(2)} < 0$ and $g_1^{(2)} \geqslant 0$ for the second mass. For a point with $g_1^{(1)} = 0$, we have $G_1^{(1,1)}(\mathbf{z}_1^{(1)}, t_{m\pm}) < 0$. However, we have $G_2^{(1,2)}(\mathbf{z}_2^{(2)}, t_{m\pm}) > 0$ for a point with $g_2^{(2)} = 0$. Thus, the sliding periodic motion is kept on the boundary forever. In Fig.6.17(d), the velocity time-histories for two masses are presented, and both of them are identical. In Fig.6.17 (e) and (f), the relative force time-histories for two masses are shown. The conditions ($g_1^{(1)} \leqslant 0$ and $g_2^{(1)} > 0$) for the first mass and the conditions ($g_2^{(2)} < 0$ and $g_1^{(2)} \geqslant 0$) for the second mass are observed.

References

Luo, A. C. J. (2005a), A theory for non-smooth dynamical systems on connectable domains, *Communication in Nonlinear Science and Numerical Simulation*, **10**, pp.1-55.

Luo, A. C. J. (2005b), Imaginary, sink and source flows in the vicinity of the separatrix of non-smooth dynamic systems, *Journal of Sound and Vibration*, **285**, pp.443-456.

Luo, A. C. J. (2006), *Singularity and Dynamics on Discontinuous Vector Fields*, Amsterdam: Elsevier.

Luo, A. C. J. (2007a), Differential geometry of flows in nonlinear dynamical systems, *Proceedings of IDECT'07*, ASME International Design Engineering Technical Conferences, September 4-7, 2007, Las Vegas, Nevada, USA. DETC2007-84754.

Luo, A. C. J. (2007b), On flow switching bifurcations in discontinuous dynamical system, *Communications in Nonlinear Science and Numerical Simulation*, **12**, pp.100-116.

Luo, A. C. J. (2008a), A theory for flow switchability in discontinuous dynamical systems, *Nonlinear Analysis: Hybrid Systems*, **2**, pp.1030-1061.

Luo, A. C. J. (2008b), *Global Transversality, Resonance and Chaotic Dynamics*, Singapore: World Scientific.

Luo, A. C. J. and Thapa, S. (2007), On nonlinear dynamics of simplified brake dynamical systems, *Proceedings of IMECE2007*, 2007 ASME International Mechanical Engineering Congress and Exposition, November 5-10, 2007, Seattle, Washington, USA. IMECE2007-42349.

Luo, A. C. J. and Thapa, S. (2009), Periodic motion in a simplified brake system with a periodic excitation, *Communication in Nonlinear Science and Numerical Simulation*, **14**, pp.2389-2414.

Chapter 7
Principles for System Interactions

In Chapter 3, the static boundary for discontinuous dynamical systems was discussed, and the sliding and transversality of a flow to the boundary were discussed. For such a system, domains for dynamical systems are invariant. In Chapter 4, a discontinuous dynamical system with time-varying boundary was discussed. In Chapter 5, the interaction of impacting chatters and stick between two-oscillators at time-varying displacement boundaries was investigated. In Chapter 6, the interaction of two systems on the time-varying velocity boundary was discussed. In this chapter, a generalized theory for the interaction of two dynamical systems will be presented under a certain condition. A general methodology will be presented to determine complex motions caused by the interaction of two systems.

7.1 Two dynamical systems

In this section, two dynamical systems with non-connected interactions will be introduced first. Based on the interaction boundary, such two systems with interaction can be discussed through a theory of discontinuous systems. To investigate the complexity of motions caused by the interaction of two systems, a resultant system will be developed.

7.1.1 Dynamical systems with interactions

Consider two dynamical systems as

$$\dot{\mathbf{x}} = \mathbf{F}(\mathbf{x},t,\mathbf{p}) \in \mathbb{R}^n, \tag{7.1}$$

and

$$\dot{\tilde{\mathbf{x}}} = \tilde{\mathbf{F}}(\tilde{\mathbf{x}},t,\tilde{\mathbf{p}}) \in \mathbb{R}^{\tilde{n}}, \tag{7.2}$$

where $\mathbf{F} = (F_1, F_2, \ldots, F_n)^\mathrm{T}$, $\mathbf{x} = (x_1, x_2, \ldots, x_n)^\mathrm{T}$ and $\mathbf{p} = (p_1, p_2, \ldots, p_k)^\mathrm{T}$; $\tilde{\mathbf{F}} = (\tilde{F}_1, \tilde{F}_2, \ldots, \tilde{F}_{\tilde{n}})^\mathrm{T}$, $\tilde{\mathbf{x}} = (\tilde{x}_1, \tilde{x}_2, \ldots, \tilde{x}_{\tilde{n}})^\mathrm{T}$ and $\tilde{\mathbf{p}} = (\tilde{p}_1, \tilde{p}_2, \ldots, \tilde{p}_{\tilde{k}})^\mathrm{T}$. The vector functions $\mathbf{F}$ and $\tilde{\mathbf{F}}$ can be time-dependent or time-independent. Consider a time interval $I_{12} \equiv (t_1, t_2) \in \mathbb{R}$ and domains $U_\mathbf{x} \subseteq \mathbb{R}^n$ and $\tilde{U}_{\tilde{\mathbf{x}}} \subseteq \mathbb{R}^{\tilde{n}}$. For initial conditions $(t_0, \mathbf{x}_0) \in I_{12} \times U_\mathbf{x}$ and $(t_0, \tilde{\mathbf{x}}_0) \in I_{12} \times \tilde{U}_{\tilde{\mathbf{x}}}$, the corresponding flows of the two systems are $\mathbf{x}(t) = \mathbf{\Phi}(t, \mathbf{x}_0, t_0, \mathbf{p})$ and $\tilde{\mathbf{x}}(t) = \tilde{\mathbf{\Phi}}(t, \tilde{\mathbf{x}}_0, t_0, \tilde{\mathbf{p}})$ for $(t, \mathbf{x}) \in I_{12} \times U_\mathbf{x}$ and $(t, \tilde{\mathbf{x}}) \in I_{12} \times \tilde{U}_{\tilde{\mathbf{x}}}$ with $\mathbf{p} \in U_\mathbf{p} \subseteq \mathbb{R}^k$ and $\tilde{\mathbf{p}} \in U_{\tilde{\mathbf{p}}} \subseteq \mathbb{R}^{\tilde{k}}$. The semi-group properties of two flows hold (i.e., $\mathbf{\Phi}(t+s, \mathbf{x}_0, t_0, \mathbf{p}) = \mathbf{\Phi}(t, \mathbf{\Phi}(s, \mathbf{x}_0, t_0, \mathbf{p}), s, \mathbf{p})$ and $\mathbf{x}(t_0) = \mathbf{\Phi}(t_0, \mathbf{x}_0, t_0, \mathbf{p})$, $\tilde{\mathbf{\Phi}}(t+s, \tilde{\mathbf{x}}_0, t_0, \tilde{\mathbf{p}}) = \tilde{\mathbf{\Phi}}(t, \tilde{\mathbf{\Phi}}(s, \tilde{\mathbf{x}}_0, t_0, \tilde{\mathbf{p}}), s, \tilde{\mathbf{p}})$ and $\tilde{\mathbf{x}}(t_0) = \tilde{\mathbf{\Phi}}(t_0, \tilde{\mathbf{x}}_0, t_0, \tilde{\mathbf{p}})$.

Definition 7.1. If two flows $\mathbf{x}(t)$ and $\tilde{\mathbf{x}}(t)$ of two systems in Eqs.(7.1) and (7.2) satisfy for time t

$$\varphi(\mathbf{x}(t), \tilde{\mathbf{x}}(t), t, \boldsymbol{\lambda}) = 0, \boldsymbol{\lambda} \in \mathbb{R}^{n_0}, \tag{7.3}$$

then, the systems in Eqs.(7.1) and (7.2) are called to be interacted under such a condition at time t.

From the foregoing definition, the interaction of two dynamical systems in Eqs.(7.1) and (7.2) occurs at boundary of $\varphi(\mathbf{x}(t), \tilde{\mathbf{x}}(t), t, \boldsymbol{\lambda}) = 0$ in Eq.(7.3). Such a condition causes the discontinuity for two dynamical systems. The time-varying domain and boundary for a dynamical system in Eq.(7.1) will be controlled by a flow of the dynamical system in Eq.(7.2) (i.e., $\tilde{\mathbf{x}}(t)$), and vice versa. In practice, not only one condition in Eq.(7.3) but also many conditions exist for two dynamical systems to interact. The following definition is given.

Definition 7.2. Consider l-non-identical functions $\varphi_j(\mathbf{x}(t), \tilde{\mathbf{x}}(t), t, \boldsymbol{\lambda}_j)$ ($j \in \mathbb{L}$ and $\mathbb{L} = \{1, 2, \ldots, l\}$). If two flows $\mathbf{x}(t)$ and $\tilde{\mathbf{x}}(t)$ of two systems in Eqs.(7.1) and (7.2) satisfy for time t

$$\varphi_j(\mathbf{x}(t), \tilde{\mathbf{x}}(t), t, \boldsymbol{\lambda}_j) = 0 \text{ for } \boldsymbol{\lambda}_j \in \mathbb{R}^{n_j} \text{ and } j \in \mathbb{L}, \tag{7.4}$$

then the systems in Eqs.(7.1) and (7.2) are called to be interacted under the jth-condition at time t.

For the foregoing definition, two dynamical systems in Eqs.(7.1) and (7.2) possess l-conditions for interactions. Therefore, the l-discontinuous boundaries will divide the corresponding phase space into many sub-domains for both of two dynamical systems, and these sub-domains change with time. As similar to Chapter 2, the boundaries and domains can be described. For each of two dynamical systems, sub-domains and boundaries generated by the l-interaction conditions are sketched in Figs.7.1~7.4. Under the l-boundaries, because of the interaction between two dynamical systems, there are $(N+1)$-pairs of sub-domains for two dynamical systems in Eqs.(7.1) and (7.2) in a pair of universal domains (i.e., $\mho \subset \mathbb{R}^n$ and $\tilde{\mho} \subset \mathbb{R}^{\tilde{n}}$), and a pair of the universal domains in phase space for two dynamical systems is divided into N-pairs of accessible sub-domains $(\mho_\alpha, \tilde{\mho}_\alpha)$ plus a pair of the inaccessible domain $(\mho_0, \tilde{\mho}_0)$. The union of all the accessible sub-domains is $\cup_{\alpha=1}^N \mho_\alpha$ and the universal domain is $\mho = \cup_{\alpha=1}^N \mho_\alpha \cup \mho_0$ for the dynamical system in Eq.(7.1).

7.1 Two dynamical systems

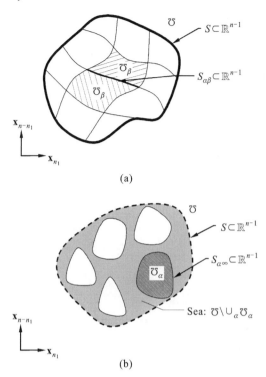

Fig. 7.1 Phase space: (a) connectable and (b) separable domains for Eq.(7.1)

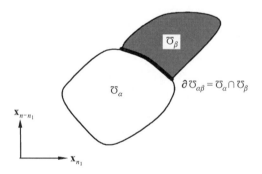

Fig. 7.2 Two adjacent sub-domains $\mathcal{U}_\alpha$ and $\mathcal{U}_\beta$, and boundary $\partial \mathcal{U}_{\alpha\beta}$ for Eq.(7.1)

However, for the dynamical system in Eq.(7.2), the union of all the accessible sub-domains is $\cup_{\alpha=1}^{N} \tilde{\mathcal{U}}_\alpha$ and the universal domain is $\tilde{\mathcal{U}} = \cup_{\alpha=1}^{N} \tilde{\mathcal{U}}_\alpha \cup \tilde{\mathcal{U}}_0$. Both $\mathcal{U}_0$ and $\tilde{\mathcal{U}}_0$ are the unions of the inaccessible domains for two systems. $\mathcal{U}_0 = \mathcal{U} \setminus \cup_{\alpha=1}^{N} \mathcal{U}_\alpha$ and $\tilde{\mathcal{U}}_0 = \tilde{\mathcal{U}} \setminus \cup_{\alpha=1}^{N} \tilde{\mathcal{U}}_\alpha$ are the complement of the union of the accessible sub-domains. From the definition of accessible and inaccessible domains in Luo (2006), a continuous dynamical system can be defined on an accessible domain in phase space. On an inaccessible domain in phase space, no any dynamical systems can be de-

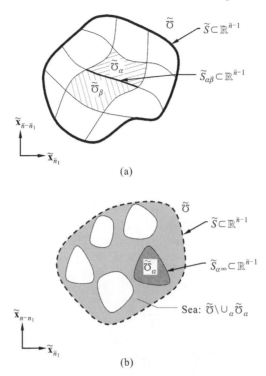

Fig. 7.3 Phase space: (a) connectable and (b) separable domains for Eq.(7.2)

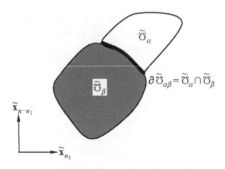

Fig. 7.4 Two adjacent sub-domains $\tilde{\mho}_\alpha$ and $\tilde{\mho}_\beta$, and boundary $\partial\tilde{\mho}_{\alpha\beta}$ for Eq.(7.2)

fined. For a dynamical system in Eq.(7.1), the boundary for two open domains $\mho_\alpha$ and $\mho_\beta$ are $\partial\mho_{\alpha\beta} = \bar{\mho}_\alpha \cap \bar{\mho}_\beta$, and for a dynamical system in Eq.(7.2), one has $\partial\tilde{\mho}_{\alpha\beta} = \bar{\tilde{\mho}}_\alpha \cap \bar{\tilde{\mho}}_\beta$. Such boundaries are formed by the intersection of the closed sub-domains. These interaction boundaries are time-varying and the corresponding domain will change with time. The reader can refer three example systems in Chapters 4~6 in order to understand how the boundaries and domains change with time.

7.1.2 Discontinuous description

Without loss of generality, to avoid the complexity of domains and boundaries generated by the l-non-identical conditions in phase space, consider the jth-condition for two dynamical systems to interact. The boundary is determined by the jth-condition in Eq.(7.4), and the corresponding domain in phase space are divided into two domains $\mho_{(\alpha_j,j)}$ and $\tilde{\mho}_{(\alpha_j,j)}$ ($\alpha_j = 1,2$) for two dynamical systems in Eqs.(7.1) and (7.2), respectively. Therefore, on the α_j^{th}-open sub-domain $\mho_{(\alpha_j,j)}$, there is a $C^{r_{\alpha_j}}$-continuous system ($r_{\alpha_j} \geq 1$) in a form of

$$\left. \begin{array}{l} \dot{\mathbf{x}}^{(\alpha_j,j)} \equiv \mathbf{F}^{(\alpha_j,j)}\left(\mathbf{x}^{(\alpha_j,j)},t,\mathbf{p}^{(\alpha_j,j)}\right) \in \mathbb{R}^n, \\ \mathbf{x}^{(\alpha_j,j)} = \left(x_1^{(\alpha_j,j)}, x_2^{(\alpha_j,j)}, \ldots, x_n^{(\alpha_j,j)}\right)^{\text{T}} \in \mho_{(\alpha_j,j)}. \end{array} \right\} \quad (7.5)$$

In a sub-domain $\mho_{(\alpha_j,j)}$, the vector field $\mathbf{F}^{(\alpha_j,j)}(\mathbf{x}^{(\alpha_j,j)},t,\mathbf{p}^{(\alpha_j,j)})$ with $\mathbf{p}^{(\alpha_j,j)} = (p_1^{(\alpha_j,j)}, p_2^{(\alpha_j,j)}, \ldots, p_k^{(\alpha_j,j)})^{\text{T}} \in \mathbb{R}^{k_j}$ is $C^{r_{\alpha_j}}$-continuous ($r_{\alpha_j} \geq 1$) in a state vector $\mathbf{x}^{(\alpha_j,j)}$ for all time t; and the continuous flow in Eq.(7.1) $\mathbf{x}^{(\alpha_j,j)}(t) = \mathbf{\Phi}^{(\alpha_j,j)}(\mathbf{x}^{(\alpha_j,j)}(t_0), t, \mathbf{p}^{(\alpha_j,j)})$ is $C^{r_{\alpha_j}+1}$-continuous for time t with $\mathbf{x}^{(\alpha_j,j)}(t_0) = \mathbf{\Phi}^{(\alpha_j,j)}(\mathbf{x}^{(\alpha_j,j)}(t_0), t_0, \mathbf{p}^{(\alpha_j,j)})$. The hypothesis (H1~H4) in Chapter 2 should hold.

The corresponding boundary relative to the jth non-connected interaction is defined as follows.

Definition 7.3. The boundary in n-dimensional phase space for dynamical system in Eq.(7.1) under the jth-interaction condition in Eq.(7.4) is defined as

$$\begin{array}{l} S_{(\alpha_j\beta_j,j)} \equiv \partial \mho_{(\alpha_j\beta_j,j)} = \bar{\mho}_{(\alpha_j,j)} \cap \bar{\mho}_{(\alpha_j,j)} \\ = \left\{ \mathbf{x}^{(0,j)} \left| \begin{array}{l} \varphi_j\left(\mathbf{x}^{(0,j)}, \tilde{\mathbf{x}}^{(0,j)}, t, \boldsymbol{\lambda}\right) = 0, \\ \varphi_j \text{ is } C^{r_j}\text{-continuous}(r_j \geq 1) \end{array} \right. \right\} \subset \mathbb{R}^{n-1}. \end{array} \quad (7.6)$$

Similarly, the discontinuous system of Eq.(7.2) caused by the interaction at the jth-condition in Eq.(7.4) can be described. On the α_j^{th}-open ($\alpha_j = 1,2$) sub-domain $\tilde{\mho}_{(\alpha_j,j)}$, there is a $C^{\tilde{r}_{\alpha_j}}$-continuous system ($\tilde{r}_{\alpha_j} \geq 1$) in a form of

$$\left. \begin{array}{l} \dot{\tilde{\mathbf{x}}}^{(\alpha_j,j)} \equiv \tilde{\mathbf{F}}^{(\alpha_j,j)}\left(\tilde{\mathbf{x}}^{(\alpha_j,j)},t,\tilde{\mathbf{p}}^{(\alpha_j,j)}\right) \in \mathbb{R}^{\tilde{n}}, \\ \tilde{\mathbf{x}}^{(\alpha_j,j)} = \left(\tilde{x}_1^{(\alpha_j,j)}, \tilde{x}_2^{(\alpha_j,j)}, \ldots, \tilde{x}_{\tilde{n}}^{(\alpha_j,j)}\right)^{\text{T}} \in \tilde{\mho}_{(\alpha_j,j)}. \end{array} \right\} \quad (7.7)$$

In an sub-domain $\tilde{\mho}_{(\alpha_j,j)}$, the vector field $\tilde{\mathbf{F}}^{(\alpha_j,j)}(\tilde{\mathbf{x}}^{(\alpha_j,j)},t,\tilde{\mathbf{p}}^{(\alpha_j,j)})$ with $\tilde{\mathbf{p}}^{(\alpha_j,j)} = (\tilde{p}_1^{(\alpha_j,j)}, \tilde{p}_2^{(\alpha_j,j)}, \ldots, \tilde{p}_{\tilde{k}}^{(\alpha_j,j)})^{\text{T}} \in \mathbb{R}^{\tilde{k}_j}$ is $C^{\tilde{r}_{\alpha_j}}$-continuous ($\tilde{r}_{\alpha_j} \geq 1$) in a state vector $\mathbf{x}^{(\alpha_j,j)}$ for all time t; and the continuous flow in Eq.(7.2) $\tilde{\mathbf{x}}^{(\alpha_j,j)}(t) = \tilde{\mathbf{\Phi}}^{(\alpha_j,j)}(\tilde{\mathbf{x}}^{(\alpha_j,j)}(t_0), t, \tilde{\mathbf{p}}^{(\alpha_j,j)})$ is $C^{\tilde{r}_{\alpha_j}+1}$-continuous for time t with $\tilde{\mathbf{x}}^{(\alpha_j,j)}(t_0) = \tilde{\mathbf{\Phi}}^{(\alpha_j,j)}(\tilde{\mathbf{x}}^{(\alpha_j,j)}(t_0), t_0, \tilde{\mathbf{p}}^{(\alpha_j,j)})$. The corresponding hypothesis (H1~H4) in Chapter 2 should also hold.

The corresponding boundary relative to non-connected interaction is defined for dynamical system in Eq.(7.2). That is,

Definition 7.4. The boundary in n-dimensional phase space for dynamical system in Eq.(7.2) is defined as

$$\tilde{S}_{(\alpha_j\beta_j,j)} \equiv \partial \tilde{\mho}_{(\alpha_j\beta_j,j)} = \bar{\tilde{\mho}}_{(\alpha_j,j)} \cap \bar{\tilde{\mho}}_{(\beta_j,j)}$$
$$= \left\{ \tilde{\mathbf{x}}^{(0,j)} \left| \begin{array}{l} \varphi_j\left(\mathbf{x}^{(0,j)},\tilde{\mathbf{x}}^{(0,j)},t,\lambda\right) = 0, \\ \varphi_j \text{ is } C^{r_j}\text{-continuous}(r_j \geqslant 1) \end{array} \right. \right\} \subset \mathbb{R}^{\tilde{n}-1}. \tag{7.8}$$

On the boundaries $\partial \mho_{\alpha_j\beta_j}$ and $\partial \tilde{\mho}_{\alpha_j\beta_j}$ with $\varphi_j(\mathbf{x}^{(\alpha_j,j)},\tilde{\mathbf{x}}^{(\alpha_j,j)},t,\lambda_j) = 0$, there is a dynamical system as

$$\left.\begin{array}{l} \dot{\mathbf{x}}^{(0,j)} = \mathbf{F}^{(0,j)}(\mathbf{x}^{(0,j)},\tilde{\mathbf{x}}^{(0,j)},t,\lambda_j), \\ \dot{\tilde{\mathbf{x}}}^{(0,j)} = \tilde{\mathbf{F}}^{(0,j)}(\mathbf{x}^{(0,j)},\tilde{\mathbf{x}}^{(0,j)},t,\lambda_j), \end{array}\right\} \tag{7.9}$$

where $\mathbf{x}^{(0,j)} = (x_1^{(0,j)},x_2^{(0,j)},\ldots,x_n^{(0,j)})^{\mathrm{T}}$ and $\tilde{\mathbf{x}}^{(0,j)} = (\tilde{x}_1^{(0,j)},\tilde{x}_2^{(0,j)},\ldots,\tilde{x}_{\tilde{n}}^{(0,j)})^{\mathrm{T}}$. The corresponding flow $\mathbf{x}^{(0,j)}$, (i.e., $\mathbf{x}^{(0,j)}(t) = \mathbf{\Phi}^{(0,j)}(\mathbf{x}^{(0,j)}(t_0),t,\lambda_j)$ with $\mathbf{x}^{(0,j)}(t_0) = \mathbf{\Phi}^{(0,j)}(\mathbf{x}^{(0,j)}(t_0),t_0,\lambda_j)$) is C^{r_j+1}-continuous for time t. The corresponding flow $\tilde{\mathbf{x}}^{(0,j)}$ (i.e., $\tilde{\mathbf{x}}^{(0,j)}(t) = \tilde{\mathbf{\Phi}}^{(0,j)}(\tilde{\mathbf{x}}^{(0,j)}(t_0),t,\lambda_j)$ with the initial condition $\tilde{\mathbf{x}}^{(0,j)}(t_0) = \tilde{\mathbf{\Phi}}^{(0,j)}(\tilde{\mathbf{x}}^{(0,j)}(t_0),t_0,\lambda_j)$) is also $C^{\tilde{r}_j+1}$-continuous.

7.1.3 Resultant dynamical systems

In the previous section, the discontinuity at the interaction boundary for two dynamical systems is described through two different systems. In this section, a resultant system will be introduced to describe such interaction between two dynamical systems. For doing so, a new vector of state variables of two dynamical systems in Eqs.(7.1) and (7.2) is introduced as

$$\mathbf{X} \equiv (\mathbf{x};\tilde{\mathbf{x}})^{\mathrm{T}} = (x_1,x_2,\ldots,x_n;\tilde{x}_1,\tilde{x}_2,\ldots,\tilde{x}_{\tilde{n}})^{\mathrm{T}} \in \mathbb{R}^{n+\tilde{n}}. \tag{7.10}$$

The notation $(\cdot;\cdot) \equiv (\cdot,\cdot)$ is for a combined vector of state vectors of two dynamical systems. From the interaction condition in Eq.(7.3) or (7.4), the interaction of two dynamical systems in Eqs.(7.1) and (7.2) can be investigated through a discontinuous dynamical system, and the corresponding domain in phase pace is separated into two sub-domains by such an interaction boundary. The interaction boundary and domains are described as follows.

Definition 7.5. A interaction boundary in an $(n+\tilde{n})$-dimensional phase space for the interaction of two dynamical systems in Eqs.(7.1) and (7.2) to the interaction condition in Eq.(7.3) is defined as

7.1 Two dynamical systems

$$\partial \Omega_{12} = \bar{\Omega}_1 \cap \bar{\Omega}_2$$
$$= \left\{ \mathbf{X}^{(0)} \middle| \begin{array}{l} \varphi\left(\mathbf{X}^{(0)}, t, \boldsymbol{\lambda}\right) \equiv \varphi\left(\mathbf{x}^{(0)}(t), \tilde{\mathbf{x}}^{(0)}(t), t, \boldsymbol{\lambda}\right) = 0, \\ \varphi \text{ is } C^r\text{-continuous}(r \geq 1) \end{array} \right\} \subset \mathbb{R}^{n+\tilde{n}-1}, \quad (7.11)$$

and two corresponding domains for a resultant system of two dynamical systems in Eqs.(7.1) and (7.2) are defined as

$$\left. \begin{array}{l} \Omega_1 = \left\{ \mathbf{X}^{(1)} \middle| \begin{array}{l} \varphi\left(\mathbf{X}^{(1)}, t, \boldsymbol{\lambda}\right) \equiv \varphi\left(\mathbf{x}^{(1)}(t), \tilde{\mathbf{x}}^{(1)}(t), t, \boldsymbol{\lambda}\right) > 0, \\ \varphi \text{ is } C^r\text{-continuous}(r \geq 1) \end{array} \right\} \subset \mathbb{R}^{n+\tilde{n}}, \\ \Omega_2 = \left\{ \mathbf{X}^{(2)} \middle| \begin{array}{l} \varphi\left(\mathbf{X}^{(2)}, t, \boldsymbol{\lambda}\right) \equiv \varphi\left(\mathbf{x}^{(2)}(t), \tilde{\mathbf{x}}^{(2)}(t), t, \boldsymbol{\lambda}\right) < 0, \\ \varphi \text{ is } C^r\text{-continuous}(r \geq 1) \end{array} \right\} \subset \mathbb{R}^{n+\tilde{n}}. \end{array} \right\} \quad (7.12)$$

From the previous section, the boundary or domains can be expressed by the direct product of two boundaries or domains

$$\left. \begin{array}{l} \Omega_\alpha = \mho_\alpha \otimes \tilde{\mho}_\alpha \text{ for } \alpha \in \{1,2\}, \\ \partial \Omega_{\alpha\beta} = \partial \mho_{\alpha\beta} \otimes \partial \tilde{\mho}_{\alpha\beta} \text{ for } \alpha, \beta \in \{1,2\}. \end{array} \right\} \quad (7.13)$$

On the two domains, a resultant system of two dynamical systems is discontinuous to the interaction boundary, defined by

$$\dot{\mathbf{X}}^{(\alpha)} = \mathbb{F}^{(\alpha)}(\mathbf{X}^{(\alpha)}, t, \boldsymbol{\pi}^{(\alpha)}) \text{ in } \Omega_\alpha(\alpha = 1, 2), \quad (7.14)$$

where $\mathbb{F}^{(\alpha)} = (\mathbf{F}^{(\alpha)}; \tilde{\mathbf{F}}^{(\alpha)})^{\mathrm{T}} = (F_1^{(\alpha)}, F_2^{(\alpha)}, \ldots, F_n^{(\alpha)}; \tilde{F}_1^{(\alpha)}, \tilde{F}_1^{(\alpha)}, \ldots, \tilde{F}_{\tilde{n}}^{(\alpha)})^{\mathrm{T}}$ and $\boldsymbol{\pi}^{(\alpha)} = (\mathbf{p}_\alpha, \tilde{\mathbf{p}}_\alpha)^{\mathrm{T}}$. Suppose there is a vector field $\mathbb{F}^{(0)}(\mathbf{X}^{(0)}, t, \boldsymbol{\lambda})$ on the interaction boundary with $\varphi(\mathbf{X}^{(0)}, t, \boldsymbol{\lambda}) = 0$, and the corresponding dynamical system on such a boundary is expressed by

$$\dot{\mathbf{X}}^{(0)} = \mathbb{F}^{(0)}(\mathbf{X}^{(0)}, t, \boldsymbol{\lambda}) \text{ on } \partial \Omega_{12}. \quad (7.15)$$

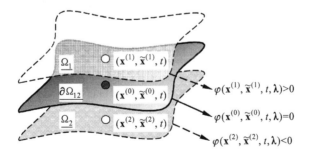

Fig. 7.5 Interaction boundary and domains in $(n+\tilde{n})$-dimensional state space

The domains Ω_α ($\alpha = 1,2$) are separated by the constraint boundary of $\partial\Omega_{12}$, as shown in Fig.7.5. For a point $(\mathbf{x}^{(1)}, \tilde{\mathbf{x}}^{(1)}) \in \Omega_1$ at time t, one obtains $\varphi(\mathbf{x}^{(1)}, \tilde{\mathbf{x}}^{(1)}, t, \boldsymbol{\lambda}) > 0$. For a point $(\mathbf{x}^{(2)}, \tilde{\mathbf{x}}^{(2)}) \in \Omega_2$ at time t, one obtains $\varphi(\mathbf{x}^{(2)}, \tilde{\mathbf{x}}^{(2)}, t, \boldsymbol{\lambda}) < 0$. However, on the boundary $(\mathbf{x}^{(0)}, \tilde{\mathbf{x}}^{(0)}) \in \partial\Omega_{12}$ at time t, the condition for interaction should be satisfied (i.e., $\varphi(\mathbf{x}^{(0)}, \tilde{\mathbf{x}}^{(0)}, t, \boldsymbol{\lambda}) = 0$). If the interaction condition is time-independent, the interaction boundary determined by the interaction condition is invariant. If the interaction condition is time-dependent, the interaction boundary determined by the interaction condition is time-varying, and the corresponding domain for the resultant system is time-varying. As in Eq.(7.4), there are many conditions for interactions of two dynamical systems. Suppose only the jth-interaction boundary occurs for two system for time t. The above definitions can be extended accordingly.

Definition 7.6. The jth-interaction boundary in an $(n+\tilde{n})$-dimensional phase space for the interaction of two dynamical systems in Eqs.(7.1) and (7.2), relative to the jth-interaction condition in Eq.(7.4), is defined as

$$\partial\Omega_{(\alpha_j\beta_j,j)} = \bar{\Omega}_{(\alpha_j,j)} \cap \bar{\Omega}_{(\beta_j,j)}$$

$$= \left\{ \mathbf{X}^{(0,j)} \middle| \begin{array}{l} \varphi_j\left(\mathbf{X}^{(0,j)}, t, \lambda_j\right) \\ \equiv \varphi_j\left(\mathbf{x}^{(0,j)}(t), \tilde{\mathbf{x}}^{(0,j)}(t), t, \lambda_j\right) = 0, \\ \varphi_j \text{ is } C^{r_j}\text{-continuous}(r_j \geq 1) \end{array} \right\} \subset \mathbb{R}^{n+\tilde{n}-1}, \quad (7.16)$$

and two domains pertaining to the jth-boundary for a resultant system of two dynamical systems in Eqs.(7.1) and (7.2) are defined as

$$\left.\begin{array}{l}\Omega_{(1,j)} = \left\{ \mathbf{X}^{(1,j)} \middle| \begin{array}{l} \varphi_j\left(\mathbf{X}^{(1,j)}, t, \lambda_j\right) \\ \equiv \varphi_j\left(\mathbf{x}^{(1,j)}(t), \tilde{\mathbf{x}}^{(1,j)}(t), t, \lambda_j\right) > 0, \\ \varphi_j \text{ is } C^{r_j}\text{-continuous}(r_j \geq 1) \end{array} \right\} \subset \mathbb{R}^{n+\tilde{n}}, \\ \\ \Omega_{(2,j)} = \left\{ \mathbf{X}^{(2,j)} \middle| \begin{array}{l} \varphi_j\left(\mathbf{X}^{(2,j)}, t, \lambda_j\right) \\ \equiv \varphi_j\left(\mathbf{x}^{(2,j)}(t), \tilde{\mathbf{x}}^{(2,j)}(t), t, \lambda_j\right) < 0, \\ \varphi_j \text{ is } C^{r_j}\text{-continuous}(r_j \geq 1) \end{array} \right\} \subset \mathbb{R}^{n+\tilde{n}}. \end{array}\right\} \quad (7.17)$$

As in Fig.7.5, such a boundary and domains for the jth-interaction condition can be sketched in Fig.7.6. From the previous section, the boundary or domains for the jth-interaction condition can be expressed by the direct product of two corresponding individual boundaries or domains

$$\left.\begin{array}{l}\Omega_{(\alpha_j,j)} = \mho_{(\alpha_j,j)} \otimes \tilde{\mho}_{(\alpha_j,j)} \text{ for } \alpha_j \in \{1,2\}, \\ \partial\Omega_{(\alpha_j\beta_j,j)} = \partial\mho_{(\alpha_j\beta_j,j)} \otimes \partial\tilde{\mho}_{(\alpha_j\beta_j,j)} \text{ for } \alpha_j, \beta_j \in \{1,2\}. \end{array}\right\} \quad (7.18)$$

On the two domains relative to the jth-interaction boundary, a discontinuous resultant system of two dynamical systems in Eqs.(7.1) and (7.2) with the jth-interaction

7.2 Fundamental interactions

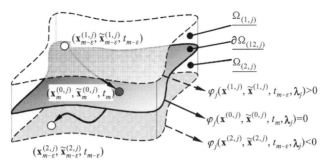

Fig. 7.6 The jth-boundary and domains of a resultant flow to the interaction boundary in $(n+\tilde{n})$-dimensional state space

in Eq.(7.4) is defined by

$$\dot{\mathbf{X}}^{(\alpha_j,j)} = \mathbb{F}^{(\alpha_j,j)}(\mathbf{X}^{(\alpha_j,j)}, t, \boldsymbol{\pi}_j^{(\alpha_j)}) \text{ in } \Omega_{(\alpha_j,j)}, \tag{7.19}$$

where $\mathbb{F}^{(\alpha_j,j)} = (\mathbf{F}^{(\alpha_j,j)}; \tilde{\mathbf{F}}^{(\alpha_j,j)})^{\mathrm{T}}$ and $\boldsymbol{\pi}_j^{(\alpha_j)} = (\mathbf{p}_j^{(\alpha_j)}, \tilde{\mathbf{p}}_j^{(\alpha_j)})^{\mathrm{T}}$. Suppose there is a vector field of $\mathbb{F}^{(0,j)}(\mathbf{X}^{(0,j)}, t, \boldsymbol{\lambda}_j)$ on the jth-interaction boundary with $\varphi_j(\mathbf{X}^{(0,j)}, t, \boldsymbol{\lambda}_j) = 0$, and the corresponding dynamical system on the jth-interaction boundary is expressed by

$$\dot{\mathbf{X}}^{(0,j)} = \mathbb{F}^{(0,j)}(\mathbf{X}^{(0,j)}, t, \boldsymbol{\lambda}_j) \text{ on } \partial \Omega_{(12,j)}. \tag{7.20}$$

Based on the description, the dynamical behaviors between two discontinuous dynamical systems with the interaction boundary can be investigated through the theory presented in Chapter 2. In next section, a generalized relative coordinate system to each interaction will be introduced.

7.2 Fundamental interactions

In this section, the interaction behaviors between two dynamical systems will be investigated in the vicinity of interaction boundary. For doing such investigation, new variables are introduced in domain $\Omega_{(\alpha_j,j)}$

$$z^{(\alpha_j,j)} = \varphi_j\left(\mathbf{x}^{(\alpha_j,j)}(t), \tilde{\mathbf{x}}^{(\alpha_j,j)}(t), t, \boldsymbol{\lambda}_j\right) \text{ for } j \in L. \tag{7.21}$$

On the boundary $\partial \Omega_{(\alpha_j \beta_j, j)}$, we obtain

$$z^{(0,j)} = \varphi_j\left(\mathbf{x}^{(0,j)}(t), \tilde{\mathbf{x}}^{(0,j)}(t), t, \boldsymbol{\lambda}_j\right) = 0 \text{ for } j \in L. \tag{7.22}$$

If the two systems do not interact each other, the new variables ($z_j \neq 0$, $j = 1, 2, \ldots, l$) will change with time t. The corresponding time-change rate is given by

$$\dot{z}^{(\alpha_j, j)} = \frac{d}{dt} \varphi_j \left(\mathbf{x}^{(\alpha_j, j)}, \tilde{\mathbf{x}}^{(\alpha_j, j)}, t, \lambda_j \right)$$
$$= \sum_{i=1}^{n} \frac{\partial \varphi_j}{\partial x_i^{(\alpha_j, j)}} \dot{x}_i^{(\alpha_j, j)} + \sum_{i=1}^{\tilde{n}} \frac{\partial \varphi_j}{\partial \tilde{x}_i^{(\alpha_j, j)}} \dot{\tilde{x}}_i^{(\alpha_j, j)} + \frac{\partial \varphi_j}{\partial t}. \quad (7.23)$$

Substitution of Eqs. (7.5) and (7.7) into Eq.(7.23) gives

$$\dot{z}^{(\alpha_j, j)} = \sum_{i=1}^{n} \frac{\partial \varphi_j}{\partial x_i^{(\alpha_j, j)}} F_i^{(\alpha_j, j)} \left(\mathbf{x}^{(\alpha_j, j)}, t, \mathbf{p}^{(\alpha_j, j)} \right)$$
$$+ \sum_{i=1}^{\tilde{n}} \frac{\partial \varphi_j}{\partial \tilde{x}_i^{(\alpha_j, j)}} \tilde{F}_i^{(\alpha_j, j)} \left(\tilde{\mathbf{x}}^{(\alpha_j, j)}, t, \tilde{\mathbf{p}}^{(\alpha_j, j)} \right) + \frac{\partial \varphi_j}{\partial t}. \quad (7.24)$$

Two new normal vectors are defined as

$$\mathbf{n}_{\varphi_j} = \left(\frac{\partial \varphi_j}{\partial x_1}, \frac{\partial \varphi_j}{\partial x_2}, \ldots, \frac{\partial \varphi_j}{\partial x_n} \right)^{\mathrm{T}} \text{ and } \tilde{\mathbf{n}}_{\varphi_j} = \left(\frac{\partial \varphi_j}{\partial \tilde{x}_1}, \frac{\partial \varphi_j}{\partial \tilde{x}_2}, \ldots, \frac{\partial \varphi_j}{\partial \tilde{x}_n} \right)^{\mathrm{T}}. \quad (7.25)$$

Using Eq.(7.25), equation (7.24) becomes

$$\dot{z}^{(\alpha_j, j)} = \mathbf{n}_{\varphi_j} \cdot \mathbf{F}^{(\alpha_j, j)} \left(\mathbf{x}^{(\alpha_j, j)}, t, \mathbf{p}^{(\alpha_j, j)} \right)$$
$$+ \tilde{\mathbf{n}}_{\varphi_j} \cdot \tilde{\mathbf{F}}^{(\alpha_j, j)} \left(\tilde{\mathbf{x}}^{(\alpha_j, j)}, t, \tilde{\mathbf{p}}^{(\alpha_j, j)} \right) + \frac{\partial \varphi_j}{\partial t}. \quad (7.26)$$

If the vector fields in different domains $\Omega_{(\alpha_j, j)}$ ($\alpha_j = 1, 2$) are distinguishing, $\dot{z}^{(\alpha_j, j)}$ is discontinuous. Similarly, for each domain $\Omega_{(\alpha_j, j)}$, we have

$$\ddot{z}^{(\alpha_j, j)} = \frac{d}{dt} \Big(\mathbf{n}_{\varphi_j} \cdot \mathbf{F}^{(\alpha_j, j)} \left(\mathbf{x}^{(\alpha_j, j)}, t, \mathbf{p}^{(\alpha_j, j)} \right)$$
$$+ \tilde{\mathbf{n}}_{\varphi_j} \cdot \tilde{\mathbf{F}}^{(\alpha_j, j)} \left(\tilde{\mathbf{x}}^{(\alpha_j, j)}, t, \tilde{\mathbf{p}}^{(\alpha_j, j)} \right) + \frac{\partial \varphi_j}{\partial t} \Big). \quad (7.27)$$

The combination of Eqs.(7.23) and (7.27) gives a dynamical system in phase space of $(z, \dot{z})$, i.e., for $j \in \mathbb{L}$

$$\left.\begin{aligned}
\dot{z}^{(\alpha_j, j)} &= g_1^{(\alpha_j, j)}(\mathbf{z}^{(\alpha_j, j)}, t) \equiv \mathbf{n}_{\varphi_j} \cdot \mathbf{F}^{(\alpha_j, j)}(\mathbf{x}^{(\alpha_j, j)}, t, \mathbf{p}^{(\alpha_j, j)}) \\
&\quad + \tilde{\mathbf{n}}_{\varphi_j} \cdot \tilde{\mathbf{F}}^{(\alpha_j, j)}(\tilde{\mathbf{x}}^{(\alpha_j, j)}, t, \tilde{\mathbf{p}}^{(\alpha_j, j)}) + \frac{\partial \varphi_j}{\partial t}, \\
\ddot{z}^{(\alpha_j, j)} &= g_2^{(\alpha_j, j)}(\mathbf{z}^{(\alpha_j, j)}, t) \equiv \frac{d}{dt} g_1^{(\alpha_j, j)}(\mathbf{z}^{(\alpha_j, j)}, t) \\
&= \frac{d}{dt} \Big(\mathbf{n}_{\varphi_j} \cdot \mathbf{F}^{(\alpha_j, j)}(\mathbf{x}^{(\alpha_j, j)}, t, \mathbf{p}^{(\alpha_j, j)}) \\
&\quad + \tilde{\mathbf{n}}_{\varphi_j} \cdot \tilde{\mathbf{F}}^{(\alpha_j, j)}(\tilde{\mathbf{x}}^{(\alpha_j, j)}, t, \tilde{\mathbf{p}}^{(\alpha_j, j)}) + \frac{\partial \varphi_j}{\partial t} \Big),
\end{aligned}\right\} \quad (7.28)$$

7.2 Fundamental interactions

where $\mathbf{z}^{(\alpha_j,j)} = (z^{(\alpha_j,j)}, \dot{z}^{(\alpha_j,j)})^{\mathrm{T}}$. Let $\mathbf{g}^{(\alpha_j,j)} = (g_1^{(\alpha_j,j)}, g_2^{(\alpha_j,j)})^{\mathrm{T}}$, one has

$$\dot{\mathbf{z}}^{(\alpha_j,j)} = \mathbf{g}^{(\alpha_j,j)}(\mathbf{z}^{(\alpha_j,j)}, t) \text{ for } j \in \mathcal{L}. \tag{7.29}$$

For a better understanding of such a discontinuous dynamical system, the boundary and domains in phase space are defined as

$$\begin{aligned}\partial \Xi_{(\alpha_j\beta_j,j)} &= \bar{\Xi}_{(\alpha_j,j)} \cap \bar{\Xi}_{(\beta_j,j)} \\ &= \left\{ (z^{(0,j)}, \dot{z}^{(0,j)}) \,\middle|\, \psi_j(z^{(0,j)}, \dot{z}^{(0,j)}) = z^{(0,j)} = 0, \right\} \subset \mathbb{R}^1, \end{aligned} \tag{7.30}$$

and

$$\left.\begin{aligned}\Xi_{(1,j)} &= \left\{ (z^{(1,j)}, \dot{z}^{(1,j)}) \,\middle|\, z^{(1,j)} > 0 \right\} \subset \mathbb{R}^2, \\ \Xi_{(2,j)} &= \left\{ (z^{(1,j)}, \dot{z}^{(1,j)}) \,\middle|\, z^{(2,j)} < 0 \right\} \subset \mathbb{R}^2. \end{aligned}\right\} \tag{7.31}$$

On the boundary, because of $\varphi_j(\mathbf{x}^{(0,j)}(t), \tilde{\mathbf{x}}^{(0,j)}(t), t, \lambda_j) = 0$, we have

$$\frac{d^s z^{(0,j)}}{dt^s} = \frac{d^s}{dt^s}\varphi_j\left(\mathbf{x}^{(0,j)}(t), \tilde{\mathbf{x}}^{(0,j)}(t), t, \lambda_j\right) = 0 \text{ for } s = 1, 2, \ldots \tag{7.32}$$

From the above discussion, the response on the boundary is given by

$$z^{(0,j)} = 0, \dot{z}^{(0,j)} = 0 \text{ for } j \in \mathcal{L} \tag{7.33}$$

The domains and boundary in phase space of $(z^{(j)}, \dot{z}^{(j)})$ are sketched in Fig.7.7, and the location for switching may not be continuous (i.e., $\mathbf{z}^{(\alpha_j,j)} \neq \mathbf{z}^{(\beta_j,j)} \neq \mathbf{z}^{(0,j)} = 0$) because the vector fields of the resultant system are discontinuous (or $\dot{z}^{(\alpha_j,j)} \neq \dot{z}^{(\beta_j,j)} \neq \dot{z}^{(0,j)} = 0$), but the boundary in such phase space is independent of time. However, the boundaries and domains in phase space of dynamical systems in Eqs.(7.1) and (7.2) are very complicated with n and $\tilde{n}$-dimensions, as shown in Fig.7.8. The boundary varying with time is presented, but switching points for a flow are continuous (i.e., $\mathbf{X}^{(\alpha_j,j)}(t_m) = \mathbf{X}^{(\beta_j,j)}(t_m) = \mathbf{X}^{(0,j)}(t_m)$). So the dynamical response will be completely determined by Eq.(7.13). However, such flows will be controlled by the vector fields $\mathbf{g}^{(1,j)}(\mathbf{z}^{(1,j)}, t)$ and $\mathbf{g}^{(2,j)}(\mathbf{z}^{(2,j)}, t)$. The dynamical systems in phase space $(z, \dot{z})$ are summarized as follows:

$$\dot{\mathbf{z}}^{(\Lambda_j,j)} = \mathbf{g}^{(\Lambda_j,j)}(\mathbf{z}^{(\Lambda_j,j)}, t) \text{ for } j \in \mathcal{L}, \Lambda_j = 0, \alpha_j, \tag{7.34}$$

where

$$\left.\begin{aligned}\mathbf{g}^{(\alpha_j,j)}(\mathbf{z}^{(\alpha_j,j)}, t) &= (g_1^{(\alpha_j,j)}(\mathbf{z}^{(\alpha_j,j)}, t), g_2^{(\alpha_j,j)}(\mathbf{z}^{(\alpha_j,j)}, t))^{\mathrm{T}} \\ &\quad \text{in} \Xi_{\alpha_j}(\alpha_j \in \{1,2\}), \\ \mathbf{g}^{(0,j)}(\mathbf{z}^{(\alpha_j,j)}, t) &\in [\mathbf{g}^{(\alpha_j,j)}(\mathbf{z}^{(\alpha_j,j)}, t), \mathbf{g}^{(\beta_j,j)}(\mathbf{z}^{(\beta_j,j)}, t)] \\ &\quad \text{on } \partial\Xi_{(\alpha_j\beta_j,j)} \text{ for non-stick}, \\ \mathbf{g}^{(0,j)}(\mathbf{z}^{(\alpha_j,j)}, t) &= (0,0)^{\mathrm{T}} \text{ on } \partial\Xi_{(\alpha_j\beta_j,j)} \text{for stick}. \end{aligned}\right\} \tag{7.35}$$

The normal vector of $\partial \Xi_{(\alpha_j\beta_j,j)}$ is computed from Eq.(7.30), i.e.,

202   7 Principles for System Interactions

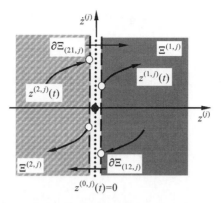

Fig. 7.7 A partition of phase space in $(z, \dot{z})$ for the jth-interaction boundary. Two dashed lines are infinitesimally close to the boundary with the dotted line

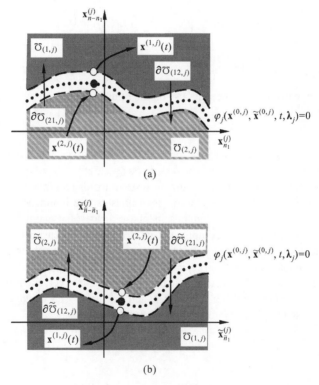

Fig. 7.8 Phase plane partitions: (a) the first system and (b) the second system. Two dashed curves are infinitesimally close to the boundary with the dotted curves

$$\mathbf{n}_{\partial \Xi_{(\alpha_j \beta_j, j)}} = (1, 0)^{\text{T}} \text{ and } D\mathbf{n}_{\partial \Xi_{(\alpha_j \beta_j, j)}} = (0, 0)^{\text{T}}, \qquad (7.36)$$

7.2 Fundamental interactions

where $D = d/dt$. From Luo (2008a,b), the corresponding two G-functions are computed by

$$\begin{aligned}
G^{(0,\alpha_j)}_{\partial\Xi_{(\alpha_j\beta_j,j)}}(\mathbf{z}^{(\alpha_j,j)},t) &= \mathbf{n}_{\partial\Xi_{(\alpha_i\beta_j,j)}} \cdot \mathbf{g}^{(\alpha_j,j)}(\mathbf{z}^{(\alpha_j,j)},t) \\
&= g_1^{(\alpha_j,j)}(\mathbf{z}^{(\alpha_j,j)},t), \\
G^{(1,\alpha_j)}_{\partial\Xi_{(\alpha_j\beta_j,j)}}(\mathbf{z}^{(\alpha_j,j)},t) &= \mathbf{n}_{\partial\Xi_{(\alpha_i\beta_j,j)}} \cdot D\mathbf{g}^{(\alpha_j,j)}(\mathbf{z}^{(\alpha_j,j)},t) \\
&= g_2^{(\alpha_j,j)}(\mathbf{z}^{(\alpha_j,j)},t).
\end{aligned} \quad (7.37)$$

With G-functions, the sufficient and necessary conditions for a passable flow at $(\mathbf{z}_m^{(0,j)}, t_m)$ with $\mathbf{z}_m^{(\alpha_j,j)} = \mathbf{z}_m^{(0,j)} = \mathbf{z}_m$ for the boundary $\partial\Xi_{(\alpha_j\beta_j,j)}$ are given by Luo (2005, 2006), i.e.,

$$\left. \begin{aligned}
G^{(0,1)}_{\partial\Xi_{(12,j)}}(\mathbf{z}_m^{(1,j)}, t_{m-}) &= g_1^{(1,j)}(\mathbf{z}_m^{(1,j)}, t_{m-}) < 0, \\
G^{(0,2)}_{\partial\Xi_{(12,j)}}(\mathbf{z}_m^{(2,j)}, t_{m+}) &= g_1^{(2,j)}(\mathbf{z}_m^{(2,j)}, t_{m+}) < 0,
\end{aligned} \right\} \text{ for } \Xi_{(1,j)} \to \Xi_{(2,j)}, \\
\left. \begin{aligned}
G^{(0,1)}_{\partial\Xi_{(12,j)}}(\mathbf{z}_m^{(1,j)}, t_{m+}) &= g_1^{(1,j)}(\mathbf{z}_m^{(1,j)}, t_{m+}) > 0, \\
G^{(0,2)}_{\partial\Xi_{(12,j)}}(\mathbf{z}_m^{(2,j)}, t_{m-}) &= g_1^{(2,j)}(\mathbf{z}_m^{(2,j)}, t_{m-}) > 0,
\end{aligned} \right\} \text{ for } \Xi_{(2,j)} \to \Xi_{(1,j)}, \quad (7.38)$$

where

$$g_1^{(\alpha_j,j)}(\mathbf{z}_m^{(\alpha_j,j)}, t_{m\pm}) = \mathbf{n}_{\varphi_j} \cdot \mathbf{F}^{(\alpha_j,j)}(\mathbf{x}_m^{(\alpha_j,j)}, t_{m\pm}, \mathbf{p}^{(\alpha_j,j)}) \\
+ \tilde{\mathbf{n}}_{\varphi_j} \cdot \tilde{\mathbf{F}}^{(\alpha_j,j)}(\tilde{\mathbf{x}}_m^{(\alpha_j,j)}, t_{m\pm}, \tilde{\mathbf{p}}^{(\alpha_j,j)}) + \frac{\partial \varphi_j}{\partial t}. \quad (7.39)$$

The foregoing condition gives the sufficient and necessary conditions for two systems to interact under the jth-interaction condition and to switch the current states through such an interaction condition. Such a flow to the boundary is called an *instantaneous interaction* between two systems.

The sufficient and necessary conditions for a stick flow (or sink flow) on the boundary $\partial\Xi_{(\alpha_j\beta_j,j)}$ are obtained from Chapter 2 (or Luo, 2005, 2006), i.e.,

$$\left. \begin{aligned}
G^{(0,1)}_{\partial\Xi_{(12,j)}}(\mathbf{z}_m^{(1,j)}, t_{m-}) &= g_1^{(1,j)}(\mathbf{z}_m^{(1,j)}, t_{m-}) < 0, \\
G^{(0,2)}_{\partial\Xi_{(12,j)}}(\mathbf{z}_m^{(2,j)}, t_{m-}) &= g_1^{(2,j)}(\mathbf{z}_m^{(2,j)}, t_{m-}) > 0,
\end{aligned} \right\} \text{ on } \partial\Xi_{(12,j)}. \quad (7.40)$$

From the foregoing condition, the two systems will stick together under the jth-interaction condition, which is called the *stick interaction*.

Similarly, the sufficient and necessary conditions for a source flow on the boundary $\partial\Xi_{(12,j)}$ are given in Luo (2005, 2006), i.e.,

$$\left. \begin{aligned}
G^{(0,1)}_{\partial\Xi_{(12,j)}}(\mathbf{z}_m^{(1,j)}, t_{m+}) &= g_1^{(1,j)}(\mathbf{z}_m^{(1,j)}, t_{m+}) > 0, \\
G^{(0,2)}_{\partial\Xi_{(12,j)}}(\mathbf{z}_m^{(2,j)}, t_{m+}) &= g_1^{(2,j)}(\mathbf{z}_m^{(2,j)}, t_{m+}) < 0,
\end{aligned} \right\} \text{ on } \partial\Xi_{(12,j)}. \quad (7.41)$$

For this case, the two dynamical systems will not interact at $(\mathbf{z}_m^{(0,j)}, t_m)$ for the boundary $\partial \Xi_{(\alpha_j \beta_j, j)}$ relative to the jth-interaction condition. The phenomenon is called the *source interaction*.

As in Chapter 2, an *L*-function is introduced to measure the above three interaction states

$$\left. \begin{aligned} L_{12}^{(j)}(t_{m\pm}) &= G_{\partial \Xi_{(\alpha_j \beta_j, j)}}^{(0,\alpha_j)}(\mathbf{z}_m^{(\alpha_j,j)}, t_{m-}) \times G_{\partial \Xi_{(\alpha_j \beta_j, j)}}^{(0,\beta_j)}(\mathbf{z}_m^{(\beta_j,j)}, t_{m+}) \\ &= g_1^{(\alpha_j,j)}(\mathbf{z}_m^{(\alpha_j,j)}, t_{m-}) \times g_1^{(\beta_j,j)}(\mathbf{z}_m^{(\beta_j,j)}, t_{m+}), \\ L_{12}^{(j)}(t_{m-}) &= G_{\partial \Xi_{(12,j)}}^{(0,1)}(\mathbf{z}_m^{(1,j)}, t_{m-}) \times G_{\partial \Xi_{(12,j)}}^{(0,2)}(\mathbf{z}_m^{(2,j)}, t_{m-}) \\ &= g_1^{(1,j)}(\mathbf{z}_m^{(1,j)}, t_{m-}) \times g_1^{(2,j)}(\mathbf{z}_m^{(2,j)}, t_{m-}), \\ L_{12}^{(j)}(t_{m+}) &= G_{\partial \Xi_{(12,j)}}^{(0,1)}(\mathbf{z}_m^{(1,j)}, t_{m+}) \times G_{\partial \Xi_{(12,j)}}^{(0,2)}(\mathbf{z}_m^{(2,j)}, t_{m+}) \\ &= g_1^{(1,j)}(\mathbf{z}_m^{(1,j)}, t_{m+}) \times g_1^{(2,j)}(\mathbf{z}_m^{(2,j)}, t_{m+}). \end{aligned} \right\} \quad (7.42)$$

No matter what interaction exists, the same quantity can be used for measuring the three interaction states. Such a quantity can be easily embedded in computer program (e.g., Luo and Gegg, 2006a,b; Luo, 2006). With Eq.(7.42), equations (7.38), (7.40) and (7.41) yield three new forms of the necessary and sufficient conditions for the three interaction states

$$\left. \begin{aligned} L_{12}^{(j)}(t_{m\pm}) &= g_1^{(\alpha_j,j)}(\mathbf{z}_m^{(\alpha_j,j)}, t_{m\mp}) \times g_1^{(\beta_j,j)}(\mathbf{z}_m^{(\beta_j,j)}, t_{m\pm}) > 0, \\ L_{12}^{(j)}(t_{m-}) &= g_1^{(1,j)}(\mathbf{z}_m^{(1,j)}, t_{m-}) \times g_1^{(2,j)}(\mathbf{z}_m^{(2,j)}, t_{m-}) < 0, \\ L_{12}^{(j)}(t_{m+}) &= g_1^{(1,j)}(\mathbf{z}_m^{(1,j)}, t_{m+}) \times g_1^{(2,j)}(\mathbf{z}_m^{(2,j)}, t_{m+}) < 0. \end{aligned} \right\} \quad (7.43)$$

The appearance and disappearance of three interaction states of the two dynamical systems to the jth-interaction condition in Eq.(7.4) can be determined from Luo (2008a,b), which are for the switching bifurcation of three states of interactions between the two dynamical systems.

(i) For appearance of the stick interaction from the instantaneous interaction, the sufficient and necessary conditions are given by

$$\left. \begin{aligned} (-1)^{\alpha_j} G_{\partial \Xi_{(\alpha_j \beta_j, j)}}^{(0,\alpha_j)}(\mathbf{z}_m^{(\alpha_j,j)}, t_{m-}) &= (-1)^{\alpha_j} g_1^{(\alpha_j,j)}(\mathbf{z}_m^{(\alpha_j,j)}, t_{m-}) > 0, \\ G_{\partial \Xi_{(\alpha_j \beta_j, j)}}^{(0,\beta_j)}(\mathbf{z}_m^{(\beta_j,j)}, t_{m\pm}) &= g_1^{(\beta_j,j)}(\mathbf{z}_m^{(\beta_j,j)}, t_{m\pm}) = 0, \\ (-1)^{\beta_j} G_{\partial \Xi_{(\alpha_j \beta_j, j)}}^{(1,\beta_j)}(\mathbf{z}_m^{(\beta_j,j)}, t_{m\pm}) &= (-1)^{\beta_j} g_2^{(\beta_j,j)}(\mathbf{z}_m^{(\beta_j,j)}, t_{m\pm}) < 0. \end{aligned} \right\} \quad (7.44)$$

For the vanishing of the stick interaction to form the instantaneous interaction on the jth-interaction boundary, the sufficient and necessary conditions are

$$\left. \begin{aligned} (-1)^{\alpha_j} G_{\partial \Xi_{(\alpha_j \beta_j, j)}}^{(0,\alpha_j)}(\mathbf{z}_m^{(\alpha_j,j)}, t_{m-}) &= (-1)^{\alpha_j} g_1^{(\alpha_j,j)}(\mathbf{z}_m^{(\alpha_j,j)}, t_{m-}) > 0, \\ G_{\partial \Xi_{(\alpha_j \beta_j, j)}}^{(0,\beta_j)}(\mathbf{z}_m^{(\beta_j,j)}, t_{m\mp}) &= g_1^{(\beta_j,j)}(\mathbf{z}_m^{(\beta_j,j)}, t_{m\mp}) = 0, \\ (-1)^{\beta_j} G_{\partial \Xi_{(\alpha_j \beta_j, j)}}^{(1,\beta_j)}(\mathbf{z}_m^{(\beta_j,j)}, t_{m\mp}) &= (-1)^{\beta_j} g_2^{(\beta_j,j)}(\mathbf{z}_m^{(\beta_j,j)}, t_{m\mp}) < 0. \end{aligned} \right\} \quad (7.45)$$

7.2 Fundamental interactions

The foregoing appearance and vanishing conditions for the stick interaction relative to the instantaneous interaction in Eq.(7.44) are also the vanishing and appearance conditions for the instantaneous interaction relative to the stick interaction, respectively. As in Chapter 2 (also see, Luo, 2006, 2008a,b), such appearance and vanishing conditions are the switching bifurcations between the stick and instantaneous interactions of the two dynamical system under the jth-interaction condition. Once the L-function in Eq.(7.42) is used, equations (7.44) and (7.45) become

$$\left.\begin{array}{l} L_{12}^{(j)}(t_{m\pm}) = g_1^{(\alpha_j,j)}(\mathbf{z}_m^{(\alpha_j,j)}, t_{m\mp}) \times g_1^{(\beta_j,j)}(\mathbf{z}_m^{(\beta_j,j)}, t_{m\pm}) = 0, \\ (-1)^{\beta_j} G_{\partial \Xi_{(\alpha_i\beta_j,j)}}^{(1,\beta_j)}(\mathbf{z}_m^{(\beta_j,j)}, t_{m\pm}) = (-1)^{\beta_j} g_2^{(\beta_j,j)}(\mathbf{z}_m^{(\beta_j,j)}, t_{m\pm}) < 0, \\ (-1)^{\alpha_j} G_{\partial \Xi_{(\alpha_i\beta_j,j)}}^{(0,\alpha_j)}(\mathbf{z}_m^{(\alpha_j,j)}, t_{m-}) = (-1)^{\alpha_j} g_1^{(\alpha_j,j)}(\mathbf{z}_m^{(\alpha_j,j)}, t_{m-}) > 0. \end{array}\right\} \quad (7.46)$$

From the foregoing equation, the L-function for such switching bifurcation is zero.

(ii) From Chapter 2 (also see, Luo, 2008a,b), the sufficient and necessary conditions for appearance and vanishing of the *source interaction*, pertaining to the instantaneous interaction are obtained. For the appearance of source interaction, we have

$$\left.\begin{array}{l} (-1)^{\alpha_j} G_{\partial \Xi_{(\alpha_i\beta_j,j)}}^{(0,\alpha_j)}(\mathbf{z}_m^{(\alpha_j,j)}, t_{m+}) = (-1)^{\alpha_j} g_1^{(\alpha_j,j)}(\mathbf{z}_m^{(\alpha_j,j)}, t_{m+}) < 0, \\ G_{\partial \Xi_{(\alpha_i\beta_j,j)}}^{(0,\beta_j)}(\mathbf{z}_m^{(\beta_j,j)}, t_{m\mp}) = g_1^{(\beta_j,j)}(\mathbf{z}_m^{(\beta_j,j)}, t_{m\mp}) = 0, \\ (-1)^{\beta_j} G_{\partial \Xi_{(\alpha_i\beta_j,j)}}^{(1,\beta_j)}(\mathbf{z}_m^{(\beta_j,j)}, t_{m\mp}) = (-1)^{\beta_j} g_2^{(\beta_j,j)}(\mathbf{z}_m^{(\beta_j,j)}, t_{m\mp}) < 0; \end{array}\right\} \quad (7.47)$$

and for the vanishing of the source interaction, one obtains

$$\left.\begin{array}{l} (-1)^{\alpha_j} G_{\partial \Xi_{(\alpha_i\beta_j,j)}}^{(0,\alpha_j)}(\mathbf{z}_m^{(\alpha_j,j)}, t_{m+}) = (-1)^{\alpha_j} g_1^{(\alpha_j,j)}(\mathbf{z}_m^{(\alpha_j,j)}, t_{m+}) < 0, \\ G_{\partial \Xi_{(\alpha_i\beta_j,j)}}^{(0,\beta_j)}(\mathbf{z}_m^{(\beta_j,j)}, t_{m\pm}) = g_1^{(\beta_j,j)}(\mathbf{z}_m^{(\beta_j,j)}, t_{m\pm}) = 0, \\ (-1)^{\beta_j} G_{\partial \Xi_{(\alpha_i\beta_j,j)}}^{(1,\beta_j)}(\mathbf{z}_m^{(\beta_j,j)}, t_{m\pm}) = (-1)^{\beta_j} g_2^{(\beta_j,j)}(\mathbf{z}_m^{(\beta_j,j)}, t_{m\pm}) < 0. \end{array}\right\} \quad (7.48)$$

Such a switching bifurcation between the non-interaction from the instantaneous interaction can be expressed as similar to Eq.(7.46).

(iii) From Chapter 2 (also see, Luo, 2008a,b), the sufficient and necessary conditions for the switching between the stick and source interactions on the jth-interaction boundary are

$$\left.\begin{array}{l} G_{\partial \Xi_{(\alpha_i\beta_j,j)}}^{(0,\alpha_j)}(\mathbf{z}_m^{(\alpha_j,j)}, t_{m\mp}) = g_1^{(\alpha_j,j)}(\mathbf{z}_m^{(\alpha_j,j)}, t_{m\mp}) = 0 \\ (-1)^{\alpha_j} G_{\partial \Xi_{(\alpha_i\beta_j,j)}}^{(1,\alpha_j)}(\mathbf{z}_m^{(\alpha_j,j)}, t_{m\mp}) = (-1)^{\alpha_j} g_2^{(\alpha_j,j)}(\mathbf{z}_m^{(\alpha_j,j)}, t_{m\mp}) < 0, \\ G_{\partial \Xi_{(\alpha_i\beta_j,j)}}^{(0,\beta_j)}(\mathbf{z}_m^{(\beta_j,j)}, t_{m\mp}) = g_1^{(\beta_j,j)}(\mathbf{z}_m^{(\beta_j,j)}, t_{m\mp}) = 0, \\ (-1)^{\beta_j} G_{\partial \Xi_{(\alpha_i\beta_j,j)}}^{(1,\beta_j)}(\mathbf{z}_m^{(\beta_j,j)}, t_{m\mp}) = (-1)^{\beta_j} g_2^{(\beta_j,j)}(\mathbf{z}_m^{(\beta_j,j)}, t_{m\mp}) < 0. \end{array}\right\} \quad (7.49)$$

Similarly, the sufficient and necessary conditions for the switching between two instantaneous states on the jth-interaction boundary for $\alpha_j \neq \beta_j$ are

$$\left.\begin{aligned}
&G^{(0,\alpha_j)}_{\partial \Xi_{(\alpha_i\beta_j,j)}}(\mathbf{z}_m^{(\alpha_j,j)},t_{m\mp}) = g_1^{(\alpha_j,j)}(\mathbf{z}_m^{(\alpha_j,j)},t_{m\mp}) = 0 \text{ for } \alpha_j \in \{1,2\},\\
&(-1)^{\alpha_j} G^{(1,\alpha_j)}_{\partial \Xi_{(\alpha_i\beta_j,j)}}(\mathbf{z}_m^{(\alpha_j,j)},t_{m\mp}) = (-1)^{\alpha_j} g_2^{(\alpha_j,j)}(\mathbf{z}_m^{(\alpha_j,j)},t_{m\mp}) < 0,\\
&G^{(0,\beta_j)}_{\partial \Xi_{(\alpha_i\beta_j,j)}}(\mathbf{z}_m^{(\beta_j,j)},t_{m\pm}) = g_1^{(\beta_j,j)}(\mathbf{z}_m^{(\beta_j,j)},t_{m\pm}) = 0 \text{ for } \beta_j \in \{1,2\},\\
&(-1)^{\beta_j} G^{(1,\beta_j)}_{\partial \Xi_{(\alpha_i\beta_j,j)}}(\mathbf{z}_m^{(\beta_j,j)},t_{m\pm}) = (-1)^{\beta_j} g_2^{(\beta_j,j)}(\mathbf{z}_m^{(\beta_j,j)},t_{m\pm}) < 0.
\end{aligned}\right\} \quad (7.50)$$

In the foregoing equation, both of the zero-order G-functions should be zero. For such two interactions, equation (7.46) becomes

$$\left.\begin{aligned}
&L_{12}^{(j)}(t_{m\pm}) = g_1^{(1,j)}(\mathbf{z}_m^{(\alpha_j,j)},t_{m\mp}) \times g_1^{(2,j)}(\mathbf{z}_m^{(\beta_j,j)},t_{m\pm}) = 0,\\
&(-1)^{\alpha_j} G^{(1,\alpha_j)}_{\partial \Xi_{(\alpha_i\beta_j,j)}}(\mathbf{z}_m^{(\alpha_j,j)},t_{m\mp}) = (-1)^{\alpha_j} g_2^{(\alpha_j,j)}(\mathbf{z}_m^{(\alpha_j,j)},t_{m\mp}) < 0,\\
&(-1)^{\beta_j} G^{(1,\beta_j)}_{\partial \Xi_{(\alpha_i\beta_j,j)}}(\mathbf{z}_m^{(\beta_j,j)},t_{m\pm}) = (-1)^{\beta_j} g_2^{(\beta_j,j)}(\mathbf{z}_m^{(\beta_j,j)},t_{m\pm}) < 0\\
&\text{for } \alpha_j, \beta_j = 1,2 \text{ and } \alpha_j \neq \beta_j.
\end{aligned}\right\} \quad (7.51)$$

In addition to the instantaneous, stick and source interactions with the corresponding singularity, a flow tangential to the boundary $\partial \Xi_{(\alpha_j\beta_j,j)}$ is another instantaneous interaction (or tangential interaction), the corresponding sufficient and necessary conditions are from Chapter 2, i.e.,

$$\left.\begin{aligned}
&G^{(0,\alpha_j)}_{\partial \Xi_{(\alpha_i\beta_j,j)}}(\mathbf{z}_m^{(\alpha_j,j)},t_{m\pm}) = g_1^{(\alpha_j,j)}(\mathbf{z}_m^{(\alpha_j,j)},t_{m\pm}) = 0 \text{ for } \alpha_j \in \{1,2\},\\
&(-1)^{\alpha_j} G^{(1,\alpha_j)}_{\partial \Xi_{(\alpha_i\beta_j,j)}}(\mathbf{z}_m^{(\alpha_j,j)},t_{m\pm}) = (-1)^{\alpha_j} g_2^{(\alpha_j,j)}(\mathbf{z}_m^{(\alpha_j,j)},t_{m\pm}) < 0.
\end{aligned}\right\} \quad (7.52)$$

7.3 Interactions with singularity

With higher-order singularity to the boundary $\partial \Xi_{(\alpha_j\beta_j,j)}$, the higher-order G-function should be defined. That is,

$$G^{(k\alpha_j,\alpha_j)}_{\partial \Xi_{(\alpha_i\beta_j,j)}}(\mathbf{z}^{(\alpha_j,j)},t)$$
$$= \mathbf{n}_{\partial \Xi_{(\alpha_i\beta_j,j)}} \cdot D^{k\alpha_j} \mathbf{g}^{(\alpha_j,j)}(\mathbf{z}^{(\alpha_j,j)},t) = g_{k\alpha_j+1}^{(\alpha_j,j)}(\mathbf{z}^{(\alpha_j,j)},t), \quad (7.53)$$

where
$$g_{k\alpha_j+1}^{(\alpha_j,j)}(\mathbf{z}^{(\alpha_j,j)},t) = D^{k\alpha_j+1} \varphi_j^{(\alpha_j,j)}(\mathbf{z}^{(\alpha_j,j)},t). \quad (7.54)$$

Using the higher order G-functions, the sufficient and necessary conditions for a $(2k_{\alpha_j} : 2k_{\beta_j})$-instantaneous interaction at $(\mathbf{z}_m^{(0,j)},t_m)$ for the boundary $\partial \Xi_{(\alpha_j\beta_j,j)}$ are

7.3 Interactions with singularity

obtained from Luo (2006, 2008a,b)

$$\left.\begin{aligned}
& G^{(s_{\alpha_j},\alpha_j)}_{\partial\Xi_{(\alpha_i\beta_j,j)}}(\mathbf{z}_m^{(\alpha_j,j)},t_{m-}) = g^{(\alpha_j,j)}_{s_{\alpha_j}+1}(\mathbf{z}_m^{(\alpha_j,j)},t_{m-}) = 0 \\
& \text{for } s_{\alpha_j} = 0,1,\ldots,2k_{\alpha_j}-1, \\
& G^{(s_{\beta_j},\beta_j)}_{\partial\Xi_{(\alpha_i\beta_j,j)}}(\mathbf{z}_m^{(\beta_j,j)},t_{m+}) = g^{(\beta_j,j)}_{s_{\beta_j}+1}(\mathbf{z}_m^{(\beta_j,j)},t_{m+}) = 0 \\
& \text{for } s_{\beta_j} = 0,1,\ldots,2k_{\beta_j}-1, \\
& (-1)^{\alpha_j} G^{(2k_{\alpha_j},\alpha_j)}_{\partial\Xi_{(\alpha_i\beta_j,j)}}(\mathbf{z}_m^{(\alpha_j,j)},t_{m-}) = (-1)^{\alpha_j} g^{(\alpha_j,j)}_{2k_{\alpha_j}+1}(\mathbf{z}_m^{(\alpha_j,j)},t_{m-}) > 0, \\
& (-1)^{\beta_j} G^{(2k_{\beta_j},\beta_j)}_{\partial\Xi_{(\alpha_i\beta_j,j)}}(\mathbf{z}_m^{(\beta_j,j)},t_{m+}) = (-1)^{\beta_j} g^{(\beta_j,j)}_{2k_{\beta_j}+1}(\mathbf{z}_m^{(\beta_j,j)},t_{m+}) < 0 \\
& \text{for}\,\Xi_{(1,j)} \to \Xi_{(2,j)},\ \alpha_j,\beta_j \in \{1,2\},\ \alpha_j \neq \beta_j.
\end{aligned}\right\} \quad (7.55)$$

From Chapter 2 (or Luo, 2006, 2008a,b), the sufficient and necessary conditions for a $(2k_{\alpha_j} : 2k_{\beta_j})$-stick interaction at $(\mathbf{z}_m^{(0,j)}, t_m)$ on the boundary $\partial\Xi_{(\alpha_j\beta_j,j)}$ are

$$\left.\begin{aligned}
& G^{(s_{\alpha_j},\alpha_j)}_{\partial\Xi_{(\alpha_i\beta_j,j)}}(\mathbf{z}_m^{(\alpha_j,j)},t_{m-}) = g^{(\alpha_j,j)}_{s_{\alpha_j}+1}(\mathbf{z}_m^{(\alpha_j,j)},t_{m-}) = 0 \\
& \text{for } s_{\alpha_j} = 0,1,\ldots,2k_{\alpha_j}-1, \\
& G^{(s_{\beta_j},\beta_j)}_{\partial\Xi_{(\alpha_i\beta_j,j)}}(\mathbf{z}_m^{(\beta_j,j)},t_{m-}) = g^{(\beta_j,j)}_{s_{\beta_j}+1}(\mathbf{z}_m^{(\beta_j,j)},t_{m-}) = 0 \\
& \text{for } s_{\beta_j} = 0,1,\ldots,2k_{\beta_j}-1, \\
& (-1)^{\alpha_j} G^{(2k_{\alpha_j},\alpha_j)}_{\partial\Xi_{(\alpha_i\beta_j,j)}}(\mathbf{z}_m^{(\alpha_j,j)},t_{m-}) = (-1)^{\alpha_j} g^{(\alpha_j,j)}_{2k_{\alpha_j}+1}(\mathbf{z}_m^{(\alpha_j,j)},t_{m-}) > 0, \\
& (-1)^{\beta_j} G^{(2k_{\beta_j},\beta_j)}_{\partial\Xi_{(\alpha_i\beta_j,j)}}(\mathbf{z}_m^{(\beta_j,j)},t_{m-}) = (-1)^{\beta_j} g^{(\beta_j,j)}_{2k_{\beta_j}+1}(\mathbf{z}_m^{(\beta_j,j)},t_{m-}) > 0 \\
& \text{for } \alpha_j,\beta_j \in \{1,2\},\ \alpha_j \neq \beta_j.
\end{aligned}\right\} \quad (7.56)$$

Similarly, the sufficient and necessary conditions for a $(2k_{\alpha_j} : 2k_{\beta_j})$-source interaction (or no any interaction) at $(\mathbf{z}_m^{(0,j)}, t_m)$ on the boundary $\partial\Xi_{(\alpha_j\beta_j,j)}$ are

$$\left.\begin{aligned}
& G^{(s_{\alpha_j},\alpha_j)}_{\partial\Xi_{(\alpha_i\beta_j,j)}}(\mathbf{z}_m^{(\alpha_j,j)},t_{m+}) = g^{(\alpha_j,j)}_{s_{\alpha_j}+1}(\mathbf{z}_m^{(\alpha_j,j)},t_{m+}) = 0 \\
& \text{for } s_{\alpha_j} = 0,1,\ldots,2k_{\alpha_j}-1, \\
& G^{(s_{\beta_j},\beta_j)}_{\partial\Xi_{(\alpha_i\beta_j,j)}}(\mathbf{z}_m^{(\beta_j,j)},t_{m+}) = g^{(\beta_j,j)}_{s_{\beta_j}+1}(\mathbf{z}_m^{(\beta_j,j)},t_{m+}) = 0 \\
& \text{for } s_{\beta_j} = 0,1,\ldots,2k_{\beta_j}-1, \\
& (-1)^{\alpha_j} G^{(2k_{\alpha_j},\alpha_j)}_{\partial\Xi_{(\alpha_i\beta_j,j)}}(\mathbf{z}_m^{(\alpha_j,j)},t_{m+}) = (-1)^{\alpha_j} g^{(\alpha_j,j)}_{2k_{\alpha_j}+1}(\mathbf{z}_m^{(\alpha_j,j)},t_{m+}) < 0, \\
& (-1)^{\beta_j} G^{(2k_{\beta_j},\beta_j)}_{\partial\Xi_{(\alpha_i\beta_j,j)}}(\mathbf{z}_m^{(\beta_j,j)},t_{m+}) = (-1)^{\beta_j} g^{(\beta_j,j)}_{2k_{\beta_j}+1}(\mathbf{z}_m^{(\beta_j,j)},t_{m+}) < 0 \\
& \text{for } \alpha_j,\beta_j \in \{1,2\},\ \alpha_j \neq \beta_j.
\end{aligned}\right\} \quad (7.57)$$

As in Chapter 2, similar to Eq.(7.42), the $(2k_{\alpha_j} : 2k_{\beta_j})$-order *L*-function is defined as follows:

$$\left.\begin{aligned}
L_{12}^{((2k_{\alpha_j}:2k_{\beta_j}),j)}(t_{m\pm}) &= G_{\partial\Xi_{(\alpha_i\beta_j,j)}}^{(2k_{\alpha_j},\alpha_j)}(\mathbf{z}_m^{(\alpha_j,j)},t_{m-}) \times G_{\partial\Xi_{(\alpha_i\beta_j,j)}}^{(2k_{\beta_j},\beta_j)}(\mathbf{z}_m^{(\beta_j,j)},t_{m+}), \\
L_{12}^{((2k_{\alpha_j}:2k_{\beta_j}),j)}(t_{m-}) &= G_{\partial\Xi_{(\alpha_i\beta_j,j)}}^{(2k_{\alpha_j},\alpha_j)}(\mathbf{z}_m^{(\alpha_j,j)},t_{m-}) \times G_{\partial\Xi_{(\alpha_i\beta_j,j)}}^{(2k_{\beta_j},\beta_j)}(\mathbf{z}_m^{(\beta_j,j)},t_{m-}), \\
L_{12}^{((2k_{\alpha_j}:2k_{\beta_j}),j)}(t_{m+}) &= G_{\partial\Xi_{(\alpha_i\beta_j,j)}}^{(2k_{\alpha_j},\alpha_j)}(\mathbf{z}_m^{(\alpha_j,j)},t_{m+}) \times G_{\partial\Xi_{(\alpha_i\beta_j,j)}}^{(2k_{\beta_j},\beta_j)}(\mathbf{z}_m^{(\beta_j,j)},t_{m+}).
\end{aligned}\right\} \quad (7.58)$$

From the $(2k_{\alpha_j} : 2k_{\beta_j})$-order L-function, the sufficient and necessary conditions for three interactions with the $(2k_{\alpha_j} : 2k_{\beta_j})$-order singularity are

$$\left.\begin{aligned}
L_{12}^{((2k_{\alpha_j}:2k_{\beta_j}),j)}(t_{m\pm}) &= g_{2k_{\alpha_j}+1}^{(\alpha_j,j)}(\mathbf{z}_m^{(\alpha_j,j)},t_{m-}) \times g_{2k_{\beta_j}+1}^{(\beta_j,j)}(\mathbf{z}_m^{(\beta_j,j)},t_{m+}) > 0, \\
L_{12}^{((2k_{\alpha_j}:2k_{\beta_j}),j)}(t_{m-}) &= g_{2k_{\alpha_j}+1}^{(\alpha_j,j)}(\mathbf{z}_m^{(\alpha_j,j)},t_{m-}) \times g_{2k_{\beta_j}+1}^{(\beta_j,j)}(\mathbf{z}_m^{(\beta_j,j)},t_{m-}) < 0, \\
L_{12}^{((2k_{\alpha_j}:2k_{\beta_j}),j)}(t_{m+}) &= g_{2k_{\alpha_j}+1}^{(\alpha_j,j)}(\mathbf{z}_m^{(\alpha_j,j)},t_{m+}) \times g_{2k_{\beta_j}+1}^{(\beta_j,j)}(\mathbf{z}_m^{(\beta_j,j)},t_{m+}) < 0.
\end{aligned}\right\} \quad (7.59)$$

The conditions for the appearance and vanishing of the $(2k_{\alpha_j} : 2k_{\beta_j})$-stick interaction relative to the $(2k_{\alpha_j} : 2k_{\beta_j})$-instantaneous interaction are

$$\left.\begin{aligned}
&G_{\partial\Xi_{(\alpha_i\beta_j,j)}}^{(s_{\alpha_j},\alpha_j)}(\mathbf{z}_m^{(\alpha_j,j)},t_{m-}) = g_{s_{\alpha_j}+1}^{(\alpha_j,j)}(\mathbf{z}_m^{(\alpha_j,j)},t_{m-}) = 0 \\
&\quad \text{for } s_{\alpha_j} = 0,1,\ldots,2k_{\alpha_j}-1, \\
&G_{\partial\Xi_{(\alpha_i\beta_j,j)}}^{(s_{\beta_j},\beta_j)}(\mathbf{z}_m^{(\beta_j,j)},t_{m\pm}) = g_{s_{\beta_j}+1}^{(\beta_j,j)}(\mathbf{z}_m^{(\beta_j,j)},t_{m\pm}) = 0 \\
&\quad \text{for } s_{\beta_j} = 0,1,\ldots,2k_{\beta_j}, \\
&(-1)^{\alpha_j} G_{\partial\Xi_{(\alpha_i\beta_j,j)}}^{(2k_{\alpha_j},\alpha_j)}(\mathbf{z}_m^{(\alpha_j,j)},t_{m-}) = (-1)^{\alpha_j} g_{2k_{\alpha_j}+1}^{(\alpha_j,j)}(\mathbf{z}_m^{(\alpha_j,j)},t_{m-}) > 0, \\
&(-1)^{\beta_j} G_{\partial\Xi_{(\alpha_i\beta_j,j)}}^{(2k_{\beta_j}+1,\beta_j)}(\mathbf{z}_m^{(\beta_j,j)},t_{m\pm}) = (-1)^{\beta_j} g_{2k_{\beta_j}+2}^{(\beta_j,j)}(\mathbf{z}_m^{(\beta_j,j)},t_{m\pm}) < 0 \\
&\quad \text{for } \alpha_j,\beta_j \in \{1,2\},\ \alpha_j \neq \beta_j,
\end{aligned}\right\} \quad (7.60)$$

and

$$\left.\begin{aligned}
&G_{\partial\Xi_{(\alpha_i\beta_j,j)}}^{(s_{\alpha_j},\alpha_j)}(\mathbf{z}_m^{(\alpha_j,j)},t_{m-}) = g_{s_{\alpha_j}+1}^{(\alpha_j,j)}(\mathbf{z}_m^{(\alpha_j,j)},t_{m-}) = 0 \\
&\quad \text{for } s_{\alpha_j} = 0,1,\ldots,2k_{\alpha_j}-1, \\
&G_{\partial\Xi_{(\alpha_i\beta_j,j)}}^{(s_{\beta_j},\beta_j)}(\mathbf{z}_m^{(\beta_j,j)},t_{m\mp}) = g_{s_{\beta_j}+1}^{(\beta_j,j)}(\mathbf{z}_m^{(\beta_j,j)},t_{m\mp}) = 0 \\
&\quad \text{for } s_{\beta_j} = 0,1,\ldots,2k_{\beta_j}, \\
&(-1)^{\alpha_j} G_{\partial\Xi_{(\alpha_i\beta_j,j)}}^{(2k_{\alpha_j},\alpha_j)}(\mathbf{z}_m^{(\alpha_j,j)},t_{m-}) = (-1)^{\alpha_j} g_{2k_{\alpha_j}+1}^{(\alpha_j,j)}(\mathbf{z}_m^{(\alpha_j,j)},t_{m-}) > 0, \\
&(-1)^{\beta_j} G_{\partial\Xi_{(\alpha_i\beta_j,j)}}^{(2k_{\beta_j}+1,\beta_j)}(\mathbf{z}_m^{(\beta_j,j)},t_{m\mp}) = (-1)^{\beta_j} g_{2k_{\beta_j}+2}^{(\beta_j,j)}(\mathbf{z}_m^{(\beta_j,j)},t_{m\mp}) < 0 \\
&\quad \text{for } \alpha_j,\beta_j \in \{1,2\},\ \alpha_j \neq \beta_j.
\end{aligned}\right\} \quad (7.61)$$

The conditions in both of two foregoing equations can be expressed by

7.3 Interactions with singularity

$$\left.\begin{array}{l}L_{12}^{((2k_{\alpha_j}:2k_{\beta_j}),j)}(t_{m\pm}) = g_{2k_{\alpha_j}+1}^{(\alpha_j,j)}(\mathbf{z}_m^{(\alpha_j,j)}, t_{m-}) \times g_{2k_{\beta_j}+1}^{(\beta_j,j)}(\mathbf{z}_m^{(\beta_j,j)}, t_{m+}) = 0, \\ (-1)^{\alpha_j} G_{\partial\Xi_{(\alpha_i\beta_j,j)}}^{(2k_{\alpha_j},\alpha_j)}(\mathbf{z}_m^{(\alpha_j,j)}, t_{m-}) = (-1)^{\alpha_j} g_{2k_{\alpha_j}+1}^{(\alpha_j,j)}(\mathbf{z}_m^{(\alpha_j,j)}, t_{m-}) > 0, \\ (-1)^{\beta_j} G_{\partial\Xi_{(\alpha_i\beta_j,j)}}^{(2k_{\beta_j}+1,\beta_j)}(\mathbf{z}_m^{(\beta_j,j)}, t_{m\pm}) = (-1)^{\beta_j} g_{2k_{\beta_j}+2}^{(\beta_j,j)}(\mathbf{z}_m^{(\beta_j,j)}, t_{m\pm}) < 0.\end{array}\right\} \quad (7.62)$$

The sufficient and necessary conditions for appearance and vanishing of the $(2k_{\alpha_j}:2k_{\beta_j})$-*source interaction* pertaining to the $(2k_{\alpha_j}:2k_{\beta_j})$-instantaneous interaction are

$$\left.\begin{array}{l}G_{\partial\Xi_{(\alpha_i\beta_j,j)}}^{(s_{\alpha_j},\alpha_j)}(\mathbf{z}_m^{(\alpha_j,j)}, t_{m+}) = g_{s_{\alpha_j}+1}^{(\alpha_j,j)}(\mathbf{z}_m^{(\alpha_j,j)}, t_{m+}) = 0 \\ \text{for } s_{\alpha_j} = 0, 1, \ldots, 2k_{\alpha_j}-1, \\ G_{\partial\Xi_{(\alpha_i\beta_j,j)}}^{(s_{\beta_j},\beta_j)}(\mathbf{z}_m^{(\beta_j,j)}, t_{m\mp}) = g_{s_{\beta_j}+1}^{(\beta_j,j)}(\mathbf{z}_m^{(\beta_j,j)}, t_{m\mp}) = 0 \\ \text{for } s_{\beta_j} = 0, 1, \ldots, 2k_{\beta_j}, \\ (-1)^{\alpha_j} G_{\partial\Xi_{(\alpha_i\beta_j,j)}}^{(2k_{\alpha_j},\alpha_j)}(\mathbf{z}_m^{(\alpha_j,j)}, t_{m+}) = (-1)^{\alpha_j} g_{2k_{\alpha_j}+1}^{(\alpha_j,j)}(\mathbf{z}_m^{(\alpha_j,j)}, t_{m+}) < 0, \\ (-1)^{\beta_j} G_{\partial\Xi_{(\alpha_i\beta_j,j)}}^{(2k_{\beta_j}+1,\beta_j)}(\mathbf{z}_m^{(\beta_j,j)}, t_{m\mp}) = (-1)^{\beta_j} g_{2k_{\beta_j}+2}^{(\beta_j,j)}(\mathbf{z}_m^{(\beta_j,j)}, t_{m\mp}) < 0 \\ \text{for } \alpha_j, \beta_j \in \{1,2\}, \alpha_j \neq \beta_j.\end{array}\right\} \quad (7.63)$$

$$\left.\begin{array}{l}G_{\partial\Xi_{(\alpha_i\beta_j,j)}}^{(s_{\alpha_j},\alpha_j)}(\mathbf{z}_m^{(\alpha_j,j)}, t_{m+}) = g_{s_{\alpha_j}+1}^{(\alpha_j,j)}(\mathbf{z}_m^{(\alpha_j,j)}, t_{m+}) = 0 \\ \text{for } s_{\alpha_j} = 0, 1, \ldots, 2k_{\alpha_j}-1, \\ G_{\partial\Xi_{(\alpha_i\beta_j,j)}}^{(s_{\beta_j},\beta_j)}(\mathbf{z}_m^{(\beta_j,j)}, t_{m\pm}) = g_{s_{\beta_j}+1}^{(\beta_j,j)}(\mathbf{z}_m^{(\beta_j,j)}, t_{m\pm}) = 0 \\ \text{for } s_{\beta_j} = 0, 1, \ldots, 2k_{\beta_j}, \\ (-1)^{\alpha_j} G_{\partial\Xi_{(\alpha_i\beta_j,j)}}^{(2k_{\alpha_j},\alpha_j)}(\mathbf{z}_m^{(\alpha_j,j)}, t_{m+}) = (-1)^{\alpha_j} g_{2k_{\alpha_j}+1}^{(\alpha_j,j)}(\mathbf{z}_m^{(\alpha_j,j)}, t_{m+}) > 0, \\ (-1)^{\beta_j} G_{\partial\Xi_{(\alpha_i\beta_j,j)}}^{(2k_{\beta_j}+1,\beta_j)}(\mathbf{z}_m^{(\beta_j,j)}, t_{m\pm}) = (-1)^{\beta_j} g_{2k_{\beta_j}+2}^{(\beta_j,j)}(\mathbf{z}_m^{(\beta_j,j)}, t_{m\pm}) < 0 \\ \text{for } \alpha_j, \beta_j \in \{1,2\}, \alpha_j \neq \beta_j.\end{array}\right\} \quad (7.64)$$

The sufficient and necessary conditions for the switching between the $(2k_{\alpha_j}:2k_{\beta_j})$-stick and source interactions on the jth-interaction boundary are

$$\left.\begin{array}{l}G_{\partial\Xi_{(\alpha_i\beta_j,j)}}^{(s_{\alpha_j},\alpha_j)}(\mathbf{z}_m^{(\alpha_j,j)}, t_{m\pm}) = g_{s_{\alpha_j}+1}^{(\alpha_j,j)}(\mathbf{z}_m^{(\alpha_j,j)}, t_{m\pm}) = 0 \\ \text{for } s_{\alpha_j} = 0, 1, \ldots, 2k_{\alpha_j} \text{ and } \alpha_j = 1, 2, \\ (-1)^{\alpha_j} G_{\partial\Xi_{(\alpha_i\beta_j,j)}}^{(2k_{\alpha_j}+1,\alpha_j)}(\mathbf{z}_m^{(\alpha_j,j)}, t_{m\pm}) = (-1)^{\alpha_j} g_{2k_{\alpha_j}+2}^{(\alpha_j,j)}(\mathbf{z}_m^{(\alpha_j,j)}, t_{m\pm}) < 0.\end{array}\right\} \quad (7.65)$$

Similarly, the sufficient and necessary conditions for the switching between two $(2k_{\alpha_j}:2k_{\beta_j})$-instantaneous states on the jth-interaction boundary are

$$\left.\begin{aligned}
& G_{\partial\Xi_{(\alpha_i\beta_j,j)}}^{(s_{\alpha_j},\alpha_j)}(\mathbf{z}_m^{(\alpha_j,j)},t_{m\mp}) = g_{s_{\alpha_j}+1}^{(\alpha_j,j)}(\mathbf{z}_m^{(\alpha_j,j)},t_{m\mp}) = 0 \\
& \quad \text{for } s_{\alpha_j} = 0,1,\ldots,2k_{\alpha_j} \text{ and } \alpha_j \in \{1,2\}, \\
& (-1)^{\alpha_j} G_{\partial\Xi_{(\alpha_i\beta_j,j)}}^{(2k_{\alpha_j}+1,\alpha_j)}(\mathbf{z}_m^{(\alpha_j,j)},t_{m\mp}) = (-1)^{\alpha_j} g_{2k_{\alpha_j}+2}^{(\alpha_j,j)}(\mathbf{z}_m^{(\alpha_j,j)},t_{m\mp}) < 0, \\
& G_{\partial\Xi_{(\alpha_i\beta_j,j)}}^{(s_{\beta_j},\beta_j)}(\mathbf{z}_m^{(\beta_j,j)},t_{m\pm}) = g_{s_{\beta_j}+1}^{(\beta_j,j)}(\mathbf{z}_m^{(\beta_j,j)},t_{m\pm}) = 0 \\
& \quad \text{for } s_{\beta_j} = 0,1,\ldots,2k_{\beta_j} \text{ and } \beta_j \in \{1,2\} \text{ and } \alpha_j \neq \beta_j, \\
& (-1)^{\beta_j} G_{\partial\Xi_{(\alpha_i\beta_j,j)}}^{(2k_{\beta_j}+1,\beta_j)}(\mathbf{z}_m^{(\beta_j,j)},t_{m\pm}) = (-1)^{\beta_j} g_{2k_{\beta_j}+2}^{(\beta_j,j)}(\mathbf{z}_m^{(\beta_j,j)},t_{m\pm}) < 0.
\end{aligned}\right\} \quad (7.66)$$

Both of the G-functions with the $(2k_{\alpha_j} : 2k_{\beta_j})$-order singularity are zero, thus, one obtains for $\alpha_j \neq \beta_j$

$$\left.\begin{aligned}
& L_{12}^{((2k_{\alpha_j}:2k_{\beta_j}),j)}(t_{m\pm}) = g_{2k_{\alpha_j}+1}^{(\alpha_j,j)}(\mathbf{z}_m^{(\alpha_j,j)},t_{m-}) \times g_{2k_{\beta_j}+1}^{(\beta_j,j)}(\mathbf{z}_m^{(\beta_j,j)},t_{m+}) = 0, \\
& (-1)^{\alpha_j} G_{\partial\Xi_{(\alpha_i\beta_j,j)}}^{(2k_{\alpha_j}+1,\alpha_j)}(\mathbf{z}_m^{(\alpha_j,j)},t_{m\mp}) = (-1)^{\alpha_j} g_{2k_{\alpha_j}+2}^{(\alpha_j,j)}(\mathbf{z}_m^{(\alpha_j,j)},t_{m\mp}) < 0, \\
& (-1)^{\beta_j} G_{\partial\Xi_{(\alpha_i\beta_j,j)}}^{(2k_{\beta_j}+1,\beta_j)}(\mathbf{z}_m^{(\beta_j,j)},t_{m\pm}) = (-1)^{\beta_j} g_{2k_{\beta_j}+2}^{(\beta_j,j)}(\mathbf{z}_m^{(\beta_j,j)},t_{m\pm}) < 0.
\end{aligned}\right\} \quad (7.67)$$

For the $(2k_{\alpha_j}+1)$-order tangential interaction to the boundary $\partial\Xi_{(\alpha_j\beta_j,j)}$, the corresponding sufficient and necessary conditions are

$$\left.\begin{aligned}
& G_{\partial\Xi_{(\alpha_i\beta_j,j)}}^{(s_{\alpha_j},\alpha_j)}(\mathbf{z}_m^{(\alpha_j,j)},t_{m\pm}) = g_{s_{\alpha_j}+1}^{(\alpha_j,j)}(\mathbf{z}_m^{(\alpha_j,j)},t_{m\pm}) = 0 \\
& \quad \text{for } s_{\alpha_j} = 0,1,\ldots,2k_{\alpha_j} \text{ and } \alpha_j \in \{1,2\}, \\
& (-1)^{\alpha_j} G_{\partial\Xi_{(\alpha_i\beta_j,j)}}^{(2k_{\alpha_j}+1,\alpha_j)}(\mathbf{z}_m^{(\alpha_j,j)},t_{m\pm}) = (-1)^{\alpha_j} g_{2k_{\alpha_j}+2}^{(\alpha_j,j)}(\mathbf{z}_m^{(\alpha_j,j)},t_{m\pm}) < 0.
\end{aligned}\right\} \quad (7.68)$$

7.4 Interactions with corner singularity

Suppose there are l_1-linear-independent interaction conditions among l -interaction conditions in Eq.(7.4). Such l_1-interaction conditions form a corner point (or singular point), which is defined in Luo (2006). At the corner point, the conditions presented in this section should be checked for all l_1-interaction boundaries through the following definitions:

$$z^{(\alpha_{j_1},j_1)} = \varphi_{j_1}\left(\mathbf{x}^{(\alpha_{j_1},j_1)}(t), \tilde{\mathbf{x}}^{(\alpha_{j_1},j_1)}(t), t, \boldsymbol{\lambda}_{j_1}\right) \text{ for all } j_1 \in \mathbb{L}_1, \quad (7.69)$$

where $\mathbb{L}_1 \subseteq \mathbb{L}$. On the boundary $\partial\Omega_{(\alpha_{j_1}\beta_{j_1},j_1)}$, one obtains

$$z^{(0,j_1)} = \varphi_{j_1}\left(\mathbf{x}^{(0,j_1)}(t), \tilde{\mathbf{x}}^{(0,j_1)}(t), t, \boldsymbol{\lambda}_{j_1}\right) = 0 \text{ for all } j_1 \in \mathbb{L}_1. \quad (7.70)$$

7.4 Interactions with corner singularity

Consider three subsets $L_{1_i} \subseteq L_1 (i=1,2,3)$ with l_{1_i}-interaction conditions among l_1-conditions and $l_{1_1}+l_{1_2}+l_{1_3}=l_1$. For $j_{1_i} \in L_{1_i}$, the corresponding conditions for l_{1_1}-stick interaction, l_{1_2}-non-interaction and l_{1_3}-instantaneous interaction are given as follows.

Theorem 7.1. *Consider two dynamical systems in Eqs.(7.1) and (7.2) with constraints in Eq.(7.4). For $\mathbf{X}_m^{(\alpha_{j_1}, j_1)} = (\mathbf{x}^{(\alpha_{j_1}, j_1)}, \tilde{\mathbf{x}}^{(\alpha_{j_1}, j_1)})^T \in \Omega_{(\alpha_{j_1}, j_1)}$ ($\alpha_{j_1} \in I$ and $j_1 \in L_1$ with $I = \{1,2\}$ and $L_1 = \{1,2,\ldots,l_1\} \subseteq L$) and $\mathbf{X}_m^{(0,j_1)} = (\mathbf{x}^{(0,j_1)}, \tilde{\mathbf{x}}^{(0,j_1)})^T \in \partial\Omega_{(12,j_1)}$ at time t_m, $\mathbf{X}_m^{(\alpha_{j_1}, j_1)} = \mathbf{X}_m^{(0,j_1)}$. For any small $\varepsilon > 0$, there is a time interval $[t_{m-\varepsilon}, t_m)$ or $(t_m, t_{m+\varepsilon}]$. At $\mathbf{X}^{(\alpha_{j_1}, j_1)} \in \Omega_{(\alpha_{j_1}, j_1)}^{\pm \varepsilon}$ for time $t \in [t_{m-\varepsilon}, t_m)$ or $(t_m, t_{m+\varepsilon}]$, $z^{(\alpha_{j_1}, j_1)}(t) = \varphi^{(\alpha_{j_1}, j_1)}(\mathbf{X}^{(\alpha_{j_1}, j_1)}(t), t, \lambda_{j_1})$ is $C^{r_{\alpha_{j_1}}}$-continuous and $\left| \dfrac{d^{r_{\alpha_{j_1}}+1}}{dt^{r_{\alpha_{j_1}}+1}} z^{(\alpha_{j_1}, j_1)}(t) \right| < \infty (r_{\alpha_{j_1}} \geq 3)$. For $\mathbf{X}^{(\alpha_{j_1}, j_1)} \in \Omega_{(\alpha_{j_1}, j_1)}$ and $\mathbf{X}^{(0, j_1)} \in \partial \Omega_{(12, j_1)}$, let $\mathbb{F}^{(\alpha_{j_1}, j_1)}(\mathbf{X}^{(\alpha_{j_1}, j_1)}, t, \pi^{(\alpha_{j_1}, j_1)}) \neq \mathbb{F}^{(0, j_1)}(\mathbf{X}^{(0, j_1)}, t, \lambda_{j_1})$ for $\mathbf{X}^{(\alpha_{j_1}, j_1)} = \mathbf{X}^{(0, j_1)}$. The two dynamical systems in Eqs.(7.1) and (7.2) to the corner points of l_1-interaction conditions in Eq.(7.4) are of the $(l_{1_1}, l_{1_2}, l_{1_3})$-stick interaction, non-interaction and instantaneous interaction for time t_m if and only if for all $j_{1_i} \in L_{1_i}$ and $L_1 = \cup_{i=1}^{3} L_{1_i}$.*

(i) *For time t_m for all $j_{1_1} \in L_{1_1}$ with $\alpha_{j_{1_1}} = 1,2$*

$$\left.\begin{array}{l} \mathbf{X}_{m-}^{(\alpha_{j_{1_1}}, j_{1_1})} = \mathbf{X}_m^{(0, j_{1_1})} \text{ or } z_{m-}^{(\alpha_{j_{1_1}}, j_{1_1})} = z_m^{(0, j_{1_1})}, \\ (-1)^{\alpha_{j_{1_1}}} G_{\partial \Xi_{(\alpha_{j_{1_1}} \beta_{j_{1_1}}, j_{1_1})}}^{(0, \alpha_{j_{1_1}})} (\mathbf{z}_m^{(\alpha_{j_{1_1}}, j_{1_1})}, t_{m-}) \\ = (-1)^{\alpha_{j_{1_1}}} g_1^{(\alpha_{j_{1_1}}, j_{1_1})} (\mathbf{z}_m^{(\alpha_{j_{1_1}}, j_{1_1})}, t_{m-}) > 0. \end{array}\right\} \quad (7.71)$$

(ii) *For time t_m for all $j_{1_2} \in L_{1_2}$ with $\alpha_{j_{1_2}} = 1,2$*

$$\left.\begin{array}{l} \mathbf{X}_{m+}^{(\alpha_{j_{1_1}}, j_{1_1})} = \mathbf{X}_m^{(0, j_{1_1})} \text{ or } z_{m+}^{(\alpha_{j_{1_2}}, j_{1_2})} = z_m^{(0, j_{1_2})}, \\ (-1)^{\alpha_{j_{1_2}}} G_{\partial \Xi_{(\alpha_{j_{1_2}} \beta_{j_{1_2}}, j_{1_2})}}^{(0, \alpha_{j_{1_2}})} (\mathbf{z}_m^{(\alpha_{j_{1_2}}, j_{1_2})}, t_{m+}) \\ = (-1)^{\alpha_{j_{1_2}}} g_1^{(\alpha_{j_{1_2}}, j_{1_2})} (\mathbf{z}_m^{(\alpha_{j_{1_2}}, j_{1_2})}, t_{m+}) < 0. \end{array}\right\} \quad (7.72)$$

(iii) *For time t_m for all $j_{1_3} \in L_{1_3}$ with $\alpha_{j_{1_3}}, \beta_{j_{1_3}} = 1,2$ and $\alpha_{j_{1_3}} \neq \beta_{j_{1_3}}$*

$$\left.\begin{array}{l} \mathbf{X}_m^{(\alpha_{j_{1_3}}, j_{1_3})} = \mathbf{X}_m^{(0, j_{1_3})} \text{ or } z_{m-}^{(\alpha_{j_{1_3}}, j_{1_2})} = z_m^{(0, j_{1_3})}, \\ (-1)^{\alpha_{j_{1_3}}} G_{\partial \Xi_{(\alpha_{j_{1_3}} \beta_{j_{1_3}}, j_{1_3})}}^{(0, \alpha_{j_{1_3}})} (\mathbf{z}_m^{(\alpha_{j_{1_3}}, j_{1_3})}, t_{m-}) \\ = (-1)^{\alpha_{j_{1_3}}} g_1^{(\alpha_{j_{1_3}}, j_{1_3})} (\mathbf{z}_m^{(\alpha_{j_{1_3}}, j_{1_3})}, t_{m-}) > 0; \end{array}\right\} \quad (7.73a)$$

$$\left.\begin{array}{l}\mathbf{X}_m^{(\beta_{j_{1_3}},j_{1_3})} = \mathbf{X}_m^{(0,j_{1_3})} \text{ or } z_{m+}^{(\beta_{j_{1_3}},j_{1_3})} = z_m^{(0,j_{1_3})},\\ (-1)^{\beta_{j_{1_3}}} G_{\partial \Xi_{(\alpha_{j_{1_3}}\beta_{j_{1_3}},j_{1_3})}}^{(0,\beta_{j_{1_3}})}(\mathbf{z}_m^{(\beta_{j_{1_3}},j_{1_3})}, t_{m+})\\ = (-1)^{\beta_{j_{1_3}}} g_1^{(\beta_{j_{1_3}},j_{1_3})}(\mathbf{z}_m^{(\beta_{j_{1_3}},j_{1_3})}, t_{m+}) < 0.\end{array}\right\} \quad (7.73b)$$

(iv) *The switching bifurcation conditions for one of three cases* (i)∼(iii) *for* $j_1 \in \{j_{1_1}, j_{1_2}, j_{1_3}\}$ *with time* $t = t_m$ *and* $\alpha_{j_1} \in \{1, 2\}$

$$\left.\begin{array}{l}\mathbf{X}_{m\pm}^{(\alpha_{j_1},j_1)} = \mathbf{X}_m^{(0,j_1)} \text{ or } z_{m\pm}^{(\alpha_{j_1},j_1)} = z_m^{(0,j_1)},\\ G_{\partial \Xi_{(\alpha_{j_1}\beta_{j_1},j_1)}}^{(0,\alpha_{j_1})}(\mathbf{z}_m^{(\alpha_{j_1},j_1)}, t_{m\pm}) = g_1^{(\alpha_{j_1},j_1)}(\mathbf{z}_m^{(\alpha_{j_1},j_1)}, t_{m\pm}) = 0,\\ (-1)^{\alpha_{j_1}} G_{\partial \Xi_{(\alpha_{j_1}\beta_{j_1},j_1)}}^{(1,\alpha_{j_1})}(\mathbf{z}_m^{(\alpha_{j_1},j_1)}, t_{m\pm})\\ = (-1)^{\alpha_{j_1}} g_2^{(\alpha_{j_1},j_1)}(\mathbf{z}_m^{(\alpha_{j_1},j_1)}, t_{m\pm}) < 0,\end{array}\right\} \quad (7.74)$$

and for $\beta_{j_1} = \{1, 2\}$ *and* $\beta_{j_1} \neq \alpha_{j_1}$

$$\left.\begin{array}{l}\mathbf{X}_{m\pm}^{(\beta_{j_1},j_1)} = \mathbf{X}_m^{(0,j_1)} \text{ or } z_{m+}^{(\beta_{j_1},j_1)} = z_m^{(0,j_1)},\\ G_{\partial \Xi_{(\alpha_{j_1}\beta_{j_1},j_1)}}^{(0,\beta_{j_1})}(\mathbf{z}_m^{(\beta_{j_1},j_1)}, t_{m\pm}) = g_1^{(\beta_{j_1},j_1)}(\mathbf{z}_m^{(\beta_{j_1},j_1)}, t_{m\pm}) \neq 0,\end{array}\right\} \quad (7.75)$$

or

$$\left.\begin{array}{l}\mathbf{X}_{m\pm}^{(\beta_{j_1},j_1)} = \mathbf{X}_m^{(0,j_1)} \text{ or } z_{m+}^{(\beta_{j_1},j_1)} = z_m^{(0,j_1)},\\ G_{\partial \Xi_{(\alpha_{j_1}\beta_{j_1},j_1)}}^{(0,\beta_{j_1})}(\mathbf{z}_m^{(\beta_{j_1},j_1)}, t_{m\pm}) = g_1^{(\beta_{j_1},j_1)}(\mathbf{z}_m^{(\beta_{j_1},j_1)}, t_{m\pm}) = 0,\\ (-1)^{\beta_{j_1}} G_{\partial \Xi_{(\alpha_{j_1}\beta_{j_1},j_1)}}^{(1,\beta_{j_1})}(\mathbf{z}_m^{(\beta_{j_1},j_1)}, t_{m\pm})\\ = (-1)^{\beta_{j_1}} g_2^{(\beta_{j_1},j_1)}(\mathbf{z}_m^{(\beta_{j_1},j_1)}, t_{m\pm}) < 0.\end{array}\right\} \quad (7.76)$$

Proof. For each interaction, the proof is similar to Luo (2008a,b).

Theorem 7.2. *Consider two dynamical systems in Eqs.(7.1) and (7.2) with interaction conditions in Eq.(7.4). For* $\mathbf{X}_m^{(\alpha_{j_1},j_1)} = (\mathbf{x}^{(\alpha_{j_1},j_1)}, \tilde{\mathbf{x}}^{(\alpha_{j_1},j_1)})^T \in \Omega_{(\alpha_{j_1},j_1)}$ *(* $\alpha_{j_1} \in I$ *and* $j_1 \in L_1$ *with* $I = \{1, 2\}$ *and* $L_1 = \{1, 2, \ldots, l_1\} \subseteq L$*) and* $\mathbf{X}_m^{(0,j_1)} = (\mathbf{x}^{(0,j_1)}, \tilde{\mathbf{x}}^{(0,j_1)})^T \in \partial \Omega_{(12,j_1)}$ *at* t_m, $\mathbf{X}_m^{(\alpha_{j_1},j_1)} = \mathbf{X}_m^{(0,j_1)}$. *For any small,* $\varepsilon > 0$, *there is a time interval* $[t_{m-\varepsilon}, t_m)$ *or* $(t_m, t_{m+\varepsilon}]$. *At* $\mathbf{X}^{(\alpha_{j_1},j_1)} \in \Omega_{(\alpha_{j_1},j_1)}^{\pm \varepsilon}$ *for* $t \in [t_{m-\varepsilon}, t_m)$ *or* $(t_m, t_{m+\varepsilon}]$, $z^{(\alpha_{j_1},j_1)}(t) = \varphi^{(\alpha_{j_1},j_1)}(\mathbf{X}^{(\alpha_{j_1},j_1)}(t), t, \lambda_{j_1})$ *is* $C^{r_{\alpha_{j_1}}}$-*continuous and* $\left| \frac{d^{r_{\alpha_{j_1}}+1}}{dt^{r_{\alpha_{j_1}}+1}} z^{(\alpha_{j_1},j_1)}(t) \right| < \infty$ *(* $r_{\alpha_{j_1}} \geq 2k_{\alpha_{j_1}} + 2$ *). For* $\mathbf{X}^{(\alpha_{j_1},j_1)} \in \Omega_{(\alpha_{j_1},j_1)}$ *and* $\mathbf{X}^{(0,j_1)} \in \partial \Omega_{(12,j_1)}$, *let* $\mathbf{F}^{(\alpha_{j_1},j_1)}(\mathbf{X}^{(\alpha_{j_1},j_1)}, t, \pi^{(\alpha_{j_1},j_1)}) \neq \mathbf{F}^{(0,j_1)}(\mathbf{X}^{(0,j_1)}, t, \lambda_{j_1})$ *for* $\mathbf{X}^{(\alpha_{j_1},j_1)} = \mathbf{X}^{(0,j_1)}$. *The two dynamical systems in Eqs.(7.1) and (7.2) to the corner points of* l_1-*interaction conditions in Eq.(7.4) are of the* (l_1, l_2, l_3)-*stick interaction, source-*

7.4 Interactions with corner singularity

interaction and instantaneous interaction of the $(2k_{\alpha_j} : 2k_{\beta_j})$-type for time t_m if and only if for all $j_{1_i} \in \mathbb{L}_{1_i}$ and $\mathbb{L}_1 = \bigcup_{i=1}^{3} \mathbb{L}_{1_i}$.

(i) *For time t_m for all $j_{1_1} \in \mathbb{L}_{1_1}$ with $\alpha_{j_{1_1}}, \beta_{j_{1_1}} \in \{1,2\}$ and $\alpha_{j_{1_1}} \neq \beta_{j_{1_1}}$*

$$\left.\begin{aligned}
& \mathbf{X}_{m-}^{(\alpha_{j_{1_1}}, j_{1_1})} = \mathbf{X}_m^{(0, j_{1_1})} \text{ or } z_{m-}^{(\alpha_{j_{1_1}}, j_{1_1})} = z_m^{(0, j_{1_1})}, \\
& G_{\partial \Xi_{(\alpha_{j_{1_1}} \beta_{j_{1_1}}, j_{1_1})}}^{(s_{\alpha_{j_{1_1}}}, \alpha_{j_{1_1}})}(\mathbf{z}_m^{(\alpha_{j_{1_1}}, j_{1_1})}, t_{m-}) = g_{s_{\alpha_{j_{1_1}}}+1}^{(\alpha_{j_{1_1}}, j_{1_1})}(\mathbf{z}_m^{(\alpha_{j_{1_1}}, j_{1_1})}, t_{m-}) = 0 \\
& \quad \text{for } s_{\alpha_{j_{1_1}}} = 0, 1, \ldots, 2k_{\alpha_{j_{1_1}}} - 1, \\
& (-1)^{\alpha_{j_{1_1}}} G_{\partial \Xi_{(\alpha_{j_{1_1}} \beta_{j_{1_1}}, j_{1_1})}}^{(2k_{\alpha_{j_{1_1}}}, \alpha_{j_{1_1}})}(\mathbf{z}_m^{(\alpha_{j_{1_1}}, j_{1_1})}, t_{m-}) \\
& \quad = (-1)^{\alpha_{j_{1_1}}} g_{2k_{\alpha_{j_{1_1}}}+1}^{(\alpha_{j_{1_1}}, j_{1_1})}(\mathbf{z}_m^{(\alpha_{j_{1_1}}, j_{1_1})}, t_{m-}) > 0, \\
& \mathbf{X}_{m-}^{(\beta_{j_{1_1}}, j_{1_1})} = \mathbf{X}_m^{(0, j_{1_1})} \text{ or } z_{m-}^{(\beta_{j_{1_1}}, j_{1_1})} = z_m^{(0, j_{1_1})}, \\
& G_{\partial \Xi_{(\alpha_{j_{1_1}} \beta_{j_{1_1}}, j_{1_1})}}^{(s_{\beta_{j_{1_1}}}, \beta_{j_{1_1}})}(\mathbf{z}_m^{(\beta_{j_{1_1}}, j_{1_1})}, t_{m-}) = g_{s_{\beta_{j_{1_1}}}+1}^{(\beta_{j_{1_1}}, j_{1_1})}(\mathbf{z}_m^{(\beta_{j_{1_1}}, j_{1_1})}, t_{m-}) = 0 \\
& \quad \text{for } s_{\beta_{j_{1_1}}} = 0, 1, \ldots, 2k_{\beta_{j_{1_1}}} - 1, \\
& (-1)^{\beta_{j_{1_1}}} G_{\partial \Xi_{(\alpha_{j_{1_1}} \beta_{j_{1_1}}, j_{1_1})}}^{(2k_{\alpha_{j_{1_1}}}, \alpha_{j_{1_1}})}(\mathbf{z}_m^{(\alpha_{j_{1_1}}, j_{1_1})}, t_{m-}) \\
& \quad = (-1)^{\beta_{j_{1_1}}} g_{2k_{\alpha_{j_{1_1}}}+1}^{(\alpha_{j_{1_1}}, j_{1_1})}(\mathbf{z}_m^{(\alpha_{j_{1_1}}, j_{1_1})}, t_{m-}) > 0.
\end{aligned}\right\} \quad (7.77)$$

(ii) *For time t_m for all $j_{1_2} \in \mathbb{L}_{1_2}$ with $\alpha_{j_{1_2}}, \beta_{j_{1_2}} \in \{1,2\}$ and $\alpha_{j_{1_2}} \neq \beta_{j_{1_2}}$*

$$\left.\begin{aligned}
& \mathbf{X}_{m+}^{(\alpha_{j_{1_2}}, j_{1_2})} = \mathbf{X}_m^{(0, j_{1_2})} \text{ or } z_{m+}^{(\alpha_{j_{1_2}}, j_{1_2})} = z_m^{(0, j_{1_2})}, \\
& G_{\partial \Xi_{(\alpha_{j_{1_2}} \beta_{j_{1_2}}, j_{1_2})}}^{(s_{\alpha_{j_{1_2}}}, \alpha_{j_{1_2}})}(\mathbf{z}_m^{(\alpha_{j_{1_2}}, j_{1_2})}, t_{m+}) = g_{s_{\alpha_{j_{1_2}}}+1}^{(\alpha_{j_{1_2}}, j_{1_2})}(\mathbf{z}_m^{(\alpha_{j_{1_2}}, j_{1_2})}, t_{m+}) = 0 \\
& \quad \text{for } s_{\alpha_{j_{1_2}}} = 0, 1, \ldots, 2k_{\alpha_{j_{1_2}}} - 1, \\
& (-1)^{\alpha_{j_{1_2}}} G_{\partial \Xi_{(\alpha_{j_{1_2}} \beta_{j_{1_2}}, j_{1_2})}}^{(2k_{\alpha_{j_{1_2}}}, \alpha_{j_{1_2}})}(\mathbf{z}_m^{(\alpha_{j_{1_2}}, j_{1_2})}, t_{m+}) \\
& \quad = (-1)^{\alpha_{j_{1_2}}} g_{2k_{\alpha_{j_{1_2}}}+1}^{(\alpha_{j_{1_2}}, j_{1_2})}(\mathbf{z}_m^{(\alpha_{j_{1_2}}, j_{1_2})}, t_{m+}) < 0; \\
& \mathbf{X}_{m+}^{(\beta_{j_{1_2}}, j_{1_2})} = \mathbf{X}_m^{(0, j_{1_2})} \text{ or } z_{m+}^{(\beta_{j_{1_2}}, j_{1_2})} = z_m^{(0, j_{1_2})}, \\
& G_{\partial \Xi_{(\alpha_{j_{1_2}} \beta_{j_{1_2}}, j_{1_2})}}^{(s_{\beta_{j_{1_2}}}, \beta_{j_{1_2}})}(\mathbf{z}_m^{(\beta_{j_{1_2}}, j_{1_2})}, t_{m+}) = g_{s_{\beta_{j_{1_2}}}+1}^{(\beta_{j_{1_2}}, j_{1_2})}(\mathbf{z}_m^{(\beta_{j_{1_2}}, j_{1_2})}, t_{m+}) = 0 \\
& \quad \text{for } s_{\beta_{j_{1_2}}} = 0, 1, \ldots, 2k_{\beta_{j_{1_2}}} - 1, \\
& (-1)^{\beta_{j_{1_2}}} G_{\partial \Xi_{(\alpha_{j_{1_2}} \beta_{j_{1_2}}, j_{1_2})}}^{(2k_{\beta_{j_{1_2}}}, \beta_{j_{1_2}})}(\mathbf{z}_m^{(\beta_{j_{1_2}}, j_{1_2})}, t_{m+}) \\
& \quad = (-1)^{\beta_{j_{1_2}}} g_{2k_{\alpha_{j_{1_2}}}+1}^{(\beta_{j_{1_2}}, j_{1_2})}(\mathbf{z}_m^{(\beta_{j_{1_2}}, j_{1_2})}, t_{m+}) < 0.
\end{aligned}\right\} \quad (7.78)$$

(iii) *For time t_m for all $j_{1_3} \in \mathbb{L}_{1_3}$ with $\alpha_{j_{1_3}}, \beta_{j_{1_3}} \in 1, 2$ and $\alpha_{j_{1_3}} \neq \beta_{j_{1_3}}$*

$$\left.\begin{aligned}
& \mathbf{X}_{m-}^{(\alpha_{j_{1_3}}, j_{1_3})} = \mathbf{X}_m^{(0,j_{1_3})} \text{ or } z_{m-}^{(\alpha_{j_{1_3}}, j_{1_3})} = z_m^{(0,j_{1_3})}, \\
& G_{\partial \Xi_{(\alpha_{j_{1_1}} \beta_{j_{1_1}}, j_{1_1})}}^{(s_{\alpha_{j_{1_3}}}, \alpha_{j_{1_3}})} (\mathbf{z}_m^{(\alpha_{j_{1_3}}, j_{1_3})}, t_{m-}) = g_{s_{\alpha_{j_{1_3}}}+1}^{(\alpha_{j_{1_3}}, j_{1_3})} (\mathbf{z}_m^{(\alpha_{j_{1_3}}, j_{1_3})}, t_{m-}) = 0 \\
& \quad for\ s_{\alpha_{j_{1_3}}} = 0, 1, \ldots, 2k_{\alpha_{j_{1_3}}} - 1, \\
& (-1)^{\alpha_{j_{1_3}}} G_{\partial \Xi_{(\alpha_{j_{1_3}} \beta_{j_{1_3}}, j_{1_3})}}^{(2k_{\alpha_{j_{1_3}}}, \alpha_{j_{1_3}})} (\mathbf{z}_m^{(\alpha_{j_{1_3}}, j_{1_3})}, t_{m-}) \\
& \quad = (-1)^{\alpha_{j_{1_3}}} g_{2k_{\alpha_{j_{1_3}}}+1}^{(\alpha_{j_{1_3}}, j_{1_3})} (\mathbf{z}_m^{(\alpha_{j_{1_3}}, j_{1_3})}, t_{m-}) > 0, \\
& \mathbf{X}_{m-}^{(\beta_{j_{1_3}}, j_{1_3})} = \mathbf{X}_m^{(0,j_{1_3})} \text{ or } z_{m+}^{(\beta_{j_{1_3}}, j_{1_3})} = z_m^{(0,j_{1_3})} \\
& G_{\partial \Xi_{(\alpha_{j_{1_3}} \beta_{j_{1_3}}, j_{1_3})}}^{(s_{\beta_{j_{1_3}}}, \beta_{j_{1_3}})} (\mathbf{z}_m^{(\beta_{j_{1_3}}, j_{1_3})}, t_{m+}) = g_{s_{\beta_{j_{1_3}}}+1}^{(\beta_{j_{1_3}}, j_{1_3})} (\mathbf{z}_m^{(\beta_{j_{1_3}}, j_{1_3})}, t_{m+}) = 0 \\
& \quad for\ s_{\beta_{j_{1_3}}} = 0, 1, \ldots, 2k_{\beta_{j_{1_3}}} - 1, \\
& (-1)^{\beta_{j_{1_3}}} G_{\partial \Xi_{(\alpha_{j_{1_3}} \beta_{j_{1_3}}, j_{1_3})}}^{(2k_{\alpha_{j_{1_3}}}, \alpha_{j_{1_3}})} (\mathbf{z}_m^{(\alpha_{j_{1_3}}, j_{1_3})}, t_{m+}) \\
& \quad = (-1)^{\beta_{j_{1_3}}} g_{2k_{\alpha_{j_{1_3}}}+1}^{(\alpha_{j_{1_3}}, j_{1_3})} (\mathbf{z}_m^{(\alpha_{j_{1_3}}, j_{1_3})}, t_{m+}) < 0.
\end{aligned}\right\} \quad (7.79)$$

(iv) *The switching bifurcation conditions for one of three cases* (i)∼(iii) *for $j_1 \in \{j_{1_1}, j_{1_2}, j_{1_3}\}$ with time $t = t_m$ and $\alpha_{j_1} \in \{1, 2\}$*

$$\left.\begin{aligned}
& \mathbf{X}_{m-}^{(\alpha_{j_1}, j_1)} = \mathbf{X}_m^{(0,j_1)} \text{ or } z_{m-}^{(\alpha_{j_1}, j_1)} = z_m^{(0,j_1)}, \\
& G_{\partial \Xi_{(\alpha_{j_{1_1}} \beta_{j_{1_1}}, j_{1_1})}}^{(s_{\alpha_{j_1}}, \alpha_{j_1})} (\mathbf{z}_m^{(\alpha_{j_1}, j_1)}, t_{m-}) = g_{s_{\alpha_{j_1}}+1}^{(\alpha_{j_1}, j_1)} (\mathbf{z}_m^{(\alpha_{j_1}, j_1)}, t_{m-}) = 0 \\
& \quad for\ s_{\alpha_{j_1}} = 0, 1, \ldots, 2k_{\alpha_{j_1}}, \\
& (-1)^{\alpha_{j_1}} G_{\partial \Xi_{(\alpha_{j_1} \beta_{j_1}, j_1)}}^{(2k_{\alpha_{j_1}}, \alpha_{j_1})} (\mathbf{z}_m^{(\alpha_{j_1}, j_1)}, t_{m-}) \\
& \quad = (-1)^{\alpha_{j_1}} g_{2k_{\alpha_{j_1}}+1}^{(\alpha_{j_1}, j_1)} (\mathbf{z}_m^{(\alpha_{j_1}, j_1)}, t_{m-}) < 0;
\end{aligned}\right\} \quad (7.80)$$

and for $\beta_{j_1} \in \{1, 2\}$ and $\beta_{j_1} \neq \alpha_{j_1}$

$$\left.\begin{aligned}
& \mathbf{X}_{m\pm}^{(\beta_{j_1}, j_1)} = \mathbf{X}_m^{(0,j_1)} \text{ or } z_{m\pm}^{(\beta_{j_1}, j_1)} = z_m^{(0,j_1)}, \\
& G_{\partial \Xi_{(\alpha_{j_1} \beta_{j_1}, j_1)}}^{(s_{\beta_{j_1}}, \beta_{j_1})} (\mathbf{z}_m^{(\beta_{j_1}, j_1)}, t_{m\pm}) = g_{s_{\beta_{j_{1_3}}}+1}^{(\beta_{j_1}, j_1)} (\mathbf{z}_m^{(\beta_{j_1}, j_1)}, t_{m\pm}) = 0 \\
& \quad for\ s_{\beta_{j_1}} = 0, 1, \ldots, 2k_{\beta_{j_1}} - 1, \\
& (-1)^{\beta_{j_1}} G_{\partial \Xi_{(\alpha_{j_1} \beta_{j_1}, j_1)}}^{(2k_{\alpha_{j_1}}, \alpha_{j_1})} (\mathbf{z}_m^{(\alpha_{j_1}, j_1)}, t_{m\pm}) \\
& \quad = (-1)^{\beta_{j_1}} g_{2k_{\alpha_{j_1}}+1}^{(\alpha_{j_1}, j_1)} (\mathbf{z}_m^{(\alpha_{j_1}, j_1)}, t_{m\pm}) \neq 0;
\end{aligned}\right\} \quad (7.81)$$

or

$$\left.\begin{array}{l}\mathbf{X}_{m\pm}^{(\beta_{j_1},j_1)} = \mathbf{X}_m^{(0,j_1)} \text{ or } z_{m\pm}^{(\beta_{j_1},j_1)} = z_m^{(0,j_1)}, \\ G_{\partial\Xi_{(\alpha_{j_1}\beta_{j_1},j_1)}}^{(s_{\beta_{j_1}},\beta_{j_1})}(\mathbf{z}_m^{(\beta_{j_1},j_1)}, t_{m\pm}) = g_{s_{\beta_{j_1}}+1}^{(\beta_{j_1},j_1)}(\mathbf{z}_m^{(\beta_{j_1},j_1)}, t_{m\pm}) = 0 \\ \text{for } s_{\beta_{j_1}} = 0, 1, \ldots, 2k_{\beta_{j_1}}, \\ (-1)^{\beta_{j_1}} G_{\partial\Xi_{(\alpha_{j_1}\beta_{j_1},j_1)}}^{(2k_{\alpha_{j_1}}+1,\alpha_{j_1})}(\mathbf{z}_m^{(\alpha_{j_1},j_1)}, t_{m\pm}) \\ = (-1)^{\beta_{j_1}} g_{2k_{\alpha_{j_1}}+2}^{(\alpha_{j_1},j_1)}(\mathbf{z}_m^{(\alpha_{j_1},j_1)}, t_{m\pm}) < 0. \end{array}\right\} \quad (7.82)$$

Proof. For each interaction, the proof is similar to Luo (2008a,b).

If the two system interaction is for the flows not to be passable, the transport laws should be used. The discussion about the transport laws and mapping dynamics for discontinuous dynamical systems can be referred to Luo (2006).

References

Luo, A. C. J. (2005), A theory for non-smooth dynamical systems on connectable domains, *Communication in Nonlinear Science and Numerical Simulation*, **10**, pp.1-55.

Luo, A. C. J. (2006), *Singularity and Dynamics on Discontinuous Vector Fields*, Amsterdam: Elsevier.

Luo, A. C. J.(2008a), A theory for flow switchability in discontinuous dynamical systems, *Nonlinear Analysis: Hybrid Systems*, **2**, pp.1030-1061.

Luo, A. C. J.(2008b),*Global Transversality, Resonance and Chaotic Dynamics*, Singapore: World Scientific.

Luo, A. C. J.and Gegg, B. C. (2006a), Periodic motions in a periodically forced oscillator moving on an oscillating belt with dry friction, ASME *Journal of Computational and Nonlinear Dynamics*, **1**, pp.212-220.

Luo, A. C. J.and Gegg, B. C. (2006b), Dynamics of a periodically excited oscillator with dry friction on a sinusoidally time-varying, traveling surface, *International Journal of Bifurcation and Chaos*,**16**, pp.3539-3566.

Appendix

A.1 Basic solution

Consider a general solution for motion equation for $j = 1, 2, \ldots$

$$\ddot{x} + 2d_j \dot{x} + c_j x = a_j \cos \Omega t + b_j. \tag{A.1}$$

With an initial condition $(t, x^{(j)}, \dot{x}^{(j)}) = (t_i, x_i^{(j)}, \dot{x}_i^{(j)})$, the general solution of Eq.(A.1) is given as follows:

CASE I: For $d_j^2 - c_j > 0$

$$\left.\begin{aligned}
x &= e^{-d_j(t-t_i)}(C_1^{(j)} e^{\lambda_d^{(j)}(t-t_i)} + C_2^{(j)} e^{-\lambda_d^{(j)}(t-t_i)}) \\
&\quad + D_1^{(j)} \cos \Omega t + D_2^{(j)} \sin \Omega t + D_0^{(j)}, \\
\dot{x} &= e^{-d_j(t-t_i)}[(\lambda_d^{(j)} - d_j)C_1^{(j)} e^{\lambda_d(t-t_i)} - (\lambda_d^{(j)} + d_j)C_2^{(j)} e^{-\lambda_d^{(j)}(t-t_i)}] \\
&\quad - D_1^{(j)} \Omega \sin \Omega t + D_2^{(j)} \Omega \cos \Omega t,
\end{aligned}\right\} \tag{A.2}$$

where

$$\left.\begin{aligned}
C_1^{(j)} &= \frac{1}{2\omega_d^{(j)}} \{\dot{x}_i - [D_1^{(j)}(d_j + \omega_d^{(j)}) + D_2^{(j)} \Omega] \cos \Omega t_i \\
&\quad + [D_1^{(j)} \Omega - D_2^{(j)}(d_j + \omega_d)] \sin \Omega t_i + (x_i - D_0^{(j)})(d_j + \omega_d^{(j)})\}, \\
C_2^{(j)} &= \frac{1}{2\omega_d^{(j)}} \{-\dot{x}_i - [D_1^{(j)} \Omega + D_2^{(j)}(-d_j + \omega_d^{(j)})] \sin \Omega t_i \\
&\quad - [D_1^{(j)}(-d_j + \omega_d^{(j)}) - D_2^{(j)} \Omega] \cos \Omega t_i + (x_i - D_0^{(j)})(-d_j + \omega_d^{(j)})\},
\end{aligned}\right\} \tag{A.3}$$

$$\left.\begin{aligned}
D_0^{(j)} &= -\frac{b_j}{c_j}, \quad D_1^{(j)} = \frac{a_j(c_j - \Omega^2)}{(c_j - \Omega^2)^2 + (2d_j\Omega)^2}, \\
D_2^{(j)} &= \frac{a_j(2d_j\Omega)}{(c_j - \Omega^2)^2 + (2d_j\Omega)^2}, \quad \lambda_d^{(j)} = \sqrt{d_j^2 - c_j}.
\end{aligned}\right\} \tag{A.4}$$

CASE II. For $d_j^2 - c_j < 0$

$$\left.\begin{aligned}
x &= e^{-d_j(t-t_i)}[C_1^{(j)} \cos \omega_d^{(j)}(t-t_i) + C_2^{(j)} \sin \omega_d^{(j)}(t-t_i)] \\
&\quad + D_1^{(j)} \cos \Omega t + D_2^{(j)} \sin \Omega t + D_0^{(j)}, \\
\dot{x} &= e^{-d_j(t-t_i)}[(C_2^{(j)} \omega_d^{(j)} - d_j C_1^{(j)}) \cos \omega_d^{(j)}(t-t_i) \\
&\quad - (C_1^{(j)} \omega_d^{(j)} + d_j C_2^{(j)}) \sin \omega_d^{(j)}(t-t_i)] \\
&\quad - D_1^{(j)} \Omega \sin \Omega t + D_2^{(j)} \Omega \cos \Omega t,
\end{aligned}\right\} \quad (A.5)$$

where

$$\left.\begin{aligned}
C_1^{(j)} &= x_i - D_1^{(j)} \cos \Omega t_i - D_2^{(j)} \sin \Omega t_i - D_0^{(j)}, \\
C_2^{(j)} &= \frac{1}{\omega_d^{(j)}}[d_j(x_i - D_1^{(j)} \cos \Omega t_i - D_2^{(j)} \sin \Omega t_i - D_0^{(j)}) \\
&\quad + \dot{x}_i + D_1^{(j)} \Omega \sin \Omega t_i - D_2^{(j)} \Omega \cos \Omega t_i)], \\
\omega_d^{(j)} &= \sqrt{c_j - d_j^2}.
\end{aligned}\right\} \quad (A.6)$$

CASE III: For $d_j^2 - c_j = 0$

$$\left.\begin{aligned}
x &= e^{-d_j(t-t_i)}[C_1^{(j)}(t-t_i) + C_2^{(j)}] + D_1^{(j)} \cos \Omega t + D_2^{(j)} \sin \Omega t + D_1^{(j)}, \\
\dot{x} &= e^{-d_j(t-t_i)}[C_1^{(j)} - C_1^{(j)} d_j(t-t_i) - d_j C_2^{(j)}] \\
&\quad - D_1^{(j)} \Omega \sin \Omega t + D_2^{(j)} \Omega \cos \Omega t,
\end{aligned}\right\} \quad (A.7)$$

where

$$\left.\begin{aligned}
C_2^{(j)} &= x_i - D_1^{(j)} \cos \Omega t_i - D_2^{(j)} \sin \Omega t_i - D_0^{(j)}, \\
C_1^{(j)} &= \dot{x}_i + \cos \Omega t_i (D_2^{(j)} \Omega - d_j D_1^{(j)}) \\
&\quad - \sin \Omega t_i (D_1^{(j)} \Omega + d_j D_2^{(j)}) - d_j D_1^{(j)}.
\end{aligned}\right\} \quad (A.8)$$

CASE IV: For $d_j \neq 0, c_j = 0$

$$\left.\begin{aligned}
x &= C_1^{(j)} e^{-2d_j(t-t_i)} + D_1 \cos \Omega t + D_2^{(j)} \sin \Omega t + D_0^{(j)} t + C_2^{(j)}, \\
\dot{x} &= -2d_j C_1^{(j)} e^{-2d_j(t-t_i)} - D_1^{(j)} \Omega \sin \Omega t + D_2^{(j)} \Omega \cos \Omega t + D_0^{(j)},
\end{aligned}\right\} \quad (A.9)$$

where

$$\left.\begin{aligned}
C_1^{(j)} &= -\frac{1}{2d_j}(\dot{x}_i + D_1^{(j)} \Omega \sin \Omega t_i - D_2^{(j)} \Omega \cos \Omega t_i - D_0^{(j)}), \\
C_2^{(j)} &= \frac{1}{2d_j}[2d_j x_i + \dot{x}_i + (D_1^{(j)} \Omega - 2d_j D_2^{(j)}) \sin \Omega t_i \\
&\quad - (2d_j D_1^{(j)} + D_2^{(j)} \Omega) \cos \Omega t_i - 2d_j D_0^{(j)} t_i - D_0^{(j)}].
\end{aligned}\right\} \quad (A.10)$$

CASE V: For $d_j = 0, c_j = 0$

$$\left.\begin{aligned}
x &= -\frac{a_j}{\Omega^2} \cos \Omega t - \frac{1}{2} b_j t^2 + C_1^{(j)} t + C_2^{(j)}, \\
\dot{x} &= \frac{a_j}{\Omega} \sin \Omega t - b_j t + C_1^{(j)},
\end{aligned}\right\} \quad (A.11)$$

A.2 Stability and bifurcation

where

$$\left.\begin{aligned} C_1^{(j)} &= \dot{x}_i - \frac{a_j}{\Omega}\sin \Omega t_i + b_j t_i, \\ C_2^{(j)} &= x_i - \dot{x}t_i + \frac{a_j}{\Omega^2}\cos \Omega t_i + \frac{a_j}{\Omega}t_i \sin \Omega t_i - \frac{1}{2}b_j t_i^2 \end{aligned}\right\} \quad (A.12)$$

A.2 Stability and bifurcation

Consider a periodic motion with a mapping of P as

$$\mathbf{x}_{i+1} = P\mathbf{x}_i, \ \mathbf{x}_i \in \mathbb{R}^n. \tag{A.13}$$

With periodicity condition $\mathbf{x}_{i+1} = \mathbf{x}_i$, the periodic solution is obtained for the foregoing mapping. The stability of the periodic solutions can be determined by the eigenvalue analysis of the corresponding $n \times n$ Jacobian matrix DP, i.e.,

$$|DP - \lambda \mathbf{I}| = 0. \tag{A.14}$$

Therefore, there are n eigenvalues. If the n eigenvalues lie inside the unit circle, then the period-1 motion is stable. If one of them lies outside the unit circle, the periodic motion is unstable. Namely, the stable, periodic motion requires the eigenvalues to be

$$|\lambda_j| < 1, \ j = 1, 2, \ldots, n. \tag{A.15}$$

If the magnitude of any eigenvalue is greater than one, ie.,

$$|\lambda_j| > 1, \ j \in \{1, 2, \ldots, n\}, \tag{A.16}$$

the periodic motion is unstable.

For $|\lambda_j| < 1$ $(j = 3, 4, \ldots, n)$ and real λ_j $(j = 1, 2)$, if

$$\max\{\lambda_j, \ j = 1, 2\} = 1, \ \min\{\lambda_j, \ j = 1, 2\} \in (-1, 1), \tag{A.17}$$

then the saddle-node (SN) bifurcation occurs; if

$$\min\{\lambda_j, j = 1, 2\} = -1, \ \max\{\lambda_j, j = 1, 2\} \in (-1, 1), \tag{A.18}$$

then the period-doubling bifurcation occurs.

For $|\lambda_j| < 1$ $(j = 3, 4, \ldots, n)$ and complex λ_j $(j = 1, 2)$, if

$$|\lambda_j| = 1, \ j = 1, 2, \tag{A.19}$$

then the Neimark bifurcation occurs.

Index

A
Accessible domains, 9

B
Border-collision bifurcation, 4

D
Discontinuous dynamical systems, 1, 9, 55, 80, 115, 161

F
Flow switchability, 9
Flow barrier, 51
Force product, 77, 91
    Initial force product condition, 94
    Final force product condition, 94
Friction oscillator, 3, 56, 80, 113

G
G-function, 11
Gear transmission systems, 1
Grazing, 127
    Grazing flow, 89, 96
    Grazing motion, 57
    Grazing bifurcation, 4

H
Half-non-passable flow, 24
    $(2k_i : 2k_j - 1)$-half-non-passable flow of the first kind, 24
    $(2k_i : 2k_j - 1)$-half-non-passable flow of the second kind, 28
Half-sink flow, 24
    $(2k_i : 2k_j - 1)$-half sink flow, 24
Half-source flow, 26
    $(2k_i : 2k_j - 1)$-half source flow, 28

I
Impact law, 6, 118
Impacting chatter, 147, 153
Inaccessible domain, 9, 192
Inaccessible tangential flow, 36
Interactions, 191
Interactions with corner singularity, 210
Instantaneous interaction, 203
    $(2k_{\alpha_j} : 2k_{\beta_j})$-instantaneous interaction, 213

N
Non-passable flow, 19
    Non-passable flow of the first kind, 24
    Non-passable flow of the $(2k_i : 2k_j)$-first kind, 44
    Non-passable flow of the second kind, 26
    Non-passable flow of the $(2k_i : 2k_j)$-second kind, 28
Non-passable motion, 82
Non-stick motion, 5
Normal vector field product, 40

P
Passable flow, 15
    $(2k_i : m_j)$-passable flows, 18
Passable motion, 82
Product of G-functions, 40
    $(m_i : m_j)$-product of G-functions, 40

R
Relative jerks, 132

S
Semi-passable, 15
Sink flow, 19
    $(2k_i : 2k_j)$-sink flows, 21
Sliding bifurcation, 42

$(2k_i : 2k_j)$-sliding bifurcation, 44
Sliding flow, 85
Sliding fragmentation bifurcation, 50
   $(2k_i : 2k_j)$-sliding fragmentation bifurcation, 50
   fragmentation condition, 94
Sliding mapping, 60
Sliding mode control, 55
Sliding motion, 55
Source flow, 3, 22
   $(2k_i : 2k_j)$-source flows, 23
Source flow bifurcation, 45
   $(2k_i : 2k_j)$-source flow bifurcation, 46
Source fragmentation bifurcation, 50
   $(2k_i : 2k_j)$-source fragmentation bifurcation, 50
Source interaction, 204
   $(2k_{\alpha_j} : 2k_{\beta_j})$-source interaction, 207
Stick interaction, 203
   $(2k_{\alpha_j} : 2k_{\beta_j})$-stick interaction, 207

Stick motion, 164
Switching bifurcation, 39
   Switching bifurcation of the $(2k_i : 2k_j)$-passable flow, 49
   Switching bifurcation of the $(2k_i : 2k_j)$-non-passable flow, 51
System interactions, 191
   Fundamental interactions, 199

T

Tangential flow, 28
   $(2k_j - 1)^{\text{th}}$-order tangential flow, 18
   $(2k_i - 1 : 2k_j)$-tangential flow, 30
   $(2k_\alpha - 1 : 2k_\beta - 1)$-tangential flow, 34
   $(2k_\alpha - 1 : 2k_\beta - 1)$-double tangential flow, 36
   $(2k_\alpha - 1 : 2k_\beta - 1)$-double inaccessible tangential flow, 39
Transversality, 55

Nonlinear Physical Science

(*Series Editors*: Albert C.J. Luo, Nail H. Ibragimov)

Nail. H. Ibragimov/ Vladimir. F. Kovalev: Approximate and Renormgroup Symmetries

Abdul-Majid Wazwaz: Partial Differential Equations and Solitary Waves Theory

Albert C.J. Luo: Discontinuous Dynamical Systems on Time-varying Domains

Anjan Biswas/ Daniela Milovic/ Matthew Edwards: Mathematical Theory of Dispersion-Managed Optical Solitons

Hanke,Wolfgang/ Kohn, Florian P.M./ Wiedemann, Meike: Self-organization and Pattern-formation in Neuronal Systems under Conditions of Variable Gravity

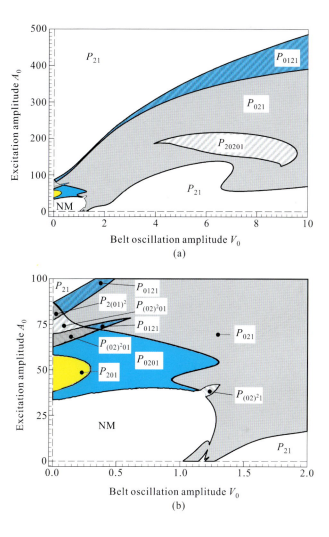

Fig. 4.18 The (V_0, A_0) parameter map for periodic motions: (a) the large range, (b) zoomed area. ($\Omega = 1, V_1 = 1, d_1 = 1, d_2 = 0, b_1 = -b_2 = 0.5, c_1 = c_2 = 30$). The acronym "NM" represents no motion interacting with the discontinuous boundary

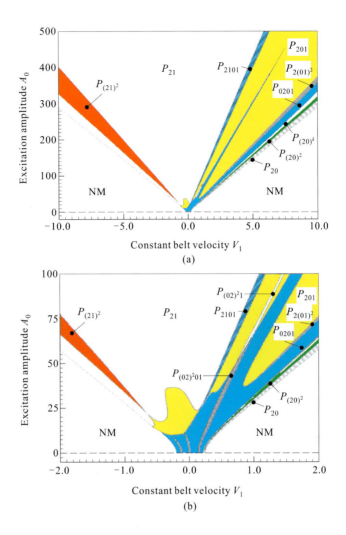

Fig. 4.19 The (V_1, A_0) parameter map for periodic motions: (a) the large range, (b) zoomed area. ($\Omega = 1, V_0 = 0.25, d_1 = 1, d_2 = 0, b_1 = -b_2 = 0.5, c_1 = c_2 = 30$). The acronym "NM" represents no motion interacting with the discontinuous boundary

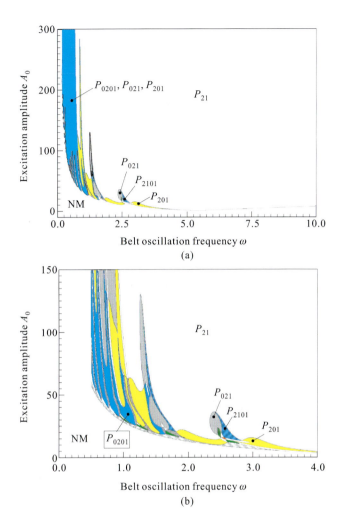

Fig. 4.20 The (A_0, ω) parameter map for periodic motions: (a) the large range, (b) zoomed area. ($\Omega = 1, V_0 = 0.25, V_1 = 1, d_1 = 1, d_2 = 0, b_1 = -b_2 = 0.5, c_1 = c_2 = 30$). The acronym "NM" represents no motion interacting with the discontinuous boundary

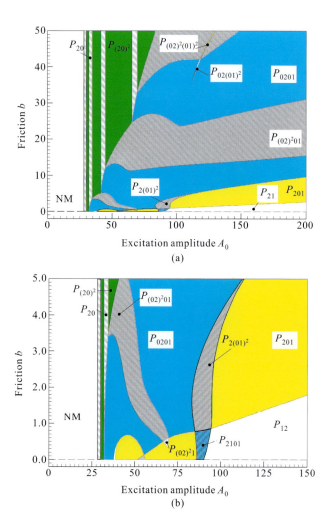

Fig. 4.21 The (A_0, b) parameter map for periodic motions: (a) the large range, (b) zoomed area. ($\Omega = 1, V_0 = 0.25, V_1 = 1, d_1 = 1, d_2 = 0, b_1 = -b_2 = b, c_1 = c_2 = 30$). The acronym "NM" represents no motion interacting with the discontinuous boundary